J. G. Kalbfleisch

Probability
and
Statistical Inference II

Springer-Verlag
New York   Heidelberg   Berlin

Dr. J. G. Kalbfleisch
Faculty of Mathematics
Department of Statistics
University of Waterloo
Ontario, Canada N2L 3G1

AMS Classifications (1980): 60-01, 62-01

With 30 illustrations

Library of Congress Cataloging in Publication Data

Kalbfleisch, J
   Probability and statistical inference.

   (Universitext)
   Includes indexes.
   1. Probabilities. 2. Mathematical Statistics.
I. Title.
QA273.K27   1979   519.2   79-22910

Printed in the United States of America.

9 8 7 6 5 4 3 2 1

ISBN 0-387-90458-1   Springer-Verlag New York Heidelberg Berlin
ISBN 3-540-90458-1   Springer-Verlag Berlin Heidelberg New York

CONTENTS—VOLUME II

CHAPTER 9. LIKELIHOOD METHODS

The first eight chapters dealt with probability models, and with mathematical methods for handling and describing them. Several of the simplest discrete and continuous probability models were considered in detail. The remainder of the book is concerned with applications of probability models in problems of data analysis and interpretation.

One important use of probability models is to provide simple mathematical descriptions of large bodies of data. For instance, we might describe a set of 1000 blood pressure measurements as being like a sample of 1000 independent values from a normal distribution whose mean μ and variance σ^2 are estimated from the data. This model gives a concise description of the data, and from it we can easily calculate the approximate proportion of blood pressure measurements which lie in any particular range. The accuracy of such calculations will, of course, depend upon how well the normal distribution model fits the data.

We shall be concerned primarily with applications of probability models in problems of statistical inference, where it is desired to draw general conclusions based on a limited amount of data. For instance, tests might be run to determine the length of life of an aircraft component prior to failure from metal fatigue. Such tests are typically very expensive and time consuming, and hence only a few specimens can be examined. Based on the small amount of data obtained, one would attempt to draw conclusions about similar components which had not been tested. The link between the observed sample and the remaining components is provided by the probability model. The data are used to check the adequacy of the model and to estimate any unknown parameters which it involves. General statements concerning this type of component are then based on the model.

Whether the model is to be used for descriptive purposes or statistical inference, it is important to check that there is good agreement between the model and the data. Methods for checking the goodness of fit of the model will be considered in Chapter 11.

Frequently, the probability model for an experiment will involve one or more unknown parameters which it is necessary to estimate from the data. We have already encountered this problem on several occasions, and have used the observed sample mean as an estimate of the mean of a Poisson or exponential distribution. Intuitively, this is a reasonable thing to do, but intuition may fail us in more complicated cases.

The method of maximum likelihood (Section 1) provides a routine procedure for obtaining estimates of unknown parameters. Section 2 discusses some computational procedures which are useful in determining maximum likelihood estimates. Section 3 considers the problem of estimating a parameter on the basis of data from two independent experiments.

In many situations, it is not enough to have merely an estimate of the parameter. Some indication of the likely accuracy of the estimate is also needed. This can be obtained by examining the relative likelihood function (Section 4), which ranks all possible values of θ according to their plausibilities in the light of the data.

Section 5 discusses likelihood methods based on continuous probability models, and the special case of censoring in lifetime experiments is considered in Section 6. Some general properties of likelihood methods are discussed in Sections 7 and 9.

In Section 8, sufficient statistics are defined. A sufficient statistic for an unknown parameter θ is a function of the data which may be said to carry all of the available information concerning θ. Sufficient statistics play an important role in determining the appropriate probability distribution to use in a test of significance; see Chapters 11 and 12.

In this chapter, we consider only the case of a single unknown parameter. Likelihood methods for the estimation of two or more unknown parameters will be discussed in Chapter 10.

9.1 The Method of Maximum Likelihood

Suppose that the probability model for an experiment involves an unknown parameter θ. The experiment is performed, and some event E is observed to occur; that is, some data are obtained. We now wish to use the data to estimate the value of θ.

Using the model and the laws of probability, the probability of the observed event E can be determined. This probability will usually be a function of the unknown parameter, $P(E;\theta)$. There will be some values of θ for which the observed event E is fairly probable, and other values of θ for which E is quite improbable. It seems reasonable that we should pick, as an estimate of θ, a value for which E is probable rather than improbable. Values of θ for which the observed event E has a relatively high probability are preferable to values of θ for which E is very improbable.

Usually, there will exist a unique value of θ which maximizes $P(E;\theta)$. This value is denoted by $\hat{\theta}$, and is called the <u>maximum likelihood estimate</u> (MLE) of θ. The MLE of θ is the value of θ for which the observed event E has the greatest probability that it can possibly have under the model.

The <u>likelihood function</u> of θ is defined as follows:

$$L(\theta) = k \cdot P(E;\theta). \qquad (9.1.1)$$

Here k is any positive constant with respect to θ; that is, k is not a function of θ, although it may be a function of the data. The <u>log likelihood function</u> is the natural logarithm of L,

$$\ell(\theta) = \log L(\theta). \qquad (9.1.2)$$

The value of θ which maximizes $P(E;\theta)$ will also maximize $L(\theta)$ and $\ell(\theta)$. The MLE $\hat{\theta}$ is thus the value of θ which maximizes the likelihood function and the log-likelihood function.

The set of all possible values of θ is called the <u>parameter space</u> and is usually denoted by Ω. In most problems with a single unknown parameter, Ω will be an interval of real values. Furthermore, the first and second derivatives

$$\ell'(\theta) = \frac{\partial}{\partial \theta} \ell(\theta); \qquad \ell''(\theta) = \frac{\partial^2}{\partial \theta^2} \ell(\theta)$$

will exist at all interior points of Ω. Then the MLE can usually be found as a root of the <u>maximum likelihood equation</u>

$$\ell'(\theta) = 0. \qquad (9.1.3)$$

In some simple examples, this equation can be solved algebraically to yield a formula for $\hat{\theta}$. In more complicated situations, it is usually necessary to solve (9.1.3) numerically on a computer; see Section 2.

A root of (9.1.3) at which $\ell''(\theta) < 0$ is a point of relative maximum. Relative minima and points of inflexion may also appear among the roots of (9.1.3). Thus it is necessary to determine the sign of the second derivative or otherwise verify that the root obtained is a relative maximum.

Situations do arise in which $\hat{\theta}$ can not be obtained by solving the maximum likelihood equation (9.1.3). For instance, the overall maximum of the likelihood function may occur on the boundary of the parameter space Ω, and then (9.1.3) need not hold at the maximum; see Examples 9.1.1 and 9.1.2. Similarly, if θ is restricted to a discrete set of values such as the integers, equation (9.1.3) does not apply; see Problems 9.1.7 and 9.1.11.

Example 9.1.1. Suppose that we wish to estimate θ, the fraction of people who have tuberculosis in a large homogeneous population. To do this, we randomly select n individuals for testing, and find that x of them have the disease. Because the population is large and homogeneous, we assume that the n individuals tested are independent, and each has probability θ of having tuberculosis. The probability of the observed event is then

$$P(E;\theta) = P(x \text{ out of } n \text{ have tuberculosis})$$
$$= \binom{n}{x}\theta^x(1-\theta)^{n-x}, \qquad (9.1.4)$$

where $0 \le \theta \le 1$. The parameter space is thus the unit interval, $\Omega = [0,1]$.

The likelihood function can be defined to be any convenient positive multiple of $P(E;\theta)$, and for simplicity we take

$$L(\theta) = \theta^x(1-\theta)^{n-x}; \quad 0 \le \theta \le 1.$$

The log likelihood function is then

$$\ell(\theta) = x \log \theta + (n-x)\log(1-\theta),$$

with first and second derivatives

$$\ell'(\theta) = \frac{x}{\theta} - \frac{n-x}{1-\theta} \; ; \quad \ell''(\theta) = -\frac{x}{\theta^2} - \frac{n-x}{(1-\theta)^2} \; .$$

If $1 \le x \le n-1$, the equation $\ell'(\theta) = 0$ has a unique solution $\theta = \frac{x}{n}$. Since $\ell''(\theta) < 0$ for $\theta = \frac{x}{n}$, this is a relative maximum. Furthermore, since $L(\theta) = 0$ for $\theta = 0$ or $\theta = 1$, this is the overall maximum, and hence $\hat{\theta} = \frac{x}{n}$. In order to maximize the probability of the data, we estimate the population fraction θ by the sample fraction $\frac{x}{n}$.

If $x = 0$, the equation $\ell'(\theta) = 0$ has no solution, and the maximum occurs on the boundary of the parameter space $[0,1]$. In this case, we have

$$P(E;\theta) = (1-\theta)^n \quad \text{for} \quad 0 \le \theta \le 1.$$

This is clearly largest when $\theta = 0$, and hence $\hat{\theta} = 0$. Similarly, $\hat{\theta} = 1$ for $x = n$, and we have $\hat{\theta} = \frac{x}{n}$ for $x = 0,1,\ldots,n$.

Example 9.1.2. Some laboratory tests are run on samples of river water in order to determine whether the water is safe for swimming. Of particular interest is the concentration of coliform bacteria in the water. The number of coliform bacteria is determined for each of n unit-volume samples of river water, giving n observed counts $x_1, x_2, \ldots, x_n$. The problem is to estimate μ, the average number of coliform bacteria per unit volume in the river.

We assume that the bacteria are randomly dispersed throughout the river water, so that the locations of the bacteria are random points in space (Section 4.4). Then the probability of finding x_i bacteria in a sample of unit volume is given by a Poisson distribution:

$$f(x_i) = \mu^{x_i} e^{-\mu}/x_i! ; \quad x_i = 0, 1, 2, \ldots$$

where $0 \le \mu < \infty$. Since disjoint volumes are independent, the probability of the n observed counts $x_1, x_2, \ldots, x_n$ is

$$P(E; \mu) = f(x_1, x_2, \ldots, x_n) = \prod_{i=1}^{n} f(x_i)$$

$$= \prod_{i=1}^{n} \frac{\mu^{x_i} e^{-\mu}}{x_i!} = \frac{\mu^{\sum x_i} e^{-n\mu}}{x_1! x_2! \ldots x_n!} .$$

The likelihood function of μ is $k \cdot P(E; \mu)$, and by a suitable choice of k we may take

$$L(\mu) = \mu^{\sum x_i} e^{-n\mu} \quad \text{for} \quad 0 \le \mu < \infty.$$

The log likelihood function and its derivatives are

$$\ell(\mu) = \sum x_i \log \mu - n\mu ; \quad \ell'(\mu) = \frac{1}{\mu} \sum x_i - n ; \quad \ell''(\mu) = -\sum x_i / \mu^2 .$$

If $\sum x_i > 0$, the maximum likelihood equation $\ell'(\mu) = 0$ has a unique solution $\mu = \frac{1}{n} \sum x_i = \bar{x}$. The second derivative is negative at this point, indicating that we do have a relative maximum. Since $L(0) = 0$ and $L(\mu) \to 0$ as $\mu \to \infty$, we have found the overall maximum. If $\sum x_i = 0$, the equation $\ell'(\mu) = 0$ has no solution, and the maximum occurs on the boundary of the parameter space: $\hat{\mu} = 0$. Thus, in either case, we have $\hat{\mu} = \bar{x}$. The probability of the sample is greatest if the population mean μ is estimated by the sample mean $\bar{x}$.

Example 9.1.3. It is usually not possible to count the number of bacteria in a sample of river water; one can only determine whether or not any are present. n test tubes each containing a volume v of river water are incubated and tested. A negative test shows that there were no bacteria present, while a positive test shows that at least one bacterium was present. If y tubes out of the n tested give negative results, what is the maximum likelihood estimate of μ?

Solution. The probability that there are x bacteria in a volume v of river water is given by a Poisson distribution with mean μv:

$$f(x) = (\mu v)^x e^{-\mu v}/x!; \qquad x = 0, 1, 2, \ldots \quad .$$

The probability of a negative reaction (no bacteria) is

$$p = f(0) = e^{-\mu v};$$

the probability of a positive reaction (at least one bacterium) is

$$1 - p = 1 - e^{-\mu v}.$$

Since disjoint volumes are independent, the n test tubes constitute independent trials. The probability of observing y negative reactions out of n is therefore

$$P(E; \mu) = \binom{n}{y} p^y (1 - p)^{n-y}$$

where $p = e^{-v\mu}$ and $0 \le \mu < \infty$.

We ignore the constant factor $\binom{n}{y}$, and define the likelihood function to be

$$L(\mu) = p^y (1 - p)^{n-y}.$$

From Example 9.1.1, this function attains its maximum value when $p = y/n$. The corresponding value of μ can be obtained by solving the equation $p = e^{-v\mu}$ to give $\mu = -\frac{1}{v} \log p$. Thus we obtain

$$\hat{\mu} = -\frac{1}{v} \log \frac{y}{n} = \frac{\log n - \log y}{v} \quad .$$

For instance, suppose that 40 test tubes each containing 10 ml. of river water are incubated. If 28 give negative tests and 12 give positive tests, then

$$\hat{\mu} = \frac{\log 40 - \log 28}{10} = 0.0357.$$

The concentration of coliform bacteria per ml. of river water is estimated to be 0.0357.

The greater the concentration of bacteria in the river, the

more probable it is that all n test tubes will give positive results.
Hence the larger the value of μ, the more probable the observation
y = 0. If we observe y = 0, the MLE of μ will be +∞. In this
case, it does not make much practical sense to give merely a single
estimate of μ. What we require is an indication of the range of μ-
values which are plausible in the light of the data, rather than a
single "most plausible" value. This can be obtained by examining the
relative likelihood function; see Section 4.

Likelihoods based on Frequency Tables

Data from n independent repetitions of an experiment are
often summarized in a frequency table:

Event or Class	A_1	A_2	$\cdots$	A_k	Total
Observed frequency	f_1	f_2		f_k	n
Expected frequency	np_1	np_2	$\cdots$	np_k	n

The sample space S for a single repetition of the experiment is
partitioned into k mutually exclusive classes or events,
$S = A_1 \cup A_2 \cup \ldots \cup A_k$. Then f_j is the number of times that A_j occurs
in n repetitions $(\sum f_n = n)$. Let p_j be the probability of event
A_j in any one repetition $(\sum p_j = 1)$. The p_j's can be determined
from the probability model. If the model involves an unknown parameter θ, the p_j's will generally be functions of θ.

The probability of observing a particular frequency table is
given by the multinomial distribution:

$$P(E;\theta) = \binom{n}{f_1\ f_2\ \ldots\ f_k} p_1^{f_1} p_2^{f_2} \ldots p_k^{f_k}.$$

The likelihood function of θ based on the frequency table is proportional to $P(E;\theta)$. Thus we may define

$$L(\theta) = c p_1^{f_1} p_2^{f_2} \ldots p_k^{f_k} \tag{9.1.5}$$

where c is any convenient positive constant. The MLE of θ is obtained by maximizing (9.1.5). Using $\hat{\theta}$, one can then compute expected frequencies for comparison with the observed frequencies.

Example 9.1.4. On each of 200 consecutive working days, ten items
were randomly selected from a production line and tested for imper-
fections, with the following results:

Number of defective items	0	1	2	3	≥4	Total
Frequency observed	133	52	12	3	0	200

The number of defective items out of 10 is thought to have a binomial distribution. Find the MLE of θ, the probability that an item is defective, and compute expected frequencies under the binomial distribution model.

Solution. According to a binomial distribution model, the probability of observing j defectives out of 10 is

$$p_j = \binom{10}{j}\theta^j(1-\theta)^{10-j}; \quad j = 0,1,2,\ldots,10.$$

The probability of observing 4 or more defectives is $p_{4+} = 1 - p_0 - p_1 - p_2 - p_3$. By (9.1.5), the likelihood function of θ is

$$L(\theta) = cp_0^{133}\, p_1^{52}\, p_2^{12}\, p_3^{3}\, p_{4+}^{0} \quad \text{for} \quad 0 \le \theta \le 1.$$

We substitute for the p_j's and make a convenient choice of c to get

$$L(\theta) = [(1-\theta)^{10}]^{133}[\theta(1-\theta)^9]^{52}[\theta^2(1-\theta)^8]^{12}[\theta^3(1-\theta)^7]^3$$

$$= \theta^{85}(1-\theta)^{1915}.$$

This likelihood function is of the form considered in Example 9.1.1, with $x = 85$ and $n = 2000$. Hence $\hat{\theta} = \frac{85}{2000} = 0.0425$.

Using this value of θ, the expected frequencies $np_j = 100p_j$ can be computed for $j = 0,1,2,3$. The expected frequency for the last class is then found by subtraction from 200.

Number of defectives	0	1	2	3	≥4	Total
Observed frequency	133	52	12	3	0	200
Expected frequency	129.54	57.50	11.48	1.36	0.12	200

The agreement between observed and expected frequencies appears to be reasonably good. The f_j's are random variables, and therefore it is natural that there will be some differences between the observed and expected frequencies. A goodness of fit test (Chapter 11) confirms that the differences here can easily be accounted for by chance variation of the f_j's, and hence the binomial distribution model seems satisfactory.

Problems for Section 9.1

†1. Suppose that diseased trees are distributed at random throughout a large forest with an average of λ per acre. The numbers of diseased trees observed in ten four-acre plots were 0,1,3,0,0,2, 2,0,1,1. Find the maximum likelihood estimate of λ.

2. Suppose that the n counts in Example 9.1.2 were summarized in a frequency table as follows:

Number of bacteria	0	1	2		Total
Frequency observed	f_0	f_1	f_2	...	n

The number of bacteria in a sample is assumed to have a Poisson distribution with mean μ. Find the likelihood function and maximum likelihood estimate of μ based on the frequency table, and show that they agree with the results obtained in Example 9.1.2.

3. Consider the following two experiments whose purpose is to estimate θ, the fraction of a large population having blood type A.

(i) Individuals are selected at random until 10 with blood type A are obtained. The total number of people examined is found to be 100.

(ii) 100 individuals are selected at random, and it is found that 10 of them have blood type A.

Show that the two experiments lead to proportional likelihood functions, and hence the same MLE for θ.

†4. According to genetic theory, blood types MM, NM, and NN should occur in a very large population with relative frequencies θ^2, $2\theta(1-\theta)$, and $(1-\theta)^2$, where θ is the (unknown) gene frequency.

(a) Suppose that, in a random sample of size n from the population, there are x_1, x_2, and x_3 of the three types. Find an expression for $\hat{\theta}$.

(b) The observed frequencies in a sample of size 100 were 32, 46, and 22, respectively. Compute $\hat{\theta}$ and the expected frequencies for the three blood types under the model.

5. A brick-shaped die (Example 1.3.2) is rolled n times, and the ith face comes up x_i times $(i = 1, 2, ..., 6)$, where $\sum x_i = n$.

(a) Show that $\hat{\theta} = (3t - 2n)/12n$, where $t = x_1 + x_2 + x_3 + x_4$.

(b) Suppose that the observed frequencies are 11, 15, 13, 15, 22, 24. Compute expected frequencies under the model.

6. A sample of n items is examined from each large batch of a mass-produced article. The number of good items in a sample has a binomial distribution with parameters n and p. The batch is accepted if all n items are good, and is rejected otherwise. Out

of m batches, x are accepted and m - x are rejected. Find the maximum likelihood estimate of p.

†7. "The enemy" has an unknown number N of tanks, which he has obligingly numbered 1,2,...,N. Spies have reported sighting 8 tanks with numbers 137, 24, 86, 33, 92, 129, 17, 111. Assume that sightings are independent, and that each of the N tanks has probability 1/N of being observed at each sighting. Show that $\hat{N} = 137$.

8. Blood samples from nk people are analysed to obtain information about θ, the fraction of the population infected with a certain disease. In order to save time, the nk samples are mixed together k at a time to give n pooled samples. The analysis of a pooled sample will be negative if the k individuals are free from the disease, and positive otherwise. Out of the n pooled samples, x give negative results and n - x give positive results. Find an expression for $\hat{\theta}$.

†9. Specimens of a new high-impact plastic are tested by repeatedly striking them with a hammer until they fracture. If the specimen has a constant probability θ of surviving a blow, independently of the number of previous blows received, the number of blows required to fracture a specimen will have a geometric distribution,

$$f(x) = \theta^{x-1}(1 - \theta) \quad \text{for} \quad x = 1,2,3,\ldots \ .$$

The results of tests on 200 specimens were as follows:

Number of blows required	1	2	3	≥4	Total
Number of specimens	112	36	22	30	200

Find the maximum likelihood estimate of θ, and compute expected frequencies.

10. The n progeny in a breeding experiment are of three types, there being x_i of the ith type (i = 1,2,3). According to a genetic model, the proportions of the three types should be $(2 + p)/4$, $(1 - p)/2$, and p/4, and progeny are independent of one another.

(a) Show that $\hat{p}$ is a root of the quadratic equation

$$np^2 + (2x_2 + x_3 - x_1)p - 2x_3 = 0.$$

(b) Suppose that $x_1 = 58$, $x_2 = 33$, and $x_3 = 9$. Find $\hat{p}$, and compute expected frequencies under the model.

11. An urn contains r red balls and b black balls, where r is known but b is unknown. Of n balls chosen at random without replacement, x were red and y were black (x + y = n).

(a) Show that L(b) is proportional to $b^{(y)}/(r + b)^{(n)}$.

(b) Show that $\dfrac{L(b+1)}{L(b)} = \dfrac{(b+1)(r+b-n+1)}{(b-y+1)(r+b+1)}$.

(c) By considering the conditions under which $L(b+1)/L(b)$ exceeds one, show that $\hat{b}$ is the smallest integer which exceeds $\dfrac{nr}{x} - (r+1)$. When is $\hat{b}$ not unique?

12. For a certain mass-produced article, the proportion of defectives is θ . It is customary to inspect a sample of 3 items from each large batch. Records are kept only for those samples which contain at least one defective item.

(a) Show that the conditional probability that a sample contains i defectives, given that it contains at least one defective, is

$$\binom{3}{i}\theta^i(1-\theta)^{3-i}/[1-(1-\theta)^3] \quad (i=1,2,3).$$

(b) Suppose that x_i samples out of n recorded contain i defectives $(i=1,2,3; \; \sum x_i = n)$. Show that $\hat{\theta}$ is the smaller root of the quadratic equation

$$t\theta^2 - 3t\theta + 3(t-n) = 0$$

where $t = x_1 + 2x_2 + 3x_3$.

9.2 Computational Methods

The maximum likelihood estimate $\hat{\theta}$ is the value of θ which maximizes the log likelihood function $\ell(\theta)$. In special cases, the maximum likelihood equation $\ell'(\theta) = 0$ can be solved algebraically to obtain a formula for $\hat{\theta}$, but more often it is necessary to determine $\hat{\theta}$ numerically. In past years, much effort has been expended in developing approximations and alternate estimation procedures in order to avoid computation. Most of this work has been made obsolete by modern computers and plotting facilities. In this section, we describe several algorithms for determining $\hat{\theta}$ numerically.

Graphical Procedure

$\ell(\theta)$ is computed at twenty or so well-spaced values of θ , and a preliminary graph is examined. The approximate location of the maximum is determined from the graph, and $\ell(\theta)$ is computed at twenty or so additional points near the maximum. A second graph is then prepared, and the procedure is repeated until the maximum is pinpointed with sufficient accuracy.

This informal procedure requires only a subroutine to evalu-

ate $\ell(\theta)$ at specified values of θ. It has the advantage that inspection of the graphs may reveal troublesome situations such as a non-
unique maximum, or a maximum near the boundary of the parameter space,
thereby avoiding confusion and wasted time later on. In fact, it is a
very good idea to prepare a graph of $\ell(\theta)$ as a preliminary to any
maximization procedure.

Repeated Bisection - I

Suppose that $\ell(\theta)$ is a continuous function of θ for
$a \leq \theta \leq b$, and that there exists a value c in the interval (a,b)
such that $\ell(c) > \ell(a)$ and $\ell(c) > \ell(b)$. Then there must exist a relative maximum in the interval $[a,b]$. This fact provides the basis
for an algorithm to determine $\hat{\theta}$. We begin with an interval $(\theta_0 - h_0,$
$\theta_0 + h_0)$ which contains $\hat{\theta}$, and compute $\ell(\theta_0)$. At each iteration,
the algorithm produces an interval only half as long which also con-
tains $\hat{\theta}$. If θ_i is the midpoint of the interval obtained from the
ith iteration, then $|\theta_i - \hat{\theta}| < h_0/2^i$.

We begin the $(i+1)$st iteration with an interval $(\theta_i - h_i,$
$\theta_i + h_i)$ of width $2h_i$ which contains $\hat{\theta}$, and the value $\ell(\theta_i)$.
Then we define $h_{i+1} = h_i/2$, and calculate $\ell(\theta_i - h_{i+1})$ and
$\ell(\theta_i + h_{i+1})$. Now θ_{i+1} is taken to be θ_i or $\theta_i - h_{i+1}$ or

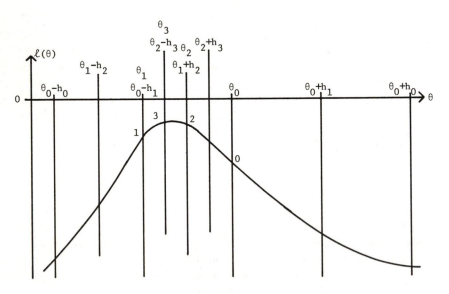

Figure 9.2.1

Maximization of $\ell(\theta)$ through Repeated Bisection

$\theta_i + h_{i+1}$, whichever gives the largest value of ℓ. The new interval $(\theta_{i+1} - h_{i+1}, \theta_{i+1} + h_{i+1})$ has width $2h_{i+1} = h_i$, which is just half the width of the preceding interval. It is easy to see that, if the old interval contains a relative maximum, so does the new one. This algorithm is illustrated in Figure 9.2.1.

The initial interval $(\theta_0 - h_0, \theta_0 + h_0)$ may be obtained from a preliminary graph, or by trial and error. Alternatively, a routine to determine a suitable starting interval can be incorporated into the algorithm. For instance, one can begin with an interval (a, b), and translate it to the left or right by steps of $(a + b)/2$ until the value of ℓ at the midpoint exceeds the value at either end.

Repeated Bisection - II

This procedure is very similar to the algorithm just described except that, instead of maximizing $\ell(\theta)$ directly, we seek a root of the maximum likelihood equation $\ell'(\theta) = 0$. It is based on the observation that, if a and b are parameter values with $a < b$ and $\ell'(a) > 0 > \ell'(b)$, and if $\ell'(\theta)$ is continuous for $a \leq \theta \leq b$, then (a, b) contains a relative maximum of $\ell(\theta)$. We begin with an interval

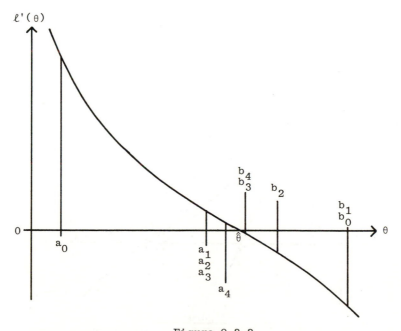

Figure 9.2.2

Solution of $\ell'(\theta) = 0$ by means of Repeated Bisection

14

(a_0, b_0) such that $\ell'(a_0) > 0 > \ell'(b_0)$. At the ith iteration we obtain an interval (a_i, b_i) of length $(b_0 - a_0)/2^i$ such that $\ell'(a_i) > 0 > \ell'(b_i)$. If θ_i is the midpoint of this interval, then $|\theta_i - \hat\theta| < (b_0 - a_0)/2^{i+1}$.

We enter the (i + 1)st interation with an interval (a_i, b_i) such that $\ell'(a_i) > 0 > \ell'(b_i)$, and calculate $\ell'(\frac{a_i + b_i}{2})$. If this value is positive, we take $a_{i+1} = (a_i + b_i)/2$ and $b_{i+1} = b_i$; if it is negative we take $a_{i+1} = a_i$ and $b_{i+1} = (a_i + b_i)/2$. The result is an interval (a_{i+1}, b_{i+1}) of length $(b_i - a_i)/2$ such that $\ell(a_{i+1}) > 0 > \ell(b_{i+1})$. This algorithm is illustrated in Figure 9.2.2.

Newton's Method

Figure 9.2.3 illustrates Newton's method for obtaining a root of the ML equation $\ell'(\theta) = 0$. We begin with an initial guess θ_0 and repeatedly improve it. If θ_i is the approximation to $\hat\theta$ which was obtained at the ith iteration, then

$$\theta_{i+1} = \theta_i - \ell'(\theta_i)/\ell''(\theta_i).$$

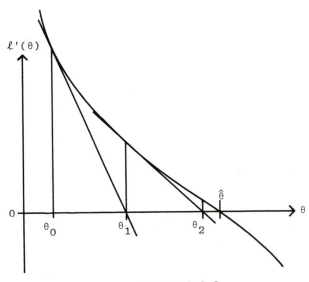

<u>Figure 9.2.3</u>
Solution of $\ell'(\theta) = 0$ by Newton's Method

As the diagram shows, θ_{i+1} is the point at which the tangent to $\ell'(\theta)$ at $\theta = \theta_i$ crosses the horizontal axis. If the initial guess θ_0 is reasonably good, this method will usually produce an accurate approximation to $\hat{\theta}$ in a few iterations.

If $\ell'(\theta) = 0$ has more than one root, Newton's method will not necessarily converge to the one desired. Difficulties can also arise if the maximum occurs at or near a boundary of the parameter space. To guard against such possibilities, a graph of $\ell(\theta)$ or $\ell'(\theta)$ should be examined before Newton's method is applied.

For an application of Newton's method, see Example 9.3.2. A generalization of this procedure, the Newton-Raphson method, is often convenient to use when there are two or more unknown parameters. See Section 10.1.

9.3 Combining Independent Likelihoods

Suppose that two independent experiments give information about the same parameter θ. In the first experiment an event E_1 is observed to occur, and the likelihood function of θ is

$$L_1(\theta) = k_1 P(E_1;\theta)$$

where k_1 is a positive constant. Similarly, an event E_2 is observed to occur in the second experiment, giving rise to the likelihood function

$$L_2(\theta) = k_2 P(E_2;\theta).$$

As in Section 3.2, we may consider the two experiments as components of a single composite experiment. In this composite experiment, the observed event is the intersection of E_1 and E_2, and the likelihood function is

$$L(\theta) = k P(E_1 E_2;\theta).$$

Because E_1 and E_2 are independent, we have

$$P(E_1 E_2:\theta) = P(E_1;\theta)P(E_2;\theta).$$

It follows that

$$L(\theta) = k' L_1(\theta) L_2(\theta)$$

where k' is a positive constant. Since k' may be chosen arbitrarily, we may write

$$L(\theta) = L_1(\theta) L_2(\theta) \qquad (9.3.1)$$

and taking the natural logarithm of both sides gives

$$\ell(\theta) = \ell_1(\theta) + \ell_2(\theta). \tag{9.3.2}$$

Therefore, to combine the information about θ from two (or more) independent experiments, we merely multiply the likelihood functions, or add the log likelihood functions.

Let us denote the maximum likelihood estimate of θ by $\hat{\theta}_1$ for the first experiment, $\hat{\theta}_2$ for the second, and $\hat{\theta}$ overall; that is $\hat{\theta}_1$ maximizes $\ell_1(\theta)$, $\hat{\theta}_2$ maximizes $\ell_2(\theta)$, and $\hat{\theta}$ maximizes $\ell(\theta)$. If $\hat{\theta}_1 = \hat{\theta}_2$, then both terms on the right hand side of (9.3.2) attain their maxima at the same point, and hence $\hat{\theta} = \hat{\theta}_1 = \hat{\theta}_2$. Otherwise, the overall maximum $\hat{\theta}$ will usually lie between $\hat{\theta}_1$ and $\hat{\theta}_2$.

Example 9.3.1. Suppose that, in Example 9.1.1, m additional people are randomly selected, and y of them are found to have tuberculosis. Find the MLE of θ based on both sets of data.

Solution. For the first experiment, the log likelihood function is

$$\ell_1(\theta) = x \log \theta + (n - x)\log(1 - \theta), \tag{9.3.3}$$

and the maximum likelihood estimate is $\hat{\theta}_1 = \frac{x}{n}$. For the second experiment, we similarly obtain

$$\ell_2(\theta) = y \log \theta + (m - y)\log(1 - \theta),$$

and $\hat{\theta}_2 = \frac{y}{m}$. Because the population is large, the two samples will be very nearly independent, and hence by (9.3.2), the log likelihood function based on both samples is

$$\ell(\theta) = \ell_1(\theta) + \ell_2(\theta)$$
$$= (x + y)\log \theta + (n + m - x - y)\log(1 - \theta). \tag{9.3.4}$$

This is of the same form as (9.3.3), and the overall MLE is

$$\hat{\theta} = \frac{x + y}{n + m} .$$

Since $x = n\hat{\theta}_1$ and $y = m\hat{\theta}_2$, we have

$$\hat{\theta} = \frac{n}{n + m} \hat{\theta}_1 + \frac{m}{n + m} \hat{\theta}_2$$

which is a weighted average of $\hat{\theta}_1$ and $\hat{\theta}_2$. For instance, if 90 individuals are examined in the first sample $(n = 90)$, and only 10 in the second (m = 10), we have

$$\hat{\theta} = 0.9\hat{\theta}_1 + 0.1\hat{\theta}_2.$$

The overall MLE will be closer to the MLE based on the large sample than to the MLE based on the small sample.

Note that the log likelihood function (9.3.4) is the same as would be obtained if we considered a single sample of n + m individuals, x + y of whom were found to have tuberculosis. The division of the results into two separate experiments is irrelevant in so far as estimation of θ is concerned.

Example 9.3.2. In performing the experiment described in Example 9.1.3, it is necessary to specify the volume v of river water which is to be placed in each test tube. If v is made too large, then all of the test tubes will contain bacteria and give a positive reaction. If v is too small, we may get only negative reactions. In either case, the experiment will be rather uninformative about μ, the concentration of bacteria in the river.

One way to guard against this difficulty is to prepare two (or more) different types of test tubes containing different volumes of river water. Suppose that 40 test tubes containing 10 ml. of river water were tested, and 28 gave negative results. Also, 40 test tubes containing 1 ml. of river water were tested, and 37 gave negative results. What is the maximum likelihood estimate of μ?

Solution. From Example 9.1.3, the likelihood function based on the 40 tubes containing 10 ml. is

$$L_1(\mu) = p_1^{28}(1 - p_1)^{12}$$

where $p_1 = e^{-10\mu}$, and the MLE of μ is $\hat{\mu}_1 = 0.0357$. The log likelihood function is

$$\ell_1(\mu) = 28 \log p_1 + 12 \log(1 - p_1).$$

Similarly, from the 40 tubes containing 1 ml. we obtain

$$\ell_2(\mu) = 37 \log p_2 + 3 \log(1 - p_2)$$

where $p_2 = e^{-\mu}$, and the MLE of μ is

$$\hat{\mu}_2 = \frac{\log n - \log y}{v} = \frac{\log 40 - \log 37}{1} = 0.078.$$

By (9.3.2), the log likelihood function based on all 80 tubes is

$$\ell(\mu) = \ell_1(\mu) + \ell_2(\mu)$$
$$= 28 \log p_1 + 12 \log(1 - p_1) + 37 \log p_2 + 3 \log(1 - p_2)$$
$$= -317\mu + 12 \log(1 - e^{-10\mu}) + 3 \log(1 - e^{-\mu}).$$

The overall maximum likelihood estimate $\hat{\mu}$ is the value of μ for which $\ell(\mu)$ is a maximum.

The derivative of ℓ with respect to μ is

$$\ell'(\mu) = -317 + \frac{120 e^{-10\mu}}{1 - e^{-10\mu}} + \frac{3 e^{-\mu}}{1 - e^{-\mu}}$$

$$= -317 + \frac{120}{e^{10\mu} - 1} + \frac{3}{e^{\mu} - 1}.$$

The maximum likelihood equation $\ell'(\mu) = 0$ cannot be solved algebraically, and $\hat{\mu}$ must be obtained numerically. Any of the procedures described in the last section may be used. We shall employ Newton's method, with starting value $\frac{1}{2}(\hat{\mu}_1 + \hat{\mu}_2) = 0.057$. For this we need the second derivative:

$$\ell''(\mu) = -\frac{1200 e^{10\mu}}{(e^{10\mu} - 1)^2} - \frac{3 e^{\mu}}{(e^{\mu} - 1)^2}.$$

The computations are summarized in Table 9.3.1. After four iterations we obtain $\hat{\mu} = 0.04005$, correct to five decimal places. Note that the second derivative is negative, indicating that a relative maximum has been obtained.

Table 9.3.1
Solution of $\ell'(\mu) = 0$ by Newton's Method

i	μ_i	$\ell'(\mu_i)$	$\ell''(\mu_i)$	$\ell'(\mu_i)/\ell''(\mu_i)$
0	0.057	-109.66	- 4518.16	0.02427
1	0.03273	83.07	-13902.58	-0.00598
2	0.03871	12.87	- 9910.74	-0.00130
3	0.04001	0.41	- 9270.86	-0.00004
4	0.04005	0.04	- 9252.15	-0.00000

Problems for Section 9.3

1. Use the four procedures described in Section 9.2 to locate the

maximum of the following log likelihood function:

$$\ell(\mu) = 100 \log \mu - 50\mu - 50 \log (1 - e^{-\mu}) \quad \text{for} \quad \mu > 0.$$

† 2. Leaves of a plant are examined for insects. The number of insects on a leaf is thought to have a Poisson distribution with mean μ, except that many leaves have no insects because they are unsuitable for feeding and not merely because of the chance variation allowed by the Poisson law. The empty leaves are therefore not counted.

   (a) Find the conditional probability that a leaf contains i insects, given that it contains at least one.

   (b) Suppose that x_i leaves are observed with i insects $(i = 1, 2, 3, \ldots)$, where $\sum x_i = n$. Show that the MLE of μ satisfies the equation

$$\hat{\mu} = \bar{x} (1 - e^{-\hat{\mu}})$$

   where $\bar{x} = \sum i x_i /n$.

   (c) Determine $\hat{\mu}$ numerically for the case $\bar{x} = 3.2$.

3. If deaths from a rare disease are spread randomly throughout the population, the number of deaths in a region of population p should have approximately a Poisson distribution with mean λp. The numbers of deaths in n regions with populations $p_1, p_2, \ldots, p_n$ were $d_1, d_2, \ldots, d_n$, respectively. Find the MLE of λ.

† 4. In a population in which the frequency of the gene for colour blindness is θ, genetic theory indicates that the probability that a male is colour blind is θ, and the probability that a female is colour blind is θ^2. A random sample of M males is found to include m colour blind, and a random sample of N females includes n colour blind. Find the likelihood function of θ based on both samples, and show that $\hat{\theta}$ can be obtained as a root of a quadratic equation.

5. In Problem 9.1.12, suppose that samples of size $k > 3$ are examined, and that x_i of those recorded contain i defectives $(i = 1, 2, \ldots, k; \sum x_i = n)$.

   (a) Show that the MLE of θ satisfies the equation

$$\bar{x}[1 - (1 - \theta)^k] - k\theta = 0$$

   where $\bar{x} = \sum i x_i /n$.

   (b) Use the binomial theorem to show that, if $\hat{\theta}$ is small, then

$$\hat{\theta} \approx 2(\bar{x} - 1)/(k - 1)\bar{x}.$$

6. Samples of river water are placed in nm test tubes, there being n tubes which contain volume v_i for $i = 1, 2, \ldots, m$. After incubation, y_i of the tubes containing volume v_i show negative reactions, indicating the absence of coliform bacteria. The remaining tubes give positive reactions, showing that at least one bacterium is present. Assuming that bacteria are randomly distributed throughout the river water, find the log likelihood function of μ, the average number of bacteria per unit volume. Show that the MLE of μ satisfies the equation

$$\sum \frac{v_i(n - y_i)}{1 - p_i} - n \sum v_i = 0 \quad \text{where} \quad p_i = e^{-\mu v_i},$$

and describe in detail how to obtain $\hat{\mu}$ by Newton's method.

9.4 Relative Likelihood

As in Section 9.1, we suppose that the probability model for an experiment involves an unknown parameter θ. The experiment is performed and some event E is observed to occur. The probability of E can be determined from the model as a function of θ, $P(E; \theta)$. The likelihood function of θ is then defined to be a constant multiple of $P(E; \theta)$,

$$L(\theta) = kP(E; \theta), \tag{9.4.1}$$

where k is positive and does not depend upon θ.

Thus far we have used the likelihood function only to determine $\hat{\theta}$, the maximum likelihood estimate of θ. This is the parameter value for which the probability of the data E is maximized. More generally, the likelihood function can be used to examine the whole range of possible parameter values, and to determine which values are plausible and which are implausible in the light of the data.

Suppose that θ_1 and θ_2 are two possible values of θ. The likelihood ratio for θ_1 versus θ_2 is defined to be

$$\frac{L(\theta_1)}{L(\theta_2)} = \frac{kP(E; \theta_1)}{kP(E; \theta_2)}$$

$$= \frac{\text{Probability of data for } \theta = \theta_1}{\text{Probability of data for } \theta = \theta_2}. \tag{9.4.2}$$

If this ratio exceeds 1, the data are more probable for $\theta = \theta_1$

than they are for $\theta = \theta_2$. We say that θ_1 is a "more plausible" or "more likely" parameter value than θ_2. The size of the ratio gives a measure of how much more likely θ_1 is than θ_2. For instance, if $L(\theta_1)/L(\theta_2) = 100$, the data are 100 times more probable for $\theta = \theta_1$ than they are for $\theta = \theta_2$. We then say that, in the light of the data, parameter value θ_1 is 100 times more likely than θ_2.

It is convenient to select one value of θ with which all other values of θ may be compared. The natural choice is $\hat{\theta}$, the most likely value of θ. Hence we define the relative likelihood function (RLF) of θ as follows:

$$R(\theta) = L(\theta)/L(\hat{\theta}). \qquad (9.4.3)$$

Because $\hat{\theta}$ is selected to maximize $L(\theta)$, we have

$$0 \le R(\theta) \le 1$$

for all possible values of θ.

If θ_1 is some particular value of θ, then $R(\theta_1)$ is the likelihood ratio for θ_1 versus $\hat{\theta}$:

$$R(\theta_1) = \frac{L(\theta_1)}{L(\hat{\theta})} = \frac{kP(E;\theta_1)}{kP(E;\hat{\theta})}$$

$$= \frac{\text{Probability of data for } \theta = \theta_1}{\text{Maximum probability of data for any value of } \theta}.$$

If $R(\theta_1)$ is small, e.g. $R(\theta_1) \le 0.1$, θ_1 is rather an implausible parameter value because there exist other values of θ for which the data are ten times as probable. However if $R(\theta_1)$ is large, e.g. $R(\theta_1) \ge 0.5$, θ_1 is a fairly plausible parameter value because it gives to the data at least 50% of the maximum probability which is possible under the model. The relative likelihood function ranks all possible parameter values according to their plausibilities in the light of the data.

In most examples which we shall encounter, $\hat{\theta}$ exists and is unique, and definition (9.4.3) applies. More generally, the relative likelihood function may be defined as the ratio of $L(\theta)$ to the supremum of $L(\theta)$ taken over all parameter values:

$$R(\theta) = L(\theta)/\sup_{\theta} L(\theta).$$

Since $L(\theta) = kP(E;\theta)$ where $P(E;\theta) \le 1$, the supremum is finite. The relative likelihood function exists and may be used to determine the plausibilities of parameter values even when $\hat{\theta}$ does not exist.

Likelihood Regions and Intervals

The set of parameter values for which $R(\theta) \geq \alpha$ is called a 100α% likelihood region for θ. In most applications with a single un-known real-valued parameter θ, the 100α% likelihood region will consist of an interval of real values, and we then refer to it as a 100α% likelihood interval (LI) for θ. We shall usually consider the 50%, 10% and 1% likelihood intervals (or regions). Values inside the 10% LI will be referred to as "plausible", and values outside this in-terval as "implausible". Similarly, we shall refer to values inside the 50% LI as "very plausible", and values outside the 1% LI as "very implausible". Of course, the choice of division points at .50, .10, and .01 is arbitrary and must not be taken too seriously.

Likelihood intervals are usually most easily obtained by plotting the log relative likelihood function,

$$r(\theta) = \log R(\theta) = \log L(\theta) - \log L(\hat{\theta}) = \ell(\theta) - \ell(\hat{\theta}). \qquad (9.4.4)$$

Since $0 \leq R(\theta) \leq 1$, we have $-\infty \leq r(\theta) \leq 0$ for all values of θ, and $r(\hat{\theta}) = 0$. For 50%, 10%, and 1% likelihood intervals (or regions), we will have $r(\theta) \geq -0.69, -2.30,$ and -4.61, respectively.

Example 9.4.1. Continuation of Example 9.1.1.

Suppose that, out of 100 people examined, three are found to have tuberculosis. On the basis of this observation, which values of θ are plausible? Compare with the results that would be obtained if 200 people were examined and six were found to have tuberculosis.

Solution. From Example 9.1.1, the log likelihood function is

$$\ell(\theta) = 3 \log \theta + 97 \log (1 - \theta),$$

and the maximum likelihood estimate is $\hat{\theta} = 0.03$. The maximum of the log likelihood is

$$\ell(\hat{\theta}) = 3 \log (.03) + 97 \log (.97) = -13.47.$$

The log relative likelihood function is thus

$$r(\theta) = \ell(\theta) - \ell(\hat{\theta}) = 3 \log \theta + 97 \log (1 - \theta) + 13.47.$$

A graph of this function is shown in Figure 9.4.1 (solid line). From

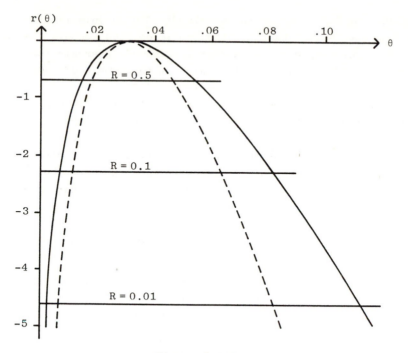

Figure 9.4.1
Log Relative Likelihood Functions from Example 9.4.1.
——————— based on 3 diseased out of 100
------- based on 6 diseased out of 200

the graph we find that r(θ) ≥ -2.30 for 0.006 ≤ θ ≤ 0.081, and this
is the 10% LI for θ. Values of θ inside this interval are fairly
plausible in the light of the data. Similarly, the 50% LI is
0.014 ≤ θ ≤ 0.054. Values within this interval are quite plausible,
because they give the data at least 50% of the maximum probability
which is possible under the model.

If we observed 6 diseased out of 200, we would have

$$\ell(\theta) = 6 \log \theta + 194 \log (1 - \theta),$$

and $\hat{\theta}$ = 0.03 as before. The maximum of the log likelihood is now

$$\ell(\hat{\theta}) = -26.95.$$

Figure 9.4.1 shows the corresponding log relative likelihood function
with a broken line. Both functions attain their maxima at $\hat{\theta}$ = 0.03.

However the log RLF based on the sample of 200 people is more sharply peaked than the log RLF based on the sample of 100 people. As a result, the larger sample gives shorter likelihood intervals for θ. For instance, the 10% LI is (.011,.063) for the sample of 200, as opposed to (.006,.081) for the sample of 100.

In general, increasing the amount of data will produce a more sharply peaked likelihood function, and shorter likelihood intervals for θ. Increasing the amount of data leads to more precise estimation of θ, in the sense that there will be a shorter range of plausible parameter values. Roughly speaking, the length of the $100\alpha\%$ likelihood interval is inversely proportional to the square root of the sample size.

Example 9.4.2. In Example 9.3.2, we considered data from two experiments with test tubes containing river water:

Observation 1: $y = 28$ negative reactions out of $n = 40$ test tubes each containing $v = 10$ ml.

Observation 2: $y = 37$ negative out of $n = 40$ tubes with $v = 1$.

Graph the log relative likelihood functions and obtain 50% likelihood intervals for μ based on the two observations taken separately, and taken together.

Solution. The log likelihood function based only on observation 1 is

$$\ell_1(\mu) = 28 \log p_1 + 12 \log (1 - p_1); \qquad p_1 = e^{-10\mu}.$$

Since $p_1 = \dfrac{y}{n} = 0.7$ at the maximum (Example 9.1.3), the maximum log likelihood is

$$\ell_1(\hat{\mu}_1) = 28 \log 0.7 + 12 \log 0.3 = -24.43.$$

The log relative likelihood function is then

$$r_1(\mu) = \ell_1(\mu) - \ell_1(\hat{\mu}_1) = -280\mu + 12 \log (1 - e^{-10\mu}) + 24.43.$$

Similarly, the log relative likelihood function based only on observation 2 is

$$r_2(\mu) = -37\mu + 3 \log (1 - e^{-\mu}) + 10.66.$$

For both observations together, the log LF is

$$\ell(\mu) = \ell_1(\mu) + \ell_2(\mu)$$
$$= -317\mu + 12 \log(1 - e^{-10\mu}) + 3 \log(1 - e^{-\mu}).$$

From Example 9.3.2, the overall MLE is $\hat{\mu} = 0.04005$, and substitution of this value gives $\ell(\hat{\mu}) = -35.71$. The log RLF based on both observations is thus

$$r(\mu) = \ell(\mu) + 35.71.$$

The three log RLF's are tabulated in Table 9.4.1 and graphed in Figure 9.4.2, with $r(\mu)$ being given by the broken line. From the graphs, the following 50% likelihood intervals may be obtained:

Observation 1 only: $0.025 \le \mu \le 0.049$

Observation 2 only: $0.036 \le \mu \le 0.144$

Both observations combined: $0.029 \le \mu \le 0.053$.

Table 9.4.1

Log Relative Likelihood Functions for Example 9.4.2

μ	$r_1(\mu)$	$r_2(\mu)$	$r(\mu)$
.005		−5.43	
.01	−6.59	−3.55	−9.51
.015	−3.42	−2.52	−5.32
.018	−2.25	−2.09	−3.71
.02	−1.66	−1.85	−2.89
.025	−0.67	−1.37	−1.42
.03	−0.17	−1.02	−0.57
.04	−0.08	−0.54	−0.00
.05	−0.76	−0.26	−0.39
.06	−1.92	−0.09	−1.39
.07	−3.40	−0.02	−2.80
.08	−5.12	−0.00	−4.50
.10		−0.10	
.20		−1.87	
.30		−4.50	

The log RLF based on observation 2 only is almost flat over a large range of μ-values, indicating that this observation provides relatively little information about μ. The combined log RLF based on all the data is very nearly the same as that based on observation 1 alone.

The combined log RLF $r(\mu)$ can be obtained directly from a table or graph of $r_1(\mu)$ and $r_2(\mu)$. We form the sum $r_1(\mu) + r_2(\mu)$, and observe the value of μ at which it is greatest. This will be the overall MLE $\hat{\mu}$. The combined log RLF is then

$$r(\mu) = r_1(\mu) + r_2(\mu) - [r_1(\hat{\mu}) + r_2(\hat{\mu})].$$

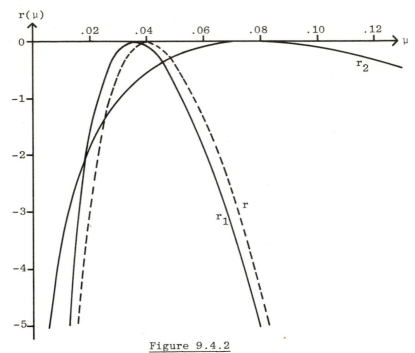

Figure 9.4.2

Combination of Log RLF's **from Independent** Experiments

If $r_1(\hat{\mu}) + r_2(\hat{\mu})$ is small (e.g. less than -2), then there exists no single value of μ which is plausible on both sets of data. The two sets of data are then in contradiction, since they point to different values for the same parameter μ. When this happens, it is generally inadvisable to combine the two data sets. Instead, the parameter should be estimated separately for each data set, and an explanation for the discrepancy should be sought.

In the present example, we find that $r_1(\hat{\mu}) + r_2(\hat{\mu}) = -0.62$. There do exist values of μ (near 0.04) which are quite plausible for both observations, and hence no contradiction is apparent. It is therefore reasonable to combine the two observations, and to base statements about μ on $r(\mu)$, the combined RLF.

Example 9.4.3. Relative likelihood when $\hat{\mu} = +\infty$.

Suppose that $n = 40$ test tubes are prepared, each containing $v = 10$ ml. of river water, and that all of them give positive results ($y = 0$). The likelihood function of μ is then

$$L(\mu) = (1 - p)^{40} = (1 - e^{-10\mu})^{40} \quad \text{for} \quad 0 \le \mu < \infty.$$

Then, as we noted at the end of Example 9.1.3, $L(\mu)$ increases as μ increases to $+\infty$. We say that $\hat{\mu} = +\infty$, although strictly speaking $\hat{\mu}$ does not exist because this value does not belong to the parameter space.

Even when $\hat{\mu}$ does not exist, the relative likelihood function is well defined and can be used to determine the range of plausible parameter values. As μ tends to $+\infty$, $L(\mu)$ increases to 1, and hence

$$\sup_{0 \le \mu < \infty} L(\mu) = 1.$$

The relative likelihood function of μ is then

$$R(\mu) = \frac{L(\mu)}{\sup L(\mu)} = (1 - e^{-10\mu})^{40} \quad \text{for} \quad 0 \le \mu < \infty.$$

The log relative likelihood function,

$$r(\mu) = 40 \log (1 - e^{-10\mu}),$$

is plotted in Figure 9.4.3. We have $r(\mu) \ge -0.69$ for $\mu > 0.41$, and hence the 50% LI for μ is $(0.41, \infty)$. Any value of μ which exceeds 0.41 is very plausible in the light of the data. Similarly, we have $r(\mu) \le -4.61$ for $\mu \le 0.22$, so that any value of μ less than 0.22 is extremely implausible.

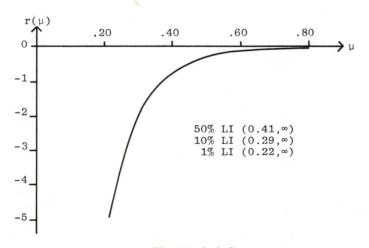

Figure 9.4.3

Log Relative Likelihood Function when $\hat{\mu} = +\infty$

Problems for Section 9.4

†1. Prepare a graph of the log RLF in Problem 9.1.1, and from it obtain 50% and 10% likelihood intervals for λ.

2. The number of west-bound vehicles which pass a fixed point on a main east-west road in 10 seconds is a Poisson variate with mean μ. The numbers passing in disjoint time intervals are independent. The following table summarizes the data from 300 ten-second intervals:

No. of vehicles in 10 sec.	0	1	2	3	4	5
Frequency observed	61	107	76	45	10	1

Plot the log RLF of μ, and from the graph obtain 50% and 10% likelihood intervals for μ.

3. A company plans to purchase either machine 1 or machine 2, and has available the following performance data:

Machine 1: 0 failures in 7800 trials

Machine 2: 4 failures in 21804 trials.

Trials are independent, and the probability of failure is θ_1 for machine 1 and θ_2 for machine 2. Plot the log RLF's of θ_1 and θ_2 on the same graph. Under what conditions would you recommend the purchase of machine 2 rather than machine 1?

†4. Find the relative likelihood of $\theta = 0$ (a balanced die) in Problem 9.1.5.

5. (a) Plot the log RLF of the gene frequency θ in Problem 9.1.4.
   (b) In a second sample of 100 humans, there were 27 with blood type MM, 52 with blood type MN, and 21 with blood type NN. Plot the log RLF of θ based on the second sample on the graph prepared in (a).
   (c) If it is appropriate to do so, obtain the log RLF for θ based on both samples, and show it on the graph prepared in (a).

6. Find 50% and 10% likelihood intervals for N in Problem 9.1.7.

†7. Suppose that $r = n = 10$ and $y = 5$ in Problem 9.1.11. Which values of b have relative likelihood 50% or more? 10% or more?

8. In Problem 9.1.10(b), graph the log RLF of p and obtain a 10% LI for p.

9. The records from 200 samples in Problem 9.1.12 showed 180 with one defective, 17 with two defectives, and 3 with three defectives. Evaluate $\hat{\theta}$, plot the log RLF of θ, and obtain a 10% likelihood interval for θ.

9.5 Likelihood for Continuous Models

Continuous probability distributions are frequently used as probability models for experiments involving the measurement of time, weight, length, etc. Suppose that X has a continuous distribution with probability density function f and cumulative distribution function F, depending upon an unknown parameter θ. The experiment is performed and values of X are observed. The problem is to use the data to estimate θ, or more generally, to determine which values of θ are plausible in the light of the data.

When X is a continuous variate, f(x) does not give the probability of observing the value x. In fact, as we noted in Section 6.1, the probability of any particular real value is zero. An actual measurement of time, weight, etc. will necessarily be made to only finitely many decimal places. An observed value x will therefore correspond to some small interval of real values $a < X \le b$, say. The probability of observing the value x is then

$$P(a < X \le b) = \int_a^b f(x)dx = F(b) - F(a). \qquad (9.5.1)$$

Suppose that, in n independent repetitions, we observe n values $x_1, x_2, \ldots, x_n$, with x_i corresponding to the real interval $(a_i, b_i]$. Because repetitions are independent, the probability of the data is obtained as a product:

$$P(E;\theta) = \prod_{i=1}^{n} P(a_i < X \le b_i) = \prod_{i=1}^{n} [F(b_i) - F(a_i)]. \qquad (9.5.2)$$

The likelihood function of θ is proportional to (9.5.2).

If the interval length $\Delta_i = b_i - a_i$ is small, then $F(b_i)$ will be close to $F(a_i)$, and computation of the difference $F(b_i) - F(a_i)$ may introduce serious roundoff errors. In this case, we make use of (6.1.7), and approximate the area under the density function between a_i and b_i by the area of a rectangle with base Δ_i and height $f(x_i)$:

$$P(a_i < X \le b_i) = F(b_i) - F(a_i) \approx f(x_i)\Delta_i. \qquad (9.5.3)$$

Some or all of the factors in (9.5.2) are approximated in this way to obtain a function which is easier to deal with computationally and mathematically.

In the most usual case, all of the measurement intervals Δ_i are small, and the approximation (9.5.3) may be applied to all of the terms in (9.5.2). This gives

$$P(E;\theta) \approx \prod_{i=1}^{n} f(x_i)\Delta_i = [\prod_{i=1}^{n} \Delta_i] \prod_{i=1}^{n} f(x_i).$$

Since the Δ_i's do not depend upon θ, <u>the likelihood function is proportional to the product of probability densities</u>,

$$L(\theta) = k \prod_{i=1}^{n} f(x_i) \qquad\qquad (9.5.4)$$

where k is any convenient positive constant. This is actually an approximation, but it will be an extremely accurate one whenever the Δ_i's are all small.

It is not necessary to replace every factor in (9.5.2) by the approximation (9.5.3). For instance, it may happen that $f(x)$ changes rapidly when x is small, in which case the original terms in (9.5.2) could be retained for small values x_i, and the approximation could be used for large x_i's. Another situation where some of the terms in (9.5.2) should be retained will be discussed in the next section.

<u>Example 9.5.1</u>. A certain type of electronic component is susceptible to instantaneous failure at any time. However, components do not deteriorate with age, and the chance of failure within a given time period does not depend upon the age of the component. From Section 6.2, the lifetime of such a component should have an exponential distribution, with probability density function

$$f(x) = \frac{1}{\theta} e^{-x/\theta} \quad \text{for} \quad x > 0,$$

where θ is the expected lifetime of such components.

Ten such components were tested independently. Their lifetimes, measured to the nearest day, were as follows:

$$70 \quad 11 \quad 66 \quad 5 \quad 20 \quad 4 \quad 35 \quad 40 \quad 29 \quad 8.$$

What values of θ are plausible in the light of the data?

<u>Solution based on (9.5.4)</u>. Each observed lifetime corresponds to an interval of length $\Delta = 1$. The average lifetime is about 30, and the exponential p.d.f. with mean $\theta = 30$ changes very little over an interval of length 1. Areas under the p.d.f. will thus be well approximated by rectangles, and (9.5.4) should give an accurate approximation. We substitute for $f(x_i)$ in (9.5.4) and take $k = 1$ to ob-

tain

$$L(\theta) = \prod_{i=1}^{n} \frac{1}{\theta} e^{-x_i/\theta} = \theta^{-n} \exp(-\frac{1}{\theta} \sum x_i).$$

The log likelihood function is

$$\ell(\theta) = -n \log \theta - \frac{1}{\theta} \sum x_i$$

with derivatives

$$\ell'(\theta) = -\frac{n}{\theta} + \frac{\sum x_i}{\theta^2} \; ; \quad \ell''(\theta) = \frac{n}{\theta^2} - \frac{2 \sum x_i}{\theta^3}.$$

We may now solve $\ell'(\theta) = 0$ to obtain $\hat{\theta} = \sum x_i/n = \bar{x}$. The value of the second derivative at this point is

$$\ell''(\hat{\theta}) = \frac{n}{\hat{\theta}^2} - \frac{2n\hat{\theta}}{\hat{\theta}^3} = -\frac{n}{\hat{\theta}^2}$$

which is negative. Hence the root obtained is a relative maximum.

The total of the $n = 10$ observed lifetimes is $\sum x_i = 288$, so that $\hat{\theta} = 28.8$ and

$$\ell(\theta) = -10 \log \theta - \frac{288}{\theta}.$$

The log relative likelihood function,

$$r(\theta) = \ell(\theta) - \ell(\hat{\theta}),$$

is plotted in Figure 9.5.1. The observations indicate a mean lifetime between 20 and 43 days (50% LI). Values of θ less than 16 days or greater than 62 days are implausible (relative likelihood less than 10%).

Exact solution based on (9.5.2).

For comparison, we shall determine the exact likelihood function based on (9.5.2). The c.d.f. of the exponential distribution with mean θ is

$$F(x) = 1 - e^{-x/\theta} \quad \text{for} \quad x > 0.$$

An observed integer value $x > 0$ corresponds to a real interval $x \pm 0.5$, with probability

$$F(x + 0.5) - F(x - 0.5) = \exp(-\frac{x - 0.5}{\theta}) - \exp(-\frac{x + 0.5}{\theta})$$

$$= [\exp(\tfrac{1}{2\theta}) - \exp(-\tfrac{1}{2\theta})]\exp(-\tfrac{x}{\theta}).$$

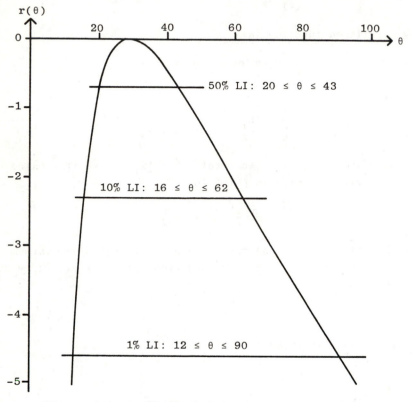

Figure 9.5.1

Log Relative Likelihood Function for the
Mean based on Ten Observations from an
Exponential Distribution

Hence by (9.5.2), the probability of observed values $x_1, x_2, \ldots, x_n$ is

$$P(E;\theta) = \prod_{i=1}^{n} [\exp(\tfrac{1}{2\theta}) - \exp(-\tfrac{1}{2\theta})]\exp(-x_i/\theta)$$

$$= [\exp(\tfrac{1}{2\theta}) - \exp(-\tfrac{1}{2\theta})]^n \exp(-\tfrac{1}{\theta}\textstyle\sum x_i).$$

The likelihood function is

$$L(\theta) = kP(E;\theta)$$

and we take $k = 1$ for convenience. The log likelihood function is

$$\ell(\theta) = n \log [\exp(\frac{1}{2\theta}) - \exp(-\frac{1}{2\theta})] - \frac{1}{\theta} \sum x_i,$$

and the solution of the equation $\ell'(\theta) = 0$ is

$$\hat{\theta} = [\log(\frac{\overline{x} + 0.5}{\overline{x} - 0.5})]^{-1}.$$

The exact log RLF is now $r(\theta) = \ell(\theta) - \ell(\hat{\theta})$.

For the ten observations given, we find that $\hat{\theta} = 28.797$, which is very close to our previous result ($\hat{\theta} = 28.800$). Table 9.5.1 compares the exact log r.l.f. with the approximate log r.l.f. which we obtained previously from (9.5.4). The agreement is extremely close over the range $12 \leq \theta \leq 100$ which includes all but the most implausible parameter values. As one might expect, the agreement becomes worse as θ becomes small; for then the p.d.f. changes more rapidly over a short interval, and the approximation (9.5.3) is less accurate.

Table 9.5.1
Comparison of Exact and Approximate Likelihoods
Based on Ten Observations from an Exponential Distribution

θ	Exact $r(\theta)$ based on (9.5.2)	Approx. $r(\theta)$ based on (9.5.4)	Difference (9.5.2)-(9.5.4)
5	-30.0745	-30.0906	+0.0161
10	- 8.2184	- 8.2221	+0.0037
12	- 5.2429	- 5.2453	+0.0024
15	- 2.6754	- 2.6767	+0.0013
20	- 0.7530	- 0.7536	+0.0006
25	- 0.1048	- 0.1050	+0.0002
40	- 0.4853	- 0.4850	-0.0003
60	- 2.1401	- 2.1397	-0.0004
80	- 3.8169	- 3.8165	-0.0004
100	- 5.3284	- 5.3279	-0.0005
200	-10.8199	-10.8194	-0.0005
300	-14.3946	-14.3941	-0.0005

More generally, if an observation x from an exponential distribution corresponds to a real interval $x \pm h$, the ratio of the exact probability (9.5.1) to the approximate probability (9.5.3) is

$$\frac{\exp(-\frac{x-h}{\theta}) - \exp(-\frac{x+h}{\theta})}{\frac{1}{\theta} \exp(-\frac{x}{\theta}) \cdot 2h} = \frac{e^c - e^{-c}}{2c} = 1 + \frac{c^2}{3!} + \frac{c^4}{5!} + \ldots,$$

where $c = \frac{h}{\theta}$ is the ratio of half the length of the measurement interval to the mean of the distribution. The approximation will be accurate whenever c is small.

Problems for Section 9.5

†1. A manufacturing process produces fibres of varying lengths. The length of a fibre is a continuous variate with p.d.f.

$$f(x) = \theta^{-2} x e^{-x/\theta} \quad \text{for} \quad x > 0$$

where $\theta > 0$ is an unknown parameter. Suppose that n randomly selected fibres have lengths $x_1, x_2, \ldots, x_n$. Find expressions for the MLE and RLF of θ.

2. Suppose that $x_1, x_2, \ldots, x_n$ are independent values from a normal distribution $N(\mu, 1)$. Find the MLE and RLF of μ.

3. Suppose that $x_1, x_2, \ldots, x_n$ are independent values from a normal distribution $N(0, \sigma^2)$. Find the MLE and RLF of σ.

4. (a) Suppose that U is a continuous variate, and that U/θ has a χ^2 distribution with n degrees of freedom. Find the p.d.f. of U, and show that $\hat{\theta} = U/n$.

   (b) Suppose that V is independent of U, and V/θ has a χ^2 distribution with m degrees of freedom. Find the joint p.d.f. of U and V, and show that the MLE of θ based on both U and V is $(U + V)/(n + m)$.

†5. The probability density function for a unit exponential distribution with guarantee time $c > 0$ is

$$f(x) = e^{c-x} \quad \text{for} \quad x \geq c.$$

Suppose that $x_1, x_2, \ldots, x_n$ are independent observations from this distribution.

   (a) Show that $\hat{c} = x_{(1)}$, the smallest observation, and find the RLF of c.

   (b) Find an expression for a $100p\%$ likelihood interval for c.

6. Suppose that $x_1, x_2, \ldots, x_n$ are independent observations from the continuous uniform distribution over the interval $[0, \theta]$. Show that the likelihood function of θ is proportional to θ^{-n} for $\theta \geq x_{(n)}$, and is zero otherwise. Hence determine the MLE and RLF of θ.

†7. Suppose that $x_1, x_2, \ldots, x_n$ are independent observations from the continuous uniform distribution over the interval $[\theta, 2\theta]$. Find the RLF of θ.

8. Suppose that X and Y are continuous variates with joint proba-
    bility density function

    $$f(x,y) = e^{-\theta x - y/\theta} \quad \text{for} \quad x > 0, \ y > 0.$$

    Find the MLE and RLF of θ on the basis of n independent pairs
    of observations (x_i, y_i), $i = 1, 2, \ldots, n$.

9. Independent measurements $x_1, x_2, \ldots, x_n$ are taken at unit time in-
    tervals. For $i = 1, 2, \ldots, \theta$ the measurements come from a stan-
    dardized normal distribution $N(0,1)$. A shift in the mean occurs
    after time θ, and for $i = \theta + 1, \theta + 2, \ldots, n$ the measurements
    come from $N(1,1)$.

    (a) Show that the likelihood function of θ is proportional to

    $$\exp\{-\sum_{i=1}^{\theta} (x_i - \tfrac{1}{2})\}.$$

    (b) Graph the log RLF for θ on the basis of the following set
         of 20 consecutive measurements:

    -1.26 -0.16 -0.64 0.56 -1.82 -0.76 -2.08 -0.58 0.14 0.94
    -0.58 0.78 1.80 0.58 0.02 0.86 2.30 1.80 0.84 -0.18

    Which values of θ have relative likelihood 10% or more?

9.6 Censoring in Lifetime Experiments

        In many experiments, the quantity of interest is the life-
time (or time to failure) of a specimen; for instance, the lifetime of
an electronic component, or the length of time until an aircraft component fails from metal fatigue, or the survival time of a cancer patient after a new treatment.

        The probability model generally assumes the lifetime X to
be a continuous variate with some particular probability density function f and cumulative distribution function F. For example, if we
thought that the chance of failure did not depend upon the age of the
specimen, we would assume an exponential distribution. Lifetime distributions for situations in which the risk of failure increases or decreases with age were considered in Section 6.4. The model will usually involve one or more unknown parameters θ which require estimation
from the data.

        Suppose that n specimens are tested independently. If the
experiment is continued sufficiently long for all of the items to have
failed, the likelihood function for θ based on the n observed lifetimes $x_1, x_2, \ldots, x_n$ can be obtained as in the last section. However,

one might wait a very long time indeed for all of the specimens to fail, and it is often desirable to analyse the data before this happens. One or two hardy specimens may tie up a laboratory for months or years without greatly adding to the information about θ, at the same time preventing other experiments from being undertaken. It often makes good practical sense to terminate the experiment before all n items have failed.

If the ith specimen has failed by the time the experiment terminates, we will know its lifetime x_i. This will actually corres-pond to a real interval $a_i < X \leq b_i$, say, with probability

$$P(a_i < X \leq b_i) = F(b_i) - F(a_i) \approx f(x_i)\Delta_i,$$

provided that the time interval $\Delta_i = b_i - a_i$ is small.

If the jth specimen has not failed when the experiment ends, we will not know its lifetime, and the lifetime is said to be censored. The censoring time T_j is the total time for which the specimen had been tested when the experiment ended. For this specimen, we know only that $T_j < X < \infty$, and the probability of this event is

$$P(T_j < X < \infty) = F(\infty) - F(T_j) = 1 - F(T_j).$$

The likelihood function of θ will be a product of n fac-tors, one for each specimen tested. Suppose that m specimens fail and $n - m$ do not, so that we have m failure times $x_1, x_2, \ldots, x_m$, and $n - m$ censoring times $T_1, T_2, \ldots, T_{n-m}$. Then the likelihood func-tion of θ will be proportional to

$$[\prod_{i=1}^{m} f(x_i)\Delta_i] \prod_{j=1}^{n-m} [1 - F(T_j)].$$

The Δ_i's do not depend upon θ and can be absorbed into the propor-tionality constant to give

$$L(\theta) = k [\prod_{i=1}^{m} f(x_i)] \prod_{j=1}^{n-m} [1 - F(T_j)], \qquad (9.6.1)$$

where k is any convenient positive constant. The maximum likelihood estimate and RLF can now be obtained.

Special Case: Exponential Distribution

If X is assumed to have an exponential distribution with

mean θ , then

$$f(x) = \frac{1}{\theta}\, e^{-x/\theta}; \quad F(x) = 1 - e^{-x/\theta} \quad \text{for} \quad x > 0.$$

In this case, (9.6.1) simplifies to give

$$L(\theta) = [\prod_{i=1}^{m} \frac{1}{\theta}\, e^{-x_i/\theta}]\prod_{j=1}^{n-m} e^{-T_j/\theta} = \theta^{-m} e^{-s/\theta}$$

where s is the total elapsed lifetime (time on test) for all n
items:

$$s = \sum_{i=1}^{m} x_i + \sum_{j=1}^{n-m} T_j.$$

The log likelihood function is

$$\ell(\theta) = -m \log \theta - \frac{s}{\theta},$$

and solving $\ell'(\theta) = 0$ gives $\hat{\theta} = \frac{s}{m}$. The log RLF is then

$$r(\theta) = \ell(\theta) - \ell(\hat{\theta}).$$

Example 9.6.1. Consider the experiment described in Example 9.5.1.
Suppose that the   n = 10   components were placed on test simultaneously,

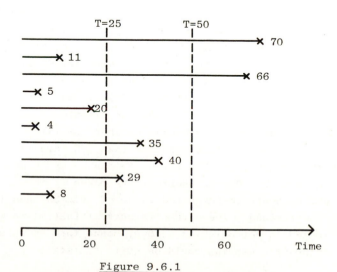

Figure 9.6.1
Diagrammatic Representation of Lifetime Data
Showing Two Possible Censoring Times

and it was decided to terminate the experiment after 50 days. The ten actual lifetimes are shown in Figure 9.6.1. If testing stopped at 50 days, everything to the right of 50 would be hidden from view, or censored. The data would then be

50+ 11 50+ 5 20 4 35 40 29 8

where 50+ indicates that the first and third lifetimes were censored at 50 days.

In the notation defined above, we have m = 8 lifetimes with total 11 + 5 + 20 + ... + 8 = 152, and n - m = 2 censoring times with total 50 + 50 = 100. The total elapsed lifetime for all 10 components is s = 152 + 100 = 252. Hence $\hat{\theta} = \frac{252}{8} = 31.5$, and

$$\ell(\theta) = -8 \log \theta - \frac{252}{\theta} .$$

If it had been decided to terminate the experiment after 25 days, the data would have been

25+ 11 25+ 5 20 4 25+ 25+ 25+ 8.

There are now m = 5 lifetimes with total 48, and n - m = 5 censoring times with total 125, giving s = 173 and $\hat{\theta} = 34.6$. The log likelihood function is now

$$\ell(\theta) = -5 \log \theta - \frac{173}{\theta}.$$

Figure 9.6.2 shows the three log relative likelihood functions resulting from (i) stopping the experiment after T = 25 days, (ii) stopping the experiment after T = 50 days, and (iii) continuing the experiment until all of the components have failed (i.e. stopping at time T > 70). The three functions agree reasonably well for $\theta \leq 30$, indicating that plausibilities of small parameter values are affected very little even when 50% of the lifetimes are censored. However, the three curves diverge considerably for large values of θ. With no censoring, values of θ greater than 62 are implausible (R < .1); with censoring at 25 days, θ can be as large as 108 before R decreases to 10%. Censoring thus makes it impossible to place as tight an upper bound on the value of θ, but has little effect on the lower bound. These results suggest that if we were primarily interested in establishing a lower bound for θ, a short experiment with heavy censoring could be quite satisfactory.

Note. In applications, the appropriate analysis will normally be that which corresponds to the pattern of censoring actually used in the ex-

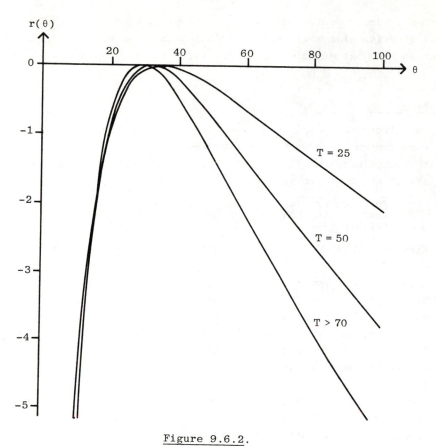

Figure 9.6.2.

Log Relative Likelihood Function for the Exponential
Mean θ under Various Levels of Censoring

periment. However, in some cases one might also wish to examine the likelihood function that would result from more severe censoring in order to see what effect a few large lifetimes have on the analysis.

Problems for Section 9.6

1. Ten electronic components with exponentially distributed lifetimes were tested for predetermined periods of time as shown. Three of the tubes survived their test periods, and the remaining seven failed at the times shown.

Tube number	1	2	3	4	5	6	7	8	9	10
Test period	81	72	70	60	41	31	31	30	29	21
Failure time	2	–	51	–	33	27	14	24	4	–

    Find the MLE and a 10% likelihood interval for the exponential mean θ.

†2. n electronic components were simultaneously placed on test. After a time T testing was stopped. It was observed that $n - k$ were still operating and that k had failed, but the times at which the failures had occurred were not known. Assuming that failure times follow an exponential distribution with mean θ, derive the maximum likelihood estimate and the relative likelihood function of θ.

3. A clinical trial was conducted to determine whether a hormone treatment benefits women who were treated previously for breast cancer. A woman entered the clinical trial when she had a recurrence. She was then treated by irradiation, and assigned to either a hormone therapy group or a control group. The observation of interest is the time until a second recurrence, which may be assumed to follow an exponential distribution with mean θ_H (hormone therapy group) or θ_C (control group). Many of the women did not have a second recurrence before the clinical trial was concluded, so that their recurrence times are censored. In the following table, a censoring time "n" means that a woman was observed for time n, and did not have a recurrence, so that her recurrence time is known to exceed n. Plot the log RLF's of θ_H and θ_C on the same graph. Is there any indication that the hormone treatment increases the mean time to recurrence?

	Hormone treated						Control					
Recurrence Times	2	4	6	9	9	9	1	4	6	7	13	24
	13	14	18	23	31	32	25	35	35	39		
	33	34	43									
Censoring Times	10	14	14	16	17	18	1	1	3	4	5	8
	18	19	20	20	21	21	10	11	13	14	14	15
	23	24	29	29	30	30	17	19	20	22	24	24
	31	31	31	33	35	37	24	25	26	26	26	28
	40	41	42	42	44	46	29	29	32	35	38	39
	48	49	51	53	54	54	40	41	44	45	47	47
	55	56					47	50	50	51		

†*4. The cumulative distribution function for the lifetime of a new
type of lightbulb is assumed to be

$$F(x) = 1 - (1 + \frac{2x}{\theta})e^{-2x/\theta} \quad \text{for} \quad x > 0.$$

(a) Find the probability density function, and show that mean of
this distribution is θ.

(b) Forty bulbs were tested and failures occurred at the following
times (in hours):

196	327	405	537	541	660	671	710	786
940	954	1004	1004	1006	1202	1459	1474	1484
1602	1662	1666	1711	1784	1796	1799		

The remaining bulbs had not failed when testing stopped at
1800 hours. Find the MLE and a 10% likelihood interval
for θ.

9.7 Invariance and Other Properties

In this section, we discuss some general properties of likelihood methods.

I. Likelihood methods are model-dependent

Throughout this chapter, we have assumed that some particular
probability model holds, and that it involves an unknown parameter θ.
Using the model, we found the probability of the observed event (data)
E as a function of θ, $P(E;\theta)$. From this, the maximum likelihood
estimate and relative likelihood function of θ were obtained.

All of these computations depend upon the particular model
which has been assumed. However, we may find (e.g. by the methods of
Chapter 11) that the probability model itself is contradicted by the
data. In this case, it often makes little sense to proceed with the
estimation of θ and determination of the range of plausible values.

Indeed, the parameter θ may no longer be a quantity of interest. Instead, we would look for a new model which was in better agreement with the data. Likelihood methods could then be used to make statements about any unspecified parameters in the new model.

II. Relative likelihoods are not additive

Relative likelihoods are similar to probabilities in that both lie between 0 and 1, and both provide objective measures of uncertainty. However, an important distinction between then is that, while probabilities of mutually exclusive events are additive, relative likelihoods of different parameter values are not additive.

Suppose that E_1 and E_2 are mutually exclusive events in the same sample space. Then from Section 3.1, the probability that one or the other of them occurs is given by

$$P(E_1 \text{ or } E_2) = P(E_1 \cup E_2) = P(E_1) + P(E_2).$$

The probability of a composite event is obtained by adding up the probabilities of its component parts.

On the other hand, suppose that θ_1 and θ_2 are two values from the parameter space. The relative likelihood of "θ_1 or θ_2" is not defined, and cannot be obtained from $R(\theta_1)$ and $R(\theta_2)$. In order to determine $R(\theta_1$ or $\theta_2)$, it would be necessary to find the probability of the data given that "$\theta = \theta_1$ or $\theta = \theta_2$", but in order to compute a probability one particular value of θ must be selected. (This is not so if the actual experiment is one of a sequence of experiments in which parameter values θ_1 and θ_2 occur with known probabilities. See Section 16.2.)

To ask for "the relative likelihood of θ_1 or θ_2" is like asking for "the height of Peter or Paul"; we cannot answer the question until we know which of them is meant.

It is generally not possible to combine relative likelihoods of different parameter values on the same data to obtain relative likelihoods for sets or intervals of parameter values. Thus, in Example 9.4.3, we may conclude that individual values of μ less than 0.29 are implausible, because each of them has relative likelihood less than 10%. However, we cannot measure how likely it is that $\mu < 0.29$. The statement that $0.41 < \mu < \infty$ is a 50% likelihood interval does not imply that there is a 50% probability (or indeed any probability) that μ is greater than 0.41. It means only that, in the light of the data, any value of μ greater than 0.41 is a reasonable guess at the value of μ.

When the likelihood function is based on a large number of independent repetitions of an experiment, it is possible to attach approximate probabilities to likelihood intervals. For instance, 50% and 10% likelihood intervals are roughly equivalent to 76% and 97% probability intervals (see Section 13.2). However, in general, there is no simple relationship between relative likelihoods and probabilities. The following example shows that in some cases the 50% LI will certainly contain the true parameter value, and in other cases it will almost certainly not.

Example 9.7.1. A deck of $n+m$ cards is made up of one card from each of $n+1$ denominations $0,1,2,\ldots,n$, and $m-1$ extra cards from some unknown demonination θ. One card is selected at random, and its denomination is found to be x. The problem is to estimate θ.

There are m cards of denomination θ, and one of each of the other n denominations. Hence the probability of drawing a card of denomination x is

$$P(x;\theta) = \begin{cases} \dfrac{m}{n+m} & \text{if } x = \theta; \\[2mm] \dfrac{1}{n+m} & \text{if } x = 0,1,\ldots,n; \quad x \neq \theta. \end{cases}$$

Since the observed value x has the greatest probability when $\theta = x$, the maximum likelihood estimate of θ is $\hat{\theta} = x$. The RLF of θ is

$$R(\theta) = \frac{P(x;\theta)}{P(x;\hat{\theta})} = \begin{cases} 1 & \text{if } \theta = x \\[2mm] \dfrac{1}{m} & \text{if } \theta = 0,1,\ldots,n; \quad \theta \neq x. \end{cases}$$

The observed denomination x is m times as plausible as any other value of θ, and is clearly the "best guess" at the value of θ, particularly when m is large. If $m \leq 2$, then $R(\theta) \geq 0.5$ for all values of θ. The 50% LI contains all possible values of θ, and hence it certainly contains the true value of θ. On the other hand, if $m > 2$, then $R(\theta) < 0.5$ except for $\theta = x$. The 50% LI contains only the single value $\theta = x$. But

$$P(X \neq \theta) = 1 - P(X = \theta) = 1 - \frac{m}{n+m} = \frac{n}{n+m} ,$$

which can be made arbitrarily close to one by taking n large. Hence it is possible to have a 50% LI which almost certainly does not contain the true parameter value. □

     The non-additivity of relative likelihoods of different para-
meter values on the same data is not necessarily a disadvantage (although it can be troublesome in the multiparameter case). Scientific
theories generally predict specific parameter values rather than intervals of values. Consequently, one is usually interested in assessing
the plausibilities of individual values of θ, and the relative likelihood function is appropriate for this purpose.

III. Likelihoods are invariant under 1-1 parameter transformations

     Suppose that the probability model for an experiment depends
upon an unknown parameter θ. The model then consists of a whole family of probability distributions, one for each value of θ in the parameter space Ω. For example, we might assume that the time to failure
of an electronic component has an exponential distribution, with probability density function

$$f(x) = \frac{1}{\theta} e^{-x/\theta} \quad \text{for} \quad 0 < x < \infty, \tag{9.7.1}$$

where θ is the expected lifetime. For each value of θ belonging to
Ω = (0,∞), we have a theoretical distribution. For instance, the distribution labelled by θ = 1 is

$$f(x) = e^{-x} \quad \text{for} \quad 0 < x < \infty, \tag{9.7.2}$$

and the distribution labelled by θ = 2 is

$$f(x) = \frac{1}{2} e^{-x/2} \quad \text{for} \quad 0 < x < \infty. \tag{9.7.3}$$

     A family of distributions can be parametrized (or labelled)
in many different ways. For instance, we could equally well write
(9.7.1) as

$$f(x) = \lambda e^{-\lambda x} \quad \text{for} \quad 0 < x < \infty$$

where $\lambda = \frac{1}{\theta}$ is the failure rate. Distributions (9.7.2) and (9.7.3)
are now labelled by λ = 1 and λ = 0.5, respectively. We have the
choice of labelling the family of exponential distributions by values
of θ, or by values of λ, or by values of any other one-to-one function of θ. We usually try to select a parametrization so that the
parameter represents some interesting characteristic of the distribu-

tion, and the mathematical expressions are fairly simple.

When we say that $\theta = 1$ is ten times as likely as $\theta = 2$, we imply that the distribution labelled by $\theta = 1$ is ten times as likely as the distribution labelled by $\theta = 2$. When we say that the maximum likelihood estimate of θ is $\hat{\theta} = 1.1$, we imply that the distribution labelled by $\theta = 1.1$ is the most likely distribution. Since the method of labelling the distributions is largely arbitrary, it would seem desirable that the plausibilities assigned to the distributions should not depend upon the particular method of labelling which has been selected. In other words, the plausibilities assigned should be _invariant_ under one-to-one transformations of the parameter.

An attractive property of the likelihood methods which we have discussed is that they are invariant under one-to-one parameter transformations. Suppose that the MLE and RLF of θ are $\hat{\theta}$ and $R(\theta)$, and define $\beta = g(\theta)$, where g is a one-to-one function. Then the MLE of β is given by

$$\hat{\beta} = g(\hat{\theta}), \qquad (9.7.4)$$

and the RLF of β is obtained by merely substituting $\theta = g^{-1}(\beta)$ in $R(\theta)$. Likelihood intervals for β can thus be obtained directly from the corresponding likelihood intervals for θ.

Example 9.7.2. In Example 9.5.1, we supposed that the lifetimes of electronic components were exponentially distributed, with mean lifetime θ. On the basis of ten observations, we found that $\hat{\theta} = 28.8$. The 50% LI for θ was $20 \le \theta \le 43$, and the 10% LI was $16 \le \theta \le 62$.

(a) Suppose that we are interested in the failure rate, $\lambda = \frac{1}{\theta}$. Then, by (9.7.4), the MLE of λ is

$$\hat{\lambda} = \frac{1}{\hat{\theta}} = \frac{1}{28.8} = 0.0347.$$

The 50% LI for λ is obtained by noting that $20 \le 1/\lambda \le 43$ if and only if $1/20 \ge \lambda \ge 1/43$. Hence the 50% LI is $0.023 \le \lambda \le 0.050$. Similarly, the 10% LI is found to be $0.016 \le \lambda \le 0.063$.

(b) Suppose that we are interested in the proportion β of such components which will last at least 25 days. Then

$$\beta = P(X \ge 25) = \int_{25}^{\infty} \frac{1}{\theta} e^{-x/\theta} dx = e^{-25/\theta}.$$

By (9.7.4), the MLE of β is

$$\hat{\beta} = e^{-25/\hat{\theta}} = 0.420.$$

Since $\theta = -25/\log\beta$, the 50% LI for β is given by

$$20 \leq -\frac{25}{\log\beta} \leq 43$$

and solving for β gives $0.287 \leq \beta \leq 0.559$. Similarly, the 10% LI is $0.210 \leq \beta \leq 0.668$.

Alternate analysis in (b). From Example 9.5.1, we see that 5 observations were greater than 25 days, and 5 were less than 25 days. Since components are independent, the probability that 5 out of 10 exceed 25 days is given by

$$P(E;\beta) = \binom{10}{5}\beta^5(1-\beta)^5; \quad 0 < \beta < 1.$$

Based on this, the MLE is $\hat{\beta} = 0.5$, and the log likelihood function of β is

$$\ell(\beta) = 5\log\beta + 5\log(1-\beta); \quad 0 < \beta < 1.$$

The log RLF can now be plotted. From this graph, we find the 50% LI to be $0.320 \leq \beta \leq 0.680$, and the 10% LI to be $0.196 \leq \beta \leq 0.804$.

The likelihood intervals obtained in this analysis are considerably wider than those found previously, indicating that the value of β is less precisely determined. The reason for this is that the second analysis is based only on the number of components which survived for 25 days or more, and not on the individual lifetimes which were observed. The second analysis thus uses only part of the available information, and the first analysis would generally be preferable. However, the second analysis does not depend upon the assumption of an exponential distribution, and it might be preferred if this assumption were in doubt.

Problems for Section 9.7

1. Let γ denote the median lifetime of electronic components in Example 9.5.1. Show that $\gamma = \theta\log 2$, and hence obtain the MLE and a 10% likelihood interval for γ.
2. We wish to estimate p, the probability of no diseased trees in a four-acre plot, in Problem 9.1.1. One approach would be to note

that 4 out of 10 plots contained no diseased trees, so that $\hat{p} = 0.4$ and $L(p) = p^4(1-p)^6$. A second approach would be to express p as a function of λ and use the invariance property of likelihood. Determine the MLE and a 10% likelihood interval for p by both methods. Under what conditions would the first method be preferable?

†3. The following table summarizes information concerning the lifetimes of one hundred V600 indicator tubes. (Ref: D.J. Davis, Journal of the American Statistical Association 47 (1952), 113-150).

Lifetime (hours)	0-100	100-200	200-300	300-400	400-600
Frequency observed	29	22	12	10	10

Lifetime (hours)	600-800	800+
Frequency observed	9	8

Suppose that the lifetimes follow an exponential distribution with mean θ.

(a) Show that the joint probability distribution of the frequencies is multinomial with probabilities

$$p_1 = P(0 < T < 100) = 1 - \beta; \quad p_2 = P(100 < T < 200) = \beta(1 - \beta); \quad \ldots;$$

$$p_7 = P(T > 800) = \beta^8, \quad \text{where} \quad \beta = e^{-100/\theta}.$$

(b) Show that $\hat{\beta}$ can be obtained as a root of a quadratic equation, and deduce the value of $\hat{\theta}$.

(c) Prepare a graph of the log RLF of β. Obtain 10% and 50% likelihood intervals for β, and transform them into likelihood intervals for θ.

*4. The arrivals of west-bound vehicles at a fixed point on an east-west road are random events in time. On the average there are μ arrivals per ten second interval. A traffic signal is to be installed a short distance beyond the observation point. It is desired that the signal remain at "STOP" for a time β such that the probability of holding up k or more vehicles is p.

(a) Show that $p = P(\chi^2_{(2k)} \leq \beta\mu/5)$. In particular, if $k = 8$ and $p = 0.05$, then $\beta = 39.8/\mu$.

(b) Assuming that $k = 8$ and $p = 0.05$, use the data in Problem 9.4.2 to determine $\hat{\beta}$ and a 10% likelihood interval for β.

9.8 Sufficient Statistics

Consider an experiment for which the probability model involves an unknown parameter θ. Let x denote a typical outcome. In most of the examples we shall consider, x will be a vector of n counts or measurements. The probability of outcome x will be a function of θ, $P(x;\theta)$, and the likelihood function $L(\theta)$ is proportional to $P(x;\theta)$. The likelihood function is defined only up to a multiplicative constant, and two likelihood functions which are proportional are considered to be the same.

A <u>statistic</u> T is a variate whose value $T(x)$ can be computed from the data x without knowledge of the value of θ. T is called a <u>sufficient statistic for</u> θ if knowledge of the observed value of T is sufficient to determine $L(\theta)$ up to a constant of proportionality. Thus, if T is a sufficient statistic, the probability of any outcome x can be written as a product,

$$P(x,\theta) = c(x) \cdot h(T(x);\theta) \tag{9.8.1}$$

where $c(x)$ is not a function of θ.

If t is a possible value of T, the probability of the event $T = t$ is obtained by summing (9.8.1) over all x such that $T(x) = t$. Since the second factor on the right hand side of (9.8.1) is constant in this sum, we obtain

$$P(T = t;\theta) = [\sum_{T(x)=t} c(x)] \cdot h(t;\theta) = d(t) \cdot h(t;\theta) \tag{9.8.2}$$

where $d(t)$ is not a function of θ. The likelihood function for θ based on (9.8.2) will be the same (up to a constant of proportionality) as that based on (9.8.1).

Now let x be an outcome such that $T(x) = t$. The ratio of (9.8.1) to (9.8.2) gives the conditional probability of outcome x given that $T(x) = t$:

$$P(x|T(x) = t) = \frac{c(x) \cdot h(t;\theta)}{d(t) \cdot h(t;\theta)} = \frac{c(x)}{d(t)}, \tag{9.8.3}$$

and this is not a function of θ. <u>The conditional distribution of outcomes given the value of a sufficient statistic for θ does not depend upon θ</u>. This property is sometimes used to define a sufficient statistic.

The likelihood function $L(\theta)$ summarizes the information provided by the data concerning θ, and a sufficient statistic T determines the likelihood function. Hence we say that <u>a sufficient statistic for θ carries all of the information about θ</u>. All that we need from the data to make inferences about θ is the observed value of T. Given T, the conditional distribution of outcomes does not depend upon θ, and thus gives no information about θ. This distribution is used for testing the goodness of fit of the model assumed; see Chapter 12.

Example 9.8.1. Suppose that n randomly chosen individuals are examined for tuberculosis (Example 9.1.1). We assume that people are independent, and that each has probability θ of having the disease. The aim is to obtain information about θ.

Define $X_i = 1$ if the ith person has tuberculosis (probability θ), and $X_i = 0$ otherwise (probability $1 - \theta$), so that

$$f(x_i) = \theta^{x_i}(1 - \theta)^{1-x_i}; \quad i = 1,2,\dots,n.$$

$$f(x_1,x_2,\dots,x_n) = \prod_{i=1}^{n} \theta^{x_i}(1 - \theta)^{1-x_i} = \theta^{\sum x_i}(1 - \theta)^{n-\sum x_i}.$$

The likelihood function for θ based on an observed sequence $(x_1,x_2,\dots,x_n)$ is thus

$$L(\theta) = \theta^{\sum x_i}(1 - \theta)^{n-\sum x_i} \quad \text{for} \quad 0 \le \theta \le 1.$$

All that we need know to write down $L(\theta)$ are n and $\sum x_i$. Since the sample size n is assumed to be known in advance, all that we require from the data is the sample total $\sum x_i$. Under the model, the variate $T \equiv \sum X_i$ is a sufficient statistic for θ, and carries all of the information about θ which is available from the sample.

Note that T is the total number of people who have the disease out of the n examined. The distribution of T is binomial:

$$P(T = t;\theta) = \binom{n}{t}\theta^t(1 - \theta)^{n-t} \quad \text{for} \quad t = 0,1,\dots,n.$$

The likelihood function for θ based on this distribution will be the same as that obtained above. In fact, we started the analysis with the binomial distribution in Example 9.1.1.

By (9.8.3), the conditional probability of a sequence $(x_1,x_2,\dots,x_n)$ given that $\sum x_i = t$ is $1/\binom{n}{t}$; that is, all $\binom{n}{t}$ possible sequences consisting of t 1's and $(n - t)$ 0's are equally probable. This conditional distribution does not depend upon θ. Given the number of 1's, no additional information about θ can be obtained by considering the order in which the 0's and 1's occurred.

Example 9.8.2. Let $X_1,X_2,\dots,X_n$ be independent Poisson variates with the same mean $\mu > 0$. Then

$$f(x_1,x_2,\ldots,x_n) = \prod_{i=1}^{n} \mu^{x_i} e^{-n\mu}/x_i! = \mu^{\sum x_i} e^{-n\mu}/(x_1!x_2!\ldots x_n!).$$

The likelihood function of μ is thus

$$L(\mu) = \mu^t e^{-n\mu} \quad \text{for} \quad \mu > 0$$

where $t = \sum x_i$. The variate $T \equiv \sum X_i$ is a sufficient statistic for μ , since knowledge of its observed value, together with n , is sufficient to determine $L(\mu)$.

By the Corollary to Example 4.5.5, T has a Poisson distribution with mean $n\mu$, and probability function

$$g(t) = (n\mu)^t e^{-n\mu}/t! \quad \text{for} \quad t = 0,1,2,\ldots .$$

The conditional distribution of the X_i 's given $T = t$ is

$$\frac{f(x_1,x_2,\ldots,x_n)}{g(t)} = \frac{t!}{x_1!x_2!\ldots x_n!} n^{-t}$$

where the x_i 's are non-negative integers and $\sum x_i = t$. This distribution does not depend upon μ . All of the information about μ is carried by the sufficient statistic T . The conditional distribution given T carries information about the adequacy of the model assumed, but not about μ .

Example 9.8.3. Suppose that $X_1,X_2,\ldots,X_n$ are independent variates having a uniform distribution on the interval $[0,\theta]$, where $\theta > 0$. Then each of the x_i 's has p.d.f.

$$f(x) = \theta^{-1} \quad \text{for} \quad 0 \le x \le \theta \tag{9.8.4}$$

and the joint p.d.f. of $X_1,X_2,\ldots,X_n$ is

$$f(x_1,x_2,\ldots,x_n) = \theta^{-n} \quad \text{for} \quad 0 \le x_i \le \theta; \quad i = 1,2,\ldots,n. \tag{9.8.5}$$

The likelihood function of θ is then

$$L(\theta) = \theta^{-n},$$

which appears not to depend upon the data at all. This would seem to imply that no information about θ can be obtained from the data - a conclusion that makes no sense intuitively, because clearly any observation x_i gives a lower bound for θ.

The difficulty here is that we have not completely specified the likelihood function, because we have not given the range of permissible values of θ. From (9.8.5), we have $\theta \geq x_i$ for $i = 1, 2, \ldots, n$, and this will be true if and only if $\theta \geq x_{(n)}$, where $x_{(n)}$ is the largest sample value. Thus

$$L(\theta) = \begin{cases} \theta^{-n} & \text{for} \quad \theta \geq x_{(n)} \\ 0 & \text{otherwise.} \end{cases} \tag{9.8.6}$$

The likelihood function depends only on the value of $X_{(n)}$, which is therefore a sufficient statistic for θ.

Sets of Sufficient Statistics

In each of the preceding three examples, there exists a single statistic which is sufficient for the unknown parameter. Example 9.8.6 considers a general family of distributions for which there is a single sufficient statistic.

In general, more than one function of the data may be needed to determine the likelihood function. If knowledge of the observed values of k statistics $T_1, T_2, \ldots, T_k$ is sufficient to determine $L(\theta)$ up to a proportionality constant, then $(T_1, T_2, \ldots, T_k)$ is called a set of sufficient statistics for θ. The likelihood function obtained from the joint distribution of $T_1, T_2, \ldots, T_k$ will be the same as that obtained from the original data. The conditional probability of the data given $T_1, T_2, \ldots, T_k$ does not depend upon θ. A set of sufficient statistics carries all of the information concerning θ, and permits the data to be condensed or reduced to k numbers without loss of information about θ.

A set of sufficient statistics which is as "small" as possible is said to be minimally sufficient. A set of minimally sufficient statistics gives the greatest possible reduction of the data such that all of the information about θ is preserved. Two different outcomes will lead to the same likelihood function if and only if they imply the same values for a set of minimally sufficient statistics.

Example 9.8.4. (a) Let $X_1, X_2, \ldots, X_n$ be independent exponential variates with the same mean θ. Their joint p.d.f. is

$$f(x_1, x_2, \ldots, x_n) = \prod_{i=1}^{n} \frac{1}{\theta} e^{-x_i/\theta} = \theta^{-n} e^{-\sum x_i/\theta}.$$

If the measurement intervals are all small (see Section 9.5), the likelihood function for θ is

$$L(\theta) = \theta^{-n}e^{-t/\theta} \quad \text{for} \quad \theta > 0$$

where $t = \sum x_i$. Assuming n to be known in advance, the total $T \equiv \sum X_i$ is a sufficient statistic for the unknown parameter θ. See Problem 6.9.5 for a derivation of the probability distribution of T.

(b) A more complicated situation was considered in Section 9.6. The lifetimes of n specimens were assumed to be independent exponential variates with mean θ, but censoring of lifetimes was permitted. The likelihood function then has the form

$$L(\theta) = \theta^{-m}e^{-s/\theta} \quad \text{for} \quad \theta > 0$$

where m is the number of specimens which fail, and s is the sum of the m failure times and $n - m$ censoring times. We would not know m until after the experiment was completed. Thus, in this case we need the observed values of two statistics, M (the number of failures), and S (the total time on test), before we can write down $L(\theta)$. Under the exponential model with censoring, we have a pair of sufficient statistics (M,S). Neither M nor S is itself a sufficient statistic.

Example 9.8.5. Let $X_1, X_2, \ldots, X_n$ be independent variates having a Cauchy distribution centred at θ. The p.d.f. of this distribution is

$$f(x) = \frac{1}{\pi[1 + (x - \theta)^2]} \quad \text{for} \quad -\infty < x < \infty.$$

The joint p.d.f. of $X_1, X_2, \ldots, X_n$ is

$$f(x_1, x_2, \ldots, x_n) = \pi^{-n} \prod_{i=1}^{n} [1 + (x_i - \theta)^2]^{-1},$$

and the likelihood function of θ is

$$L(\theta) = \prod_{i=1}^{n} [1 + (x_i - \theta)^2]^{-1} \quad \text{for} \quad -\infty < \theta < \infty.$$

We can write down the likelihood function if we know the set of observed values $\{x_i\}$; it is not necessary to know the order in which they occurred. Let $X_{(i)}$ denote the ith smallest of $X_1, X_2, \ldots, X_n$, so that

$$X_{(1)} \le X_{(2)} \le X_{(3)} \le \cdots \le X_{(n)}.$$

The order statistics $X_{(1)}, X_{(2)}, \ldots, X_{(n)}$ form a set of sufficient statistics for θ. It can be shown that no further reduction of the data is possible in this case. The best that we can do is an n-dimensional set of sufficient statistics.

Example 9.8.6.   The Exponential Family

Suppose that X is a variate whose p.f. or p.d.f. depends upon a single parameter θ, and is of the form

$$f(x;\theta) = A(\theta) \cdot B(x) \cdot e^{c(\theta) \cdot d(x)} \quad \text{for} \quad -\infty < x < \infty,$$

where A, B, c, and d are known functions. The distribution of X is then said to belong to the exponential family of distributions.

Several of the one-parameter distributions which we have considered are members of the exponential family. For example, the binomial p.f. can be written

$$f(x;\theta) = \binom{n}{x}\theta^x(1-\theta)^{n-x} = (1-\theta)^n\binom{n}{x}e^{x \log\{\theta/(1-\theta)\}},$$

which is of the exponential form with

$$A(\theta) = (1-\theta)^n; \quad c(\theta) = \log\frac{\theta}{1-\theta}; \quad d(x) = x;$$

$$B(x) = \binom{n}{x} \quad \text{for} \quad x = 0, 1, \ldots, n \quad \text{and} \quad B(x) = 0 \quad \text{otherwise.}$$

Similarly, the Poisson, exponential, and χ^2 distributions are members of the exponential family.

If $X_1, X_2, \ldots, X_n$ are independent and identically distributed variates having a distribution which belongs to the exponential family, their joint p.f. or p.d.f. is

$$f(x_1, \ldots, x_n) = [A(\theta)]^n [\prod_{i=1}^{n} B(x_i)] \cdot \exp\{c(\theta) \sum_{i=1}^{n} d(x_i)\}.$$

The likelihood function is then

$$L(\theta) = [A(\theta)]^n \exp\{c(\theta) \sum_{i=1}^{n} d(x_i)\},$$

and hence there exists a single sufficient statistic $T \equiv \sum d(X_i)$. Because of this, problems of statistical inference are easier for distributions belonging to the exponential family.

The definition of the exponential family can be extended to include distributions which depend upon several parameters $\theta_1, \theta_2, \ldots, \theta_r$. Details may be found in Chapter 2 of *Theoretical Statistics* by D.R. Cox and D.V. Hinkley.

One-to-one transformations

Let $(T_1, T_2, \ldots, T_k)$ be a set of sufficient statistics for θ. Suppose that $U_1, U_2, \ldots, U_k$ are functions of the T_i's, and that the transformation $(T_1, T_2, \ldots, T_k) \rightarrow (U_1, U_2, \ldots, U_k)$ is one-to-one. Because the transformation is reversible, the information content of the U_i's will be exactly the same as that of the T_i's. If we know the values of the U_i's, we can deduce the values of the T_i's, and hence determine $L(\theta)$. Therefore, the U_i's also form a set of sufficient statistics for θ. If we apply a one-to-one transformation to a set of sufficient statistics, we get another set of sufficient statistics.

In Example 9.8.1, the MLE of $\hat{\theta}$ is $\hat{\theta} \equiv T/n$. This is a one-to-one function of T, and is therefore also a sufficient statistic for θ. Since n is given, the values of T and $\hat{\theta}$ can be deduced from one another, and they have the same information content. Note that

$$L(\theta) = \theta^{n\hat{\theta}}(1 - \theta)^{n(1-\hat{\theta})} \quad \text{for} \quad 0 \le \theta \le 1,$$

so that knowledge of $\hat{\theta}$ is sufficient to determine $L(\theta)$. Similarly, the MLE is a sufficient statistic for the unknown parameter in Examples 9.8.2, 9.8.3, and 9.8.4(a).

In Example 9.8.4(b), we considered exponential lifetimes with censoring. The MLE of θ is then $\hat{\theta} \equiv S/M$, and

$$L(\theta) = \theta^{-M} e^{-M\hat{\theta}/\theta} \quad \text{for} \quad \theta > 0.$$

Hence $(\hat{\theta}, M)$ is a pair of sufficient statistics for θ. However, note that $\hat{\theta}$ by itself is not a sufficient statistic for θ.

Any set of sufficient statistics determines $L(\theta)$ and hence $\hat{\theta}$. Thus $\hat{\theta}$ is a function of any set of sufficient statistics for θ. In several examples, we found that $\hat{\theta}$ was itself a sufficient statistic, but Examples 9.8.4(b) and 9.8.5 show that this is not always the case. However, one can transform any set of sufficient statistics into a set $(\hat{\theta}, T_2, \ldots, T_k)$ which includes $\hat{\theta}$. We can think of $\hat{\theta}$ as the primary source of information concerning θ, while $T_2, \ldots, T_k$ provide supplementary information; see Section 12.8. In large samples, this supplementary information becomes negligible, and then $\hat{\theta}$ is approximately sufficient; see Section 13.3.

Problems for Section 9.8

1. Suppose that X has a binomial distribution with parameters (n,θ), and that Y is independent of X and has a binomial distribution with parameters (m,θ), where m and n are known. Show that $T \equiv X + Y$ is a sufficient statistic for θ, and verify that the conditional distribution of X and Y given T does not depend upon θ.

2. The number of deaths from lung cancer over a one-year period among men aged 50-54 years was recorded for each of n Ontario counties. The number D_j of deaths in the jth county is assumed to have a Poisson distribution with mean $\mu_j = \theta p_j$, where p_j is the (known) population of the county, and θ is an unknown constant. Show that $D \equiv \sum D_j$ is a sufficient statistic for θ, and find its probability distribution.

†3. A manufacturing process produces fibres of varying lengths. The length X of such a fibre is assumed to be a continuous variate with p.d.f.

$$f(x) = 2\lambda x e^{-\lambda x^2} \quad \text{for} \quad x > 0$$

where $\lambda > 0$. Suppose that n fibres are selected at random and their lengths $X_1, X_2, \ldots, X_n$ are determined.

(a) Show that $T \equiv \sum X_i^2$ is a sufficient statistic for λ.

(b) Show that $2\lambda X^2$ has a χ^2 distribution with 2 degrees of freedom, and hence that $2\lambda T \sim \chi^2_{(2n)}$. Find the p.d.f. of T, and show that it gives rise to the same likelihood function for λ as the original sample.

4. Show that $\overline{X}$ is a sufficient statistic for the Poisson mean μ in Problem 9.3.2.

5. Let $X_1, X_2, \ldots, X_n$ be independent variates having a continuous uniform distribution on the interval $(\theta, \theta+1)$. Show that $X_{(1)}$ and $X_{(n)}$ form a pair of sufficient statistics for θ.

†6. Let $X_1, X_2, \ldots, X_n$ be independent variates having a continuous uniform distribution on the interval $(-\theta, \theta)$. Find a sufficient statistic for θ.

7. Show that the Poisson, exponential, and χ^2 distributions are members of the exponential family.

8. Show that the normal distributions $N(0, \sigma^2)$ and $N(\mu, 1)$ are members of the exponential family.

9. Suppose that the distribution of X belongs to the exponential family, and Y is a one-to-one function of X. Show that the distribution of Y also belongs to the exponential family.

†10. Suppose that the distribution of X belongs to the exponential family. The parameter $\phi = c(\theta)$ is called the natural parameter of the distribution. Find the natural parameter for the binomial, Poisson, and exponential distributions.

*9.9 Long-run Properties

At the beginning of this chapter, we considered the method of maximum likelihood for obtaining an estimate of an unknown parameter θ from the experimental data. The maximum likelihood estimate $\hat{\theta}$ is the value of θ which best explains the data, in the sense that it maximizes the probability of what has been observed. Because of this direct interpretation in terms of probability, the method of maximum likelihood has considerable intuitive appeal, and the invariance property discussed in the last section is also attractive.

A great deal of attention has been given in the statistical literature to the long-run properties of parameter estimates. One imagines that the experiment is to be repeated over and over, and that each time an estimate of θ is to be computed in the same way. Even if the value of θ is the same in all repetitions, the estimate obtained will change owing to random variation in the data. The estimate t which we compute in one repetition of the experiment is thus just one possible value of a random variable T. The variate T is sometimes called an estimator.

If the method of estimation has good long-run properties, the estimates obtained in many repetitions of the experiment should be clustered tightly about the value of the parameter θ. This implies that the probability distribution of the estimator T should be centred near θ, and should have a small spread.

A convenient measure of the "centre" of a distribution is the mean, $E(T)$. The difference between $E(T)$ and θ is called the bias of the estimator T:

$$\text{Bias} = E(T) - \theta. \tag{9.9.1}$$

In particular, T is said to be an unbiased estimator of θ if

---

* This section may be omitted on first reading.

$E(T) = \theta$ for all θ.

The spread of a distribution is usually measured by the second moment or variance. The second moment of the variate $T - \theta$ is called the <u>mean squared error</u> of the estimator T:

$$MSE = E\{(T - \theta)^2\}. \qquad (9.9.2)$$

If this is small, the estimates obtained in many repetitions of the experiment will be clustered near θ. It is sometimes suggested that one should select an estimator T such that the mean squared error is minimized.

If T is an unbiassed estimator, then $E(T) = \theta$, and hence the mean squared error is equal to the variance of T. An estimator T which is unbiassed for θ and has the smallest possible variance is called <u>MVU</u> (minimum variance unbiassed).

We now give several examples to illustrate some general results concerning unbiassed estimation.

<u>Example 9.9.1</u>. If X is the number of successes obtained in n independent trials with success probability θ, the estimator of θ obtained by the method of maximum likelihood is $T \equiv X/n$ (Example 9.1.1). Since $E(X) = n\theta$, we have

$$E(T) = \frac{1}{n} E(X) = \frac{1}{n}(n\theta) = \theta,$$

and hence T is an unbiassed estimator of θ.

By the invariance property (Section 7), the maximum likelihood estimator of θ^2 is $T^2 \equiv X^2/n^2$, with expected value

$$E(T^2) = E(X^2)/n^2 = [var(X) + E(X)^2]/n^2$$

by (5.2.3). Since $var(X) = n\theta(1 - \theta)$, we have

$$E(T^2) = [n\theta(1 - \theta) + n^2\theta^2]/n^2 = \theta^2 + \frac{\theta(1 - \theta)}{n} .$$

Hence T^2 is not an unbiassed estimator of θ^2. The bias is

$$E(T^2) - \theta^2 = \frac{\theta(1 - \theta)}{n}$$

which is positive for $0 < \theta < 1$, and tends to zero as $n \to \infty$.

<u>Example 9.9.2</u>. If $X_1, X_2, \ldots, X_n$ are independent Poisson variates with mean μ, the maximum likelihood estimator of μ is $\overline{X} \equiv \sum X_i/n$

(Example 9.1.2). Since $E(X_i) = \mu$, we have

$$E(\overline{X}) = \frac{1}{n} \sum E(X_i) = \frac{1}{n}(n\mu) = \mu,$$

and hence $\overline{X}$ is an unbiassed estimator of μ.

By the invariance property, the maximum likelihood estimator of $\beta = e^{-\mu}$ is $e^{-\overline{X}}$. Since $T \equiv \sum X_i$ has a Poisson distribution with mean $n\mu$ (corollary to Example 4.5.5), the expected value of $e^{-\overline{X}}$ is

$$E(e^{-T/n}) = \sum_{t=0}^{\infty} e^{-t/n}(n\mu)^t e^{-n\mu}/t! = e^{-n\mu}\sum (n\mu e^{-1/n})^t/t!$$

$$= e^{-n\mu} \cdot e^{n\mu e^{-1/n}} = e^{-\mu(n-ne^{-1/n})} = \beta^{n(1-e^{-1/n})}.$$

Hence $e^{-\overline{X}}$ is not an unbiassed estimator of β. The bias is

$$\beta^{n(1-e^{-1/n})} - \beta$$

which is always positive, and tends to zero as $n \to \infty$.

Example 9.9.3. Suppose that $X_1, X_2, \ldots, X_n$ are independent variates with the same mean μ and variance σ^2. A linear estimator of μ is a linear combination of the X_i's, $T \equiv \sum a_i X_i$, where the a_i's are constants. Show that the sample mean $\overline{X}$ is the unique MVU linear estimator of μ.

Solution. By (5.4.11) and (5.4.12), the mean and variance of T are

$$E(T) = \mu \sum a_i; \quad \text{var}(T) = \sigma^2 \sum a_i^2.$$

Thus T is unbiassed for μ if and only if $\sum a_i = 1$. Now it is easy to show that

$$\sum a_i^2 = \sum (a_i - \overline{a})^2 + n\overline{a}^2$$

where $\overline{a} = \frac{1}{n} \sum a_i$. If $\sum a_i = 1$, then $\overline{a} = \frac{1}{n}$ and $\sum a_i^2$ is minimized for $a_1 = a_2 = \ldots = a_n = \frac{1}{n}$. Hence the unbiassed linear estimator of μ with smallest variance is

$$T \equiv \frac{1}{n}X_1 + \frac{1}{n}X_2 + \ldots + \frac{1}{n}X_n \equiv \frac{1}{n}\sum X_i \equiv \overline{X}.$$

Example 9.9.4. Let X be the number of successes before the first failure in independent trials with success probability θ. Define $T(x) = 0$ for $x = 0$, and $T(x) = 1$ for $x \geq 1$. Show that T is the unique unbiassed estimator of θ.

Solution. The distribution of X is geometric:

$$f(x) = \theta^x(1 - \theta) \quad \text{for} \quad x = 0,1,2,\ldots \quad .$$

The expected value of T is

$$
\begin{aligned}
E(T) &= 0 \cdot P(T = 0) + 1 \cdot P(T = 1) \\
&= P(X \geq 1) = 1 - f(0) = \theta.
\end{aligned}
$$

Hence T is an unbiassed estimator of θ.

Now suppose that T' is another unbiassed estimator, and define $U(x) = T'(x) - T(x)$ for $x = 0,1,\ldots$. Then

$$E(U) = E(T') - E(T) = \theta - \theta = 0 \quad \text{for all} \quad \theta.$$

Also by (5.1.3) we have

$$E(U) = \sum U(x)\theta^x(1 - \theta).$$

It follows that

$$U(0) + U(1)\theta + U(2)\theta^2 + U(3)\theta^3 + \ldots = 0$$

for all values of θ between 0 and 1. This will be true if and only if $U(0) = U(1) = U(2) = \ldots = 0$. Hence $T(x) = T'(x)$ for $x = 0,1,2,\ldots$, and T is the unique unbiassed estimator of θ.

Discussion

Examples 9.9.1 and 9.9.2 show that the criterion of unbiassedness is not invariant under one-to-one parameter changes. If T is an unbiassed estimator of θ, then $g(T)$ will generally not be an unbiassed estimator of $g(\theta)$ unless g is a linear transformation. It is not possible to require both unbiassedness and invariance.

Since maximum likelihood estimators are invariant under one-to-one parameter changes, they cannot in general be unbiassed. However, it can be shown that under suitable regularity conditions, the bias of maximum likelihood estimators tends to zero as $n \to \infty$.

The only way to achieve unbiassedness in Example 9.9.4 is to estimate the success probability to be 0 if a failure occurs on the first trial, and to be 1 if a success occurs on the first trial. This is not a very sensible estimation procedure. If we observe a success followed by a failure, it would be reasonable to guess that θ

was near 0.5, but to achieve the correct long-run average, we must estimate θ to be 1.

It may make sense to require an estimator with small bias in some applications. However, the requirement of unbiassedness is too strong, and as Example 9.9.4 shows, this requirement may eliminate all "sensible" estimation procedures. There is an extensive literature on the theory of unbiassed estimators. Although this is of mathematical interest, it does not seem terribly relevant from a practical point of view.

The problem of selecting a "best estimator" T for a parameter θ is not well-defined unless we know the precise use which is to be made of T, and can specify the loss which would result from underestimating or overestimating θ by various amounts. In minimizing the mean squared error (9.9.2), we are treating overestimation and underestimation as being equally serious. There are many situations where these are not equally important, and it would not be appropriate to minimize (9.9.2). Furthermore, one would not necessarily be interested in minimizing losses over a sequence of repetitions of the experiment in which θ was fixed. The choice of a "best estimator" is really a problem in decision theory rather than statistical inference, and we shall not consider it further here.

Problems for Section 9.9

1. Suppose that $X_1, X_2, \ldots, X_n$ are independent variates with the same mean μ and variance σ^2. Show that

$$\sum (X_i - \overline{X})^2 \equiv \sum X_i^2 - n\overline{X}^2$$

and hence verify that

$$s^2 \equiv \frac{1}{n-1} \sum (X_i - \overline{X})^2$$

is an unbiassed estimator of σ^2. Is S an unbiassed estimator of σ?

†2. Suppose that $X_1, X_2, \ldots, X_n$ are independent variates with the same mean μ but with different variances $\sigma_1^2, \sigma_2^2, \ldots, \sigma_n^2$. Find the minimum variance linear unbiassed estimator of μ.

CHAPTER 10. TWO-PARAMETER LIKELIHOODS

In this chapter we discuss likelihood methods for the case in which the probability model involves more than one unknown parameter. Section 1 gives the basic definitions, while Sections 2, 3, and 4 discuss specific examples from life testing, dosage mortality trials, and learning theory. Methods of eliminating unwanted parameters from the likelihood function are discussed in Section 5. Although attention is restricted primarily to models with two parameters, much of the discussion also applies when there are three or more.

10.1 Introduction

Suppose that the probability model for an experiment involves two unknown parameters α and β, say. The probability of the observed event (data) E will then generally be a function of both α and β, $P(E;\alpha,\beta)$. The (joint) <u>likelihood function</u> of α and β is proportional to this probability:

$$L(\alpha,\beta) = kP(E;\alpha,\beta),$$

where k is positive and does not depend upon α or β.

The (joint) <u>maximum likelihood estimate</u> of α and β is the pair of parameter values $(\hat{\alpha},\hat{\beta})$ which maximizes $L(\alpha,\beta)$. It can usually be found by solving the simultaneous equations

$$\frac{\partial \ell}{\partial \alpha} = \frac{\partial \ell}{\partial \beta} = 0, \tag{10.1.1}$$

where $\ell(\alpha,\beta) = \log L(\alpha,\beta)$ is the log likelihood function. The maximum likelihood equations (10.1.1) hold only at a point of relative maximum. Other methods are needed to determine a maximum point which occurs on a boundary of the parameter space, or when one of the parameters is restricted to a discrete set of values.

The (joint) relative likelihood function of α and β is defined by

$$R(\alpha,\beta) = L(\alpha,\beta)/L(\hat{\alpha},\hat{\beta}),$$

so that $0 \leq R(\alpha,\beta) \leq 1$ for all pairs of parameter values (α,β). As in the one-parameter case, we have

$$R(\alpha_1, \beta_1) = \frac{\text{Probability of data when } \alpha = \alpha_1 \text{ and } \beta = \beta_1}{\text{Maximum probability of data for any } \alpha, \beta}.$$

If this is small, then the pair of parameter values (α_1, β_1) is im-plausible, because there exist other pairs of parameter values such that the data are much more probable. This does not mean that α_1 is an implausible value of α, or that β_1 is an implausible value of β, but only that it is unreasonable to <u>simultaneously</u> estimate α by α_1 and β by β_1. The joint RLF is appropriate for making statements about both parameters together. In Section 5 we shall consider the problem of eliminating one of the parameters in order that statements can be made about the other parameter by itself.

As an aid to the interpretation of $R(\alpha, \beta)$, we may graph contours of constant relative likelihood in the (α, β) plane; that is, we plot the curve $R(\alpha, \beta) = c$ for selected values of c (e.g. $c = .01, .1, .5, 1$). This produces a contour map of the "mountain" of relative likelihood. The contours usually form a nested set of closed curves, roughly elliptical in shape. Several examples of contour maps are given in the following sections.

Sufficient Statistics

A set of statistics $T_1, T_2, \ldots, T_k$ is said to be sufficient for the parameters α and β if knowledge of the observed values of $T_1, T_2, \ldots, T_k$ is sufficient to determine $L(\alpha, \beta)$ up to a proportiona-lity constant. If we take θ to be a vector of parameters, $\theta = (\alpha, \beta)$, then the discussion of Section 9.8 carries over directly to the two-parameter case. In particular, the MLE $(\hat{\alpha}, \hat{\beta})$ may be determined from any set of sufficient statistics. In some simple situations, $(\hat{\alpha}, \hat{\beta})$ is itself sufficient for (α, β); see Example 10.1.1. In more compli-cated problems, it is necessary to supplement $(\hat{\alpha}, \hat{\beta})$ with additional functions of the data in order to obtain a set of sufficient statistics.

Example 10.1.1. Sample of size n from a normal distribution

Suppose that n measurements $x_1, x_2, \ldots, x_n$ are assumed to be independent values from a normal distribution with unknown mean μ and standard deviation σ. The p.d.f. of $N(\mu, \sigma^2)$ is

$$f(x) = \frac{1}{\sqrt{2\pi}\,\sigma} \exp\{-(x-\mu)^2/2\sigma^2\} \quad \text{for } -\infty < x < \infty.$$

If all of the measurement intervals are small (see Section 9.5), the likelihood function can be taken to be proportional to the product of probability densities,

$$L(\mu,\sigma) = k \prod_{i=1}^{n} f(x_i) = k \prod_{i=1}^{n} \frac{1}{\sqrt{2\pi}\,\sigma} \exp\{-(x_i - \mu)^2/2\sigma^2\},$$

where $\sigma > 0$ and $-\infty < \mu < \infty$. Hence, by a convenient choice of k,

$$L(\mu,\sigma) = \sigma^{-n} \exp\{-\textstyle\sum(x_i - \mu)^2/2\sigma^2\},$$

and the log likelihood function is

$$\ell(\mu,\sigma) = -n \log \sigma - \frac{1}{2\sigma^2} \sum(x_i - \mu)^2.$$

The first partial derivatives are

$$\frac{\partial \ell}{\partial \mu} = -\frac{1}{2\sigma^2} \sum 2(x_i - \mu)(-1) = \frac{1}{\sigma^2} \sum(x_i - \mu) = \frac{1}{\sigma^2} (\textstyle\sum x_i - n\mu);$$

$$\frac{\partial \ell}{\partial \sigma} = -\frac{n}{\sigma} + \frac{1}{\sigma^3} \sum(x_i - \mu)^2 = \frac{1}{\sigma^3} [\textstyle\sum(x_i - \mu)^2 - n\sigma^2].$$

The maximum likelihood estimate $(\hat{\mu},\hat{\sigma})$ satisfies the equations

$$\sum x_i - n\hat{\mu} = 0; \qquad \sum(x_i - \hat{\mu})^2 - n\hat{\sigma}^2 = 0,$$

and solving these gives

$$\hat{\mu} = \frac{\sum x_i}{n} = \bar{x}; \qquad \hat{\sigma}^2 = \frac{1}{n} \sum(x_i - \hat{\mu})^2 = \frac{1}{n} \sum(x_i - \bar{x})^2.$$

We therefore have

$$\hat{\mu} = \bar{x}; \qquad \hat{\sigma} = \sqrt{\frac{1}{n} \sum(x_i - \bar{x})^2}.$$

At first glance, it appears that the likelihood function depends upon all of the individual observed values $x_1, x_2, \ldots, x_n$. We now show that, in fact, only two functions of the data are needed. Since

$$(x_i - \mu)^2 = (x_i - \bar{x} + \bar{x} - \mu)^2$$
$$= (x_i - \bar{x})^2 + (\bar{x} - \mu)^2 + 2(x_i - \bar{x})(\bar{x} - \mu),$$

it follows that

$$\sum (x_i - \mu)^2 = \sum (x_i - \overline{x})^2 + n(\overline{x} - \mu)^2 + 2(\overline{x} - \mu)\sum (x_i - \overline{x}).$$

Since $\overline{x} = \frac{1}{n} \sum x_i$, we have

$$\sum (x_i - \overline{x}) = \sum x_i - n\overline{x} = 0$$

and therefore

$$\sum (x_i - \mu)^2 = \sum (x_i - \overline{x})^2 + n(\overline{x} - \mu)^2 = n\hat{\sigma}^2 + n(\hat{\mu} - \mu)^2. \qquad (10.1.2)$$

Hence the likelihood function may be written

$$L(\mu,\sigma) = \sigma^{-n} \exp\{- \frac{n}{2\sigma^2}[\hat{\sigma}^2 + (\hat{\mu} - \mu)^2]\}$$

for $-\infty < \mu < \infty$ and $\sigma > 0$. Assuming that n is known in advance, knowledge of $\hat{\mu}$ and $\hat{\sigma}$ is sufficient to determine the likelihood function. Hence $(\hat{\mu},\hat{\sigma})$ is a pair of sufficient statistics for (μ,σ).

The maximum of the likelihood function is

$$L(\hat{\mu},\hat{\sigma}) = \hat{\sigma}^{-n} \exp\{- \frac{n}{2}\} \qquad (10.1.3)$$

and hence the joint relative likelihood function of μ and σ is

$$R(\mu,\sigma) = \frac{L(\mu,\sigma)}{L(\hat{\mu},\hat{\sigma})} = (\frac{\hat{\sigma}}{\sigma})^n \exp\{\frac{n}{2}[1 - (\frac{\hat{\sigma}}{\sigma})^2 - (\frac{\hat{\mu} - \mu}{\sigma})^2]\}. \qquad (10.1.4)$$

One could plot contours of constant relative likelihood in order to determine regions of plausible values for (μ,σ). However, more convenient methods for making inferences about μ and σ are available, and these will be discussed in Chapter 14.

Computation

In some examples, such as the one just considered, it is possible to obtain an algebraic solution to the maximum likelihood equations (10.1.1). In other cases, it may be possible to eliminate one parameter from the equations, and then use the methods of Section 9.2 to solve the remaining equation. Sometimes a one-to-one transformation of the parameters will help to put the equations in a more manageable form.

If the equations do not simplify, one can often use the Newton-Raphson method, which is a generalization of Newton's method (Section 9.2). We shall describe this in matrix notation, so that the extension to three or more parameters is immediate. We begin with an initial approximation (α_0, β_0), and repeatedly improve it. If (α_i, β_i) is the approximation to $(\hat{\alpha}, \hat{\beta})$ which was obtained at the ith iteration, then

$$(\alpha_{i+1}, \beta_{i+1}) = (\alpha_i, \beta_i) - (\frac{\partial \ell}{\partial \alpha}, \frac{\partial \ell}{\partial \beta})C^{-1}$$

where $C = (c_{jk})$ is the matrix of second order partial derivatives:

$$c_{11} = \frac{\partial^2 \ell}{\partial \alpha^2} ; \quad c_{12} = c_{21} = \frac{\partial^2 \ell}{\partial \alpha \partial \beta} ; \quad c_{22} = \frac{\partial^2 \ell}{\partial \beta^2} .$$

All derivatives are evaluated at $\alpha = \alpha_i$ and $\beta = \beta_i$.

When the derivatives are too complicated, or the likelihood function is not well behaved near the maximum, $(\hat{\alpha}, \hat{\beta})$ can be evaluated by a double application of the methods described in Section 9.2. Note that

$$\max_{\alpha, \beta} \ell(\alpha, \beta) = \max_{\alpha} \{ \max_{\beta} \ell(\alpha, \beta) \} = \max_{\alpha} g(\alpha)$$

where g is a function of α only:

$$g(\alpha) = \max_{\beta} \ell(\alpha, \beta).$$

We first write a computer programme to evaluate $g(\alpha)$ for any given value of α. This involves maximizing over β only. Once we are able to evaluate $g(\alpha)$, we can obtain its maximum graphically or by repeated bisection.

Three or more parameters

The MLE and RLF can be defined in a similar way when there are more than two parameters. However, it is more difficult to display and interpret the relative likelihood function.

If there are three parameters α, β, γ, there will generally be one of them, say γ, which is of little interest. We can then choose several values $\gamma_1, \gamma_2, \ldots$ in the neighbourhood of $\hat{\gamma}$ and plot a contour map of $R(\alpha, \beta, \gamma_i)$ for each γ_i. Such a procedure would have been prohibitively time-consuming only a few years ago, but is quite feasible now with high-speed computers and plotting facilities.

A single scientific experiment may give information about many unknown parameters $\alpha, \beta, \gamma, \delta, \ldots$. Indeed, the great advantage of using a proper experimental design is that it will permit the effects of several factors (e.g. temperature, pressure, catalyst) to be investigated simultaneously. Nevertheless, in interpreting the experimental results, we rarely focus on more than one or two parameters at a time.

Suppose that, for the moment, we wish to consider only α and β. The remaining parameters $\gamma, \delta, \ldots$ are then called <u>nuisance parameters</u>. The likelihood function will generally be a function of $\gamma, \delta, \ldots$ as well as α, β. We wish to eliminate $\gamma, \delta, \ldots$ from the likelihood function in order that we may take statements about α, β only. If the number of observations is large in comparison with the number of nuisance parameters, we may eliminate $\gamma, \delta, \ldots$ by maximizing over them to obtain

$$R_{max}(\alpha, \beta) = \underset{\gamma, \delta, \ldots}{\text{maximum}} R(\alpha, \beta, \gamma, \delta, \ldots),$$

which is called the <u>maximum relative likelihood function of α and β</u>. This and other methods of eliminating nuisance parameters will be discussed in Section 5.

Problems for Section 10.1

1. Suppose that $X_1, X_2, \ldots, X_n$ are independent $N(\mu, \sigma^2)$ variates.

   (a) Show that $\bar{X}$ is a sufficient statistic for μ when σ is known.

   (b) Show that $\sum(X_i - \mu)^2$ is a sufficient statistic for σ when μ is known.

†2. A scientist makes n measurements $X_1, X_2, \ldots, X_n$ of a constant μ using an apparatus of known variance σ^2, and m additional measurements $Y_1, Y_2, \ldots, Y_m$ of μ using a second apparatus of known variance $k\sigma^2$. Assume that all measurements are independent and normally distributed. Show that $T \equiv nk\bar{X} + m\bar{Y}$ is a sufficient statistic for μ, and find its distribution.

3. Suppose that $X_1, X_2, \ldots, X_n$ are $N(\mu_1, \sigma^2)$ and $Y_1, Y_2, \ldots, Y_m$ are independent $N(\mu_2, \sigma^2)$, all independent. Show that $\bar{X}, \bar{Y}$, and $V \equiv \sum(X_i - \bar{X})^2 + \sum(Y_i - \bar{Y})^2$ form a set of sufficient statistics for μ_1, μ_2, and σ.

4. Suppose that $X_1, X_2, \ldots, X_n$ are independent normal variates with the same mean μ, but with different variances,

$$X_i \sim N(\mu, \sigma^2/a_i) \quad \text{for} \quad i = 1, 2, \ldots, n,$$

where $a_1, a_2, \ldots, a_n$ are known constants. Find expressions for the MLE of μ and σ^2.

5. Suppose that $X_1, X_2, \ldots, X_n$ are independent normal variates with the same variance σ^2, but with different means,

$$X_i \sim N(\mu b_i, \sigma^2) \quad \text{for} \quad i = 1, 2, \ldots, n$$

where $b_1, b_2, \ldots, b_n$ are known constants. Find expressions for the MLE of μ and σ^2.

†6. A set of scales gives measurements which are normally distributed about the true weight with variance 1. Three independent weighings are made, one of weight μ_1, one of weight μ_2, and one of weight $\mu_1 + \mu_2$, giving measurements x_1, x_2, and x_3. Find the MLE of μ_1 and μ_2.

7. The lengths of the gestation periods for 1000 females are summarized in the following table:

Interval (days)	Frequency	Interval (days)	Frequency
259.5-264.5	6	284.5-289.5	176
264.5-269.5	27	289.5-294.5	135
269.5-274.5	107	294.5-299.5	34
274.5-279.5	198	299.5-304.5	4
279.5-284.5	312	304.5-309.5	1

Suppose that the length of the gestation period is normally distributed with mean μ and variance σ^2.

(a) Obtain approximate values of $\hat{\mu}$ and $\hat{\sigma}^2$ by taking all times to be at the midpoints of their intervals (6 observations of 262, 27 observations of 267, etc.), and compute expected frequencies.

(b) Write down the exact likelihood function of μ and σ in terms of the $N(0,1)$ probability integral, and indicate how $\hat{\mu}$ and $\hat{\sigma}$ could be determined exactly.

†8. The probability density function for an exponential distribution with guarantee time c is

$$f(x) = \lambda e^{-\lambda(x-c)} \quad \text{for} \quad x \geq c.$$

(a) Find the MLE and RLF of λ and c based on independent observations $x_1, x_2, \ldots, x_n$ from this distribution.

(b) Find a pair of sufficient statistics for (λ, c).

9. Consider the situation described in Problem 9.1.9. It is suggested that, while the geometric distribution applies to most specimens,

a fraction $1 - \lambda$ of them have flaws and therefore always fracture on the first blow.

(a) Show that the proportions of specimens fracturing after one, two, three, and four or more blows are respectively

$$1 - \lambda\theta, \quad \lambda\theta(1 - \theta), \quad \lambda\theta^2(1 - \theta), \quad \lambda\theta^3.$$

(b) If x_i specimens are observed in the ith category ($i = 1, 2, 3, 4$; $\sum x_i = n$), show that

$$\hat{\theta} = \frac{x_3 + 2x_4}{x_2 + 2x_3 + 2x_4} \; ; \qquad \hat{\lambda} = \frac{n - x_1}{n\hat{\theta}} \; .$$

(c) Compute expected frequencies for the data given in Problem 9.1.9 and comment on the fit of the model.

10. n individuals are randomly selected. Blood serum from each is mixed with a certain chemical compound and observed for a time T in order to record the time at which a certain colour change occurs. It is observed that m individuals respond at times $t_1, t_2, \ldots, t_m$, and that the remaining $n - m$ have shown no response at the end of the observation period T. The situation is thought to be describable by a probability density function $\lambda e^{-\lambda t}$ ($t > 0$) for a fraction p of the population, and complete immunity to the reaction in the remaining fraction $1 - p$. Find the maximum likelihood equations, and indicate how these can be solved for $\hat{p}$ and $\hat{\lambda}$.

†11. The number of particles emitted in unit time from a radioactive source has a Poisson distribution. The strength of the source is decaying exponentially with time, and the mean of the Poisson distribution on the jth day is $\mu_j = \alpha\beta^j$ ($j = 0, 1, \ldots, n$). Independent counts $x_0, x_1, \ldots, x_n$ of the number of emissions in unit time are obtained on these $n + 1$ days. Find the maximum likelihood equations and indicate how these may be solved for $\hat{\alpha}$ and $\hat{\beta}$.

12. A zoologist wishes to investigate the survival of fish in an isolated pond during the winter. The population may change by death during the period, but not by birth, immigration, or emigration. He catches n fish, marks them, and returns them. On two subsequent occasions he takes a sample of fish from the pond, observes which of his marked fish are in the sample, and returns them. He finds that x_1 of the marked fish are caught in the first sample only, x_2 in the second sample only, x_3 in both samples, and x_4 in neither sample. He assumes that each individual independently has a probability ϕ of survival between sampling periods,

and a probability p of being caught in any sample if it is alive at the time of the sample.

(a) Show that the probabilities of the four classes of recapture are $\alpha(1-\alpha)$, $\alpha\beta$, α^2, and $1-\alpha-\alpha\beta$, respectively, where $\alpha = \phi p$ and $\beta = \phi(1-p)$.

(b) Show that

$$\hat{\beta} = \frac{x_2}{x_2 + x_4} \left(\frac{1-\hat{\alpha}}{\hat{\alpha}}\right); \qquad \hat{\alpha} = \frac{x_1 + 2x_3}{n + x_1 + x_3}.$$

(c) Suppose that the observed frequencies are 22, 7, 14, and 21, respectively. Find the MLE of ϕ and p, and compute expected frequencies.

*10.2 An Example from Life-Testing

In Section 9.6 we considered the likelihood analysis of data from lifetime experiments when there is censoring. The analysis was illustrated for an exponential distribution model, involving a single unknown parameter. This model is appropriate when the specimens being tested are subject to a constant risk of failure which does not change with age.

In many real-life situations, deterioration with age does take place, and an item which has already operated for some time is likely to fail sooner than a new item. There are also some instances in which the risk of failure decreases with age.

Several lifetime distributions were discussed in Section 6.4. One of these, the Weibull distribution, is often used as a model in life-testing. It is particularly convenient because its cumulative distribution function has a simple form,

$$F(x) = 1 - \exp\{-\lambda x^\beta\} \quad \text{for} \quad x \geq 0. \tag{10.2.1}$$

This simplifies the calculations in the analysis of lifetime data when there is censoring.

The Weibull distribution has two parameters, λ and β, both positive. β is called the shape parameter. For $\beta = 1$, the Weibull distribution simplifies to an exponential distribution (no ageing). There is positive ageing (deterioration) for $\beta > 1$ and negative ageing (improvement) for $0 < \beta < 1$.

Since λ does not represent a quantity of interest, we shall

---

* This section may be omitted on first reading.

usually replace the parameter pair (β, λ) by the parameter pair (β, θ), where $\lambda = \theta^{-\beta}$. Then

$$F(x) = 1 - \exp\{-(x/\theta)^{\beta}\} \quad \text{for} \quad x > 0. \tag{10.2.2}$$

The p.d.f. of the Weibull distribution is

$$f(x) = \lambda \beta x^{\beta-1} \exp\{-\lambda x^{\beta}\} = \beta \theta^{-\beta} x^{\beta-1} \exp\{-(x/\theta)^{\beta}\}. \tag{10.2.3}$$

Note that

$$P(X \leq \theta) = F(\theta) = 1 - e^{-1} = 0.63,$$

so that θ represents the .63-quantile of the distribution. When $\beta = 1$, the distribution becomes exponential with mean θ.

Likelihood analysis under a Weibull distribution model

Suppose that n specimens are tested, and that the survival distribution is assumed to be Weibull with parameters (β, λ). If m specimens fail at times $x_1, x_2, \ldots, x_m$, and $n - m$ lifetimes are censored at times $T_1, T_2, \ldots, T_{n-m}$, then, by (9.6.1), the likelihood function is

$$L(\beta, \lambda) = k \left[\prod_{i=1}^{m} f(x_i) \right] \prod_{j=1}^{n-m} [1 - F(T_j)].$$

Substitution from (10.2.1) and (10.2.3) gives

$$L(\beta, \lambda) = k \left[\prod_{i=1}^{m} \lambda \beta x_i^{\beta-1} \exp\{-\lambda x_i^{\beta}\} \right] \prod_{j=1}^{n-m} \exp\{-\lambda T_j^{\beta}\}$$

$$= k \lambda^m \beta^m \left[\prod_{i=1}^{m} x_i \right]^{\beta-1} \exp\left\{-\lambda \left[\sum_{i=1}^{m} x_i^{\beta} + \sum_{j=1}^{n-m} T_j^{\beta} \right] \right\}.$$

With $k = 1$ for convenience, the log likelihood function is

$$\ell(\beta, \lambda) = m \log \lambda + m \log \beta + (\beta - 1) \sum \log x_i - \lambda \left[\sum x_i^{\beta} + \sum T_j^{\beta} \right].$$

We wish to determine the pair of values $(\hat{\beta}, \hat{\lambda})$ for which this is maximized.

Suppose, for the moment, that a value of β is given, and consider the maximization of $\ell(\beta, \lambda)$ with respect to λ only. At the maximum, we will have

$$0 = \frac{\partial \ell}{\partial \lambda} = \frac{m}{\lambda} - [\sum x_i^{\beta} + \sum T_j^{\beta}].$$

Hence, for the given value of β, the MLE of λ is

$$\hat{\lambda}(\beta) = m/[\sum x_i^{\beta} + \sum T_j^{\beta}], \qquad (10.2.4)$$

and the maximum of the log likelihood function is

$$\ell(\beta, \hat{\lambda}(\beta)) = m \log \hat{\lambda}(\beta) + m \log \beta + (\beta - 1)\sum \log x_i - m. \qquad (10.2.5)$$

To determine the joint MLE $(\hat{\beta}, \hat{\lambda})$, we must now maximize (10.2.5) over β. This may be done numerically using any of the methods described in Section 9.2. Once $\hat{\beta}$ has been found, we obtain

$$\hat{\lambda} = \hat{\lambda}(\hat{\beta}) = m/[\sum x_i^{\hat{\beta}} + \sum T_j^{\hat{\beta}}].$$

The relative likelihood function is then

$$R(\beta, \lambda) = \frac{L(\beta, \lambda)}{L(\hat{\beta}, \hat{\lambda})} = \frac{(\lambda \beta)^m (\Pi x_i)^{\beta - 1} \exp\{-\lambda [\sum x_i^{\beta} + \sum T_j^{\beta}]\}}{(\hat{\lambda} \hat{\beta})^m (\Pi x_i)^{\hat{\beta} - 1} \exp\{-\hat{\lambda} [\sum x_i^{\hat{\beta}} + \sum T_j^{\hat{\beta}}]\}}$$

$$= (\frac{\lambda \beta}{\hat{\lambda} \hat{\beta}})^m (\Pi x_i)^{\beta - \hat{\beta}} \exp\{m - \lambda [\sum x_i^{\beta} + \sum T_j^{\beta}]\}. \qquad (10.2.6)$$

If we wish, we may now make a one-to-one parameter transformation from (β, λ) to (β, θ), where $\lambda = \theta^{-\beta}$. The MLE of θ is then $\hat{\theta} = \hat{\lambda}^{-1/\hat{\beta}}$, and the RLF of (β, θ) is obtained by substituting $\lambda = \theta^{-\beta}$ in (10.2.6).

In the special case that all n items are observed to failure (no censoring), we have $m = n$, so that i ranges from 1 to n, and the T_j's disappear from the above expressions.

Example 10.2.1. The following are the results, in millions of revolutions to failure, of endurance tests for 23 deep-groove ball bearings:

17.88	28.92	33.00	41.52	42.12	45.60
48.48	51.84	51.96	54.12	55.56	67.80
68.64	68.64	68.88	84.12	93.12	98.64
105.12	105.84	127.92	128.04	173.40	

(The data are from page 286 of a paper by J. Lieblein and M. Zelen in

J. Res. National Bureau of Standards [1956].) As a result of testing thousands of ball bearings, it is known that, to a good approximation,
their lifetimes have a Weibull distribution. We wish to estimate the
parameters (β, θ).

Since there is no censoring, we have $m = n = 23$, and the
T_j's disappear from (10.2.4) and (10.2.6). We obtain $\hat{\beta}$ by maximizing (10.2.5), with $\hat{\lambda}(\beta) = 23/\sum x_i^{\beta}$. The procedure illustrated in
Table 10.2.1 is essentially repeated bisection, and yields $\hat{\beta} = 2.10$
correct to two decimal places. If the procedure is continued for a
few more steps, we obtain $\hat{\beta} = 2.102$, correct to three decimal places.
Then (10.2.4) gives

$$\hat{\lambda} = \hat{\lambda}(\hat{\beta}) = 9.518 \times 10^{-5},$$

and hence $\hat{\theta} = \hat{\lambda}^{-1/\hat{\beta}} = 81.88$.

Table 10.2.1
Evaluation of $\hat{\beta}$ by Repeated Bisection

β	$\hat{\lambda}(\hat{\beta}) \times 10^5$	$-\ell(\beta, \hat{\lambda}(\beta))$	Range containing $\hat{\beta}$
1.0	1384.57	121.43488	
2.0	15.24	113.74061	
3.0	0.14	116.83305	1.0 - 3.0
2.5	1.49	114.36666	1.0 - 2.5
2.2	6.04	113.73476	2.0 - 2.5
2.1	9.61	113.69131	2.0 - 2.2
2.15	7.62	113.70182	2.0 - 2.15
2.05	12.10	113.70398	2.05 - 2.15
2.08	10.54	113.69355	2.08 - 2.15
2.12	8.76	113.69278	2.08 - 2.12
2.11	9.17	113.69158	2.08 - 2.11
2.09	10.06	113.69197	2.09 - 2.11

The next step is to evaluate $R(\beta, \theta)$ over a lattice of
(β, θ) values so that a contour map can be prepared. Table 10.2.2
gives values of $R(\beta, \theta)$ near the maximum, and the curve along which
$R(\beta, \theta) = 0.5$ is sketched in. This is the innermost curve on the
contour map (Figure 10.2.1). The 0.1 and 0.01 contours can be obtained in a similar fashion from a tabulation of $R(\beta, \theta)$ over a larger region.

Table 10.2.2
Relative Likelihood Function $R(\beta,\theta)$

	$\theta = 72$	75	78	81	84	87	90	93
$\beta = 2.6$	.019	.066	.155	.261	.338	.351	.306	.230
2.5	.047	.136	.275	.418	.501	.495	.416	.307
2.4	.100	.245	.437	.605	.679	.641	.525	.383
2.3	.184	.387	.619	.791	.839	.764	.613	.443
2.2	.291	.539	.783	.934	.945	.835	.660	.474
2.1	.400	.661	.885	.994	.967	.835	.653	.469
2.0	.477	.715	.890	.952	.897	.761	.591	.427
1.9	.493	.679	.796	.817	.750	.628	.487	.354
1.8	.441	.565	.630	.625	.563	.468	.364	.267
1.7	.341	.411	.439	.424	.377	.312	.244	.181

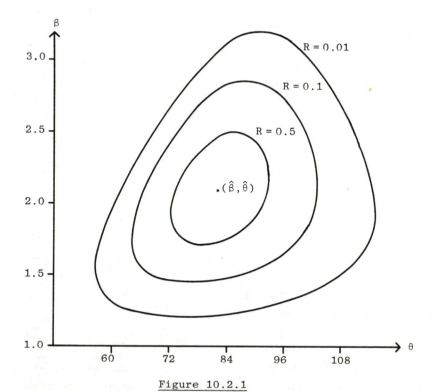

Figure 10.2.1
Contours of constant relative likelihood for the Weibull distribution parameters in Example 10.2.1 (no censoring)

The plausibility of any pair of values (β, θ) can be determined from its location on the contour map. Values inside the .5-contour are quite plausible, while values outside the .01-contour are very implausible.

The value $\beta = 1$ is of special interest, since for $\beta = 1$ the Weibull distribution simplifies to an exponential distribution. Note that the line $\beta = 1$ lies entirely outside the .01-contour in Figure 10.2.1. If $\beta = 1$, there exists no value of θ for which $R(\beta, \theta) \geq 0.01$; in fact, the maximum of $R(1, \theta)$ is 0.0004. It is therefore highly unlikely that $\beta = 1$, and the simpler exponential distribution model is not suitable. (The agreement of an exponential distribution with the data could also be checked by means of a quantile plot; see Chapter 11.)

If $\beta < 1.2$ or $\beta > 3.2$, there does not exist any value of θ for which $R(\beta, \theta) \geq 0.01$. For values of β outside the interval [1.2, 3.2], the maximum of the relative likelihood function is less than 0.01, and consequently such values of β are judged to be quite implausible. Similarly, any value of θ outside the interval [57, 115] is quite implausible because $R(\beta, \theta) < 0.01$ for all β. For further discussion, see Section 5.

Contour maps can usually be prepared with sufficient accuracy from tabulations of $R(\beta, \theta)$ such as that in Table 10.2.2. Alternatively, one can write a computer programme to solve the equation $R(\beta, \theta) = k$ for a given value of k, thereby giving points on the k-contour. This leads to a more accurate contour map, but usually at the expense of increased computer time and programming complexity.

Example 10.2.2. In Example 10.2.1, suppose that testing had stopped after 75 million revolutions. The last 8 lifetimes would then have been censored. In this case we would have $m = 15$ failure times and 8 equal censoring times:

$$x_1 = 17.88, \quad x_2 = 28.92, \ldots, \quad x_{15} = 68.88; \quad T_1 = T_2 = \ldots = T_8 = 75.$$

Proceeding as in Example 10.2.1, we obtain $\hat{\beta} = 2.763$, correct to three decimal places. Now (10.2.4) gives

$$\hat{\lambda} = \hat{\lambda}(\hat{\beta}) = 7.156 \times 10^{-6},$$

and hence $\hat{\theta} = \hat{\lambda}^{-1/\hat{\beta}} = 72.82$. After tabulating the relative likelihood function, the contours in Figure 10.2.2 may be drawn. The primary effect of the censoring has been to stretch the contours in the β-direction. The .01-contour now includes some pairs (β, θ) for

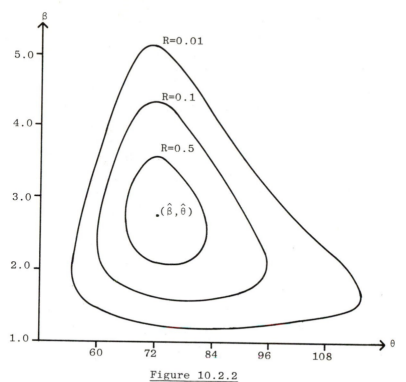

<u>Figure 10.2.2</u>

Contours of constant relative likelihood for the Weibull
distribution parameters in Example 10.2.2 (censoring at 75)

$1.2 \leq \beta \leq 5.1$, whereas previously the range of β was $1.2 \leq \beta \leq 3.2$.
Without observations from the right-hand tail of the distribution, it
is not possible to put as precise an upper bound on the value of the
shape parameter β. On the other hand, the censoring has had very
little effect on statements about θ. Within the .01-contour we now
have $55 \leq \theta \leq 115$, as opposed to $57 \leq \theta \leq 115$ with no censoring.

<u>Problems for Section 10.2</u>

1. Eighteen identical ball bearings were placed in test machines and
   subjected to a fixed radial load. The following are the numbers
   of hours the individual bearings endured at 2000 rpm:

   183   355   538   618   697   834   862   887   1056
   1147   1351   1506   1578   1607   1683   1710   2020   2410

   Assuming a Weibull model, obtain the maximum likelihood estimates
   of β and θ. Plot contours of constant relative likelihood, and

obtain intervals for β and θ .

*10.3 Dosage Response Models

Consider an experiment in which a lethal drug is administered to subjects in various doses $d_1, d_2, \ldots$. (Dosage is usually measured by log concentration of the toxin.) The probability p that a subject dies will depend upon the dose d of the drug which he receives. We first consider two theoretical models which relate p and d , and then illustrate the use of likelihood methods to estimate the parameters of one of these models.

The Logistic Model

In attempting to set up a simple relationship between p and d , the first thing which comes to mind is to express p as a linear function of d , $p = \alpha + \beta d$. However, this is generally, unsatisfactory because d (measured by log concentration) can take any real value, whereas p must lie between 0 and 1 . To overcome this difficulty, we replace p by some function $g(p)$ which can take any real value. The simplest function which maps the interval (a,b) onto the real line $(-\infty, \infty)$ is

$$g(p) = \log(\frac{p - a}{b - p}).$$

For $a = 0$ and $b = 1$, $g(p)$ is the log-odds,

$$g(p) = \log(\frac{p}{1 - p});$$

that is, $g(p)$ is logarithm of the odds ratio for death versus survival. We now assume that the log-odds is a linear function of the dose:

$$\log \frac{p}{1 - p} = \alpha + \beta d$$

where $\beta > 0$ for a harmful drug. Solving for p gives

$$p(d) = \frac{e^{\alpha + \beta d}}{1 + e^{\alpha + \beta d}} = \frac{AB^d}{1 + AB^d} \tag{10.3.1}$$

where $A = e^{\alpha}$ and $B = e^{\beta}$. This gives a hypothetical model relating

---

* This section may be omitted on first reading.

p and d which is, in a sense, the simplest possible. It is called
the logistic model, because p(d) is the cumulative distribution function of a continuous probability distribution called the logistic
distribution. There are two parameters α and β (or A and B).

The Probit Model

The probability of death, p(d), is assumed to be a smooth
non-decreasing function of d. We assume that the probability of
death from causes other than the drug is zero. A dose of -∞ (zero
concentration) will not cause death, and hence p(-∞) = 0. Furthermore, it is reasonable to assume that a very large dose will certainly
cause death, so that p(∞) = 1. These conditions will be satisfied if
p(d) is taken to be the cumulative distribution function of any continuous probability distribution.

This result can also be obtained as follows. We imagine
that different members of the population have different tolerances to
the drug. Let D represent the minimum does required to kill a randomly chosen population member, and suppose that D has a continuous
distribution with probability density function f and cumulative distribution function F. A dose d will produce death if and only if
the tolerance of the individual is at most d. Hence the probability
of death when dose d is administered is

$$p(d) = P(D \le d) = F(d).$$

The logistic model (10.3.1) corresponds to the assumption
that D has a logistic distribution. The probit model corresponds
to the assumption that the minimum lethal dose is normally distributed,
so that

$$p(d) = \int_{-\infty}^{d} \frac{1}{\sqrt{2\pi}\,\sigma} \exp\{-\frac{1}{2}(\frac{t-\mu}{\sigma})^2\}dt. \qquad (10.3.2)$$

There are two parameters, μ and σ.

Both the logistic model and the probit model are used in
analysing data from dosage response tests. The main advantage of the
logistic model is its mathematical simplicity. As a result, less computation is required than with the probit model. On the other hand,
the assumption that tolerances are normally distributed (probit model)
has some intuitive appeal because of the Central Limit Theorem. In
fact, the two models lead to quite similar results, and it would be
difficult to choose between them without a very large amount of data.

Likelihood Analysis for the Logistic Model

The remainder of this section deals with estimation of the parameters in the logistic model on the basis of some observed survival data. A similar analysis is possible for the probit model (10.3.2).

Consider a "three point assay", in which the drug is administered at three doses $d-h$, d, and $d+h$ in arithmetic progression. Then

$$\alpha + \beta d_1 = \alpha + \beta(d-h) = \alpha' + \beta'(-1),$$

$$\alpha + \beta d_2 = \alpha + \beta d \qquad = \alpha' + \beta'(0),$$

$$\alpha + \beta d_3 = \alpha + \beta(d+h) = \alpha' + \beta'(1),$$

where $\beta' = \beta h$ and $\alpha' = \alpha + \beta d$. Hence we shall take the three dose levels to be $-1, 0$, and 1. A linear transformation of the parameters will then yield the corresponding results for any three dose levels in arithmetic progression. According to the logistic model (10.3.1), the probabilities of death at doses $-1, 0, 1$ are

$$p(-1) = \frac{A/B}{1 + A/B} = \frac{A}{A+B}; \qquad p(0) = \frac{A}{1+A}; \qquad p(+1) = \frac{AB}{1+AB}.$$

Suppose that there are $3n$ subjects, with each dose being administered to n of them, and that x deaths are observed at dose -1, y deaths at dose 0, and z deaths at dose $+1$. The probability of observing x deaths out of n at dose -1 is

$$\binom{n}{x}[p(-1)]^x[1-p(-1)]^{n-x} = \binom{n}{x}\left[\frac{A}{A+B}\right]^x\left[\frac{B}{A+B}\right]^{n-x} = \binom{n}{x}\frac{A^x B^{n-x}}{(A+B)^n}.$$

Similarly, we may find the probability of observing y deaths out of n at dose 0, and z out of n at dose $+1$. The joint probability of the three observed frequencies x, y, z is then

$$\binom{n}{x}\binom{n}{y}\binom{n}{z} \frac{A^x B^{n-x} A^y (AB)^z}{(A+B)^n (1+A)^n (1+AB)^n}.$$

The likelihood function of (A,B) is proportional to this:

$$L(A,B) = \frac{A^{x+y+z} B^{n+z-x}}{(A+B)^n (1+A)^n (1+AB)^n}. \qquad (10.3.3)$$

Note that the two statistics $T_1 \equiv X + Y + Z$ and $T_2 \equiv Z - X$ are together sufficient for the pair of parameters (A, B).

To simplify the expressions which follow, we define

$$na = 2x + y; \qquad nb = y + 2z.$$

Since x, y, and z lie between 0 and n, we have $0 \le a \le 3$ and $0 \le b \le 3$. Furthermore,

$$n(a + b) = (2x + y) + (y + 2z) = 2(x + y + z);$$

$$n(b - a) = (y + 2z) - (2x + y) = 2(z - x).$$

With this substitution, the log likelihood function becomes

$$\ell(A, B) = n\left[\frac{a+b}{2} \log A + \frac{b-a+2}{2} \log B - \log(A + B) - \log(1 + A) - \log(1 + AB)\right],$$

and the maximum likelihood equations are

$$\frac{a + b}{2A} - \frac{1}{A + B} - \frac{1}{1 + A} - \frac{B}{1 + AB} = 0; \qquad (10.3.4)$$

$$\frac{b - a + 2}{2B} - \frac{1}{A + B} - \frac{A}{1 + AB} = 0. \qquad (10.3.5)$$

Adding A times the first equation to B times the second gives

$$\frac{2AB}{1 + AB} = b + 1 - \frac{A + B}{A + B} - \frac{A}{1 + A} = \frac{b(1 + A) - A}{1 + A}.$$

Solving this equation for B gives

$$B = \frac{b(1 + A) - A}{A[A + (2 - b)(1 + A)]}. \qquad (10.3.6)$$

Substitution for B in (10.3.5) now gives a polynomial equation of the fourth degree in A. It can be verified that $A = -1$ is a root, and when the factor $1 + A$ is divided out, one obtains the following cubic equation

$$(3 - a)(3 - b)A^3 + [(2 - a)(2 - b) - 1]A^2 - [(1 - a)(1 - b) - 1]A - ab = 0. \qquad (10.3.7)$$

We can obtain $\hat{A}$ as a positive real root of (10.3.7), and then substitute $A = \hat{A}$ in (10.3.6) to obtain $\hat{B}$. The relative likelihood function

$$R(A,B) = L(A,B)/L(\hat{A},\hat{B})$$

can then be found.

Example 10.3.1. Suppose that a drug was administered to n = 10 animals at each of the three doses -1,0,+1, and the observed numbers of deaths were x = 2, y = 2, z = 9. Then a = 0.6 and b = 2.0, and (10.3.7) becomes

$$2.4A^3 - A^2 + 1.4A - 1.2 = 0.$$

This equation has a unique real root $\hat{A} = 2/3$, and substituting this value into (10.3.6) gives $\hat{B} = 6$. The relative likelihood function of A and B is then found to be

$$R(A,B) = \frac{5^{30}A^{13}B^{17}}{2^{10}3^{24}(A+B)^{10}(1+A)^{10}(1+AB)^{10}} . \tag{10.3.8}$$

The estimates of the original parameters (α,β) are

$$\hat{\alpha} = \log \hat{A} = -0.4055; \quad \hat{\beta} = \log \hat{B} = 1.792.$$

The realtive likelihood function of α and β is obtained by substituting $A = e^{\alpha}$ and $B = e^{\beta}$ in (10.3.8). We may now proceed, as in Example 10.2.1, to tabulate the RLF and prepare a contour map (Figure 10.3.1) which shows the region of plausible parameter values.

The probabilities of death at the three doses -1,0,+1 are estimated to be

$$\hat{p}(-1) = \frac{\hat{A}/\hat{B}}{1 + \hat{A}/\hat{B}} = 0.1; \quad \hat{p}(0) = \frac{\hat{A}}{1 + \hat{A}} = 0.4; \quad \hat{p}(+1) = \frac{\hat{A}\hat{B}}{1 + \hat{A}\hat{B}} = 0.8.$$

Since there are ten subjects at each dose, the expected numbers of deaths at doses -1,0,+1 are 1,4,8, respectively. In order to check the fit of the logistic model to the data, we may compare these expected frequencies with the observed frequencies 2,2,9 in a goodness of fit test; see Chapter 11.

A quantity which is often considered in such experiments is the "ED50", which is the dose required to produce a 50% probability of death. If γ is the ED50, then (10.3.1) gives

$$\frac{1}{2} = p(\gamma) = \frac{AB^{\gamma}}{1 + AB^{\gamma}} .$$

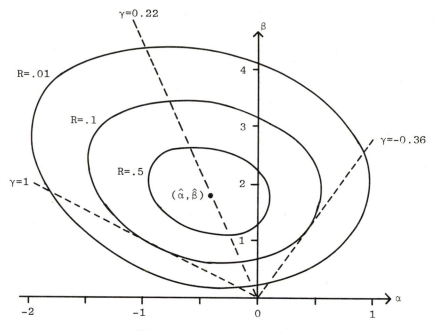

Figure 10.3.1

Contours of constant relative likelihood for
parameters of the logistic model (Example 10.3.1)

It follows that

$$1 = AB^\gamma = e^{\alpha + \beta\gamma}$$

and hence that $\alpha + \beta\gamma = 0$, so that $\gamma = -\dfrac{\alpha}{\beta}$.

To determine whether a particular value of γ is plausible,
we can draw in the line $\alpha + \beta\gamma = 0$ on the contour map. The three
lines shown in Figure 10.3.1 correspond to $\gamma = 1$, $\gamma = 0.22$, and
$\gamma = -0.36$. If γ is near 0.22, the line $\alpha + \beta\gamma = 0$ cuts through
the region of highest relative likelihood, and such values of γ are
quite plausible. On the other hand, if $\gamma < -0.36$ or $\gamma > 1$, the
line $\alpha + \beta\gamma = 0$ lies entirely outside the .1-contour, and hence such
values of γ are rather implausible. For further discussion, see
Section 5.

A Special Case

It can happen that no deaths are observed at either of the
two smallest doses, so that $x = y = 0$. Then $a = 0$ and $b \le 2$. In this

case, the constant term vanishes in (10.3.7), and hence (10.3.7) has a
root   A = 0.   It can be shown that this is the only non-negative root
of (10.3.7). By (10.3.6),   B → +∞   as   A → 0,   and it appears that
the maximum occurs for   (A,B) = (0,+∞);   that is, for $(\alpha,\beta) = (-\infty,+\infty)$.
When $x = y = 0$,   (10.3.3) becomes

$$L(A,B) = \frac{A^z B^{n+z}}{(A + B)^n (1 + A)^n (1 + AB)^n} = \frac{A^{n+z} B^{n+z}}{(A^2 + AB)^n (1 + A)^n (1 + AB)^n}$$

$$= \frac{C^{n+z}}{(A^2 + C)^n (1 + A)^n (1 + C)^n}$$

where   C = AB.   Now consider pairs of values   (A,B)   such that the
product   C   is fixed.   Then as   A → 0,   L(A,B)   increases to a maximum
of

$$\frac{C^{n+z}}{C^n (1 + C)^n} = C^z (1 + C)^{-n}.$$

This function of   C   attains its maximum value when $C = \frac{z}{n - z}$.   Hence
the overall maximum of the likelihood function is

$$(\frac{z}{n - z})^z (1 + \frac{z}{n - z})^{-n} = z^z (n - z)^{n-z} n^{-n},$$

and the realtive likelihood function is

$$R(A,B) = \frac{A^z B^{n+z} n^n}{(A + B)^n (1 + A)^n (1 + AB)^n z^z (n - z)^{n-z}}. \qquad (10.3.9)$$

The relative likelihood function of (α,β) is now obtained by substituting $A = e^{\alpha}$ and $B = e^{\beta}$.   We can now proceed to plot contours of constant relative likelihood, and hence determine the region of plausible
parameter values.
The situation is similar when   x = y = n,   when   y = z = 0,
or when   y = z = n.   In each case, $\hat{\alpha}$ and $\hat{\beta}$ are both infinite.   However, the relative likelihood function can still be used to examine the
plausibilities of parameter values.

Example 10.3.2.   Suppose that a drug was administered to   n = 10   animals at each of the three doses   -1,0,+1,   and the observed numbers of
deaths where   x = 0,   y = 0,   z = 5.   Then   a = 0,   and (10.3.9) gives

$$R(\alpha,\beta) = \frac{A^5 B^{15} 2^{10}}{(A+B)^{10}(1+A)^{10}(1+AB)^{10}} \; ,$$

where $A = e^{\alpha}$ and $B = e^{\beta}$. Contours of constant relative likelihood are shown in Figure 10.3.2. The data indicate large negative values of α and large positive values of β, with α/β near -1. The range of plausible values for β now depends very markedly on the value of α, and conversely. For example, if α is near -3, the likely values of β are between 2 and 4, while if α is near -6, the likely values of β are between 5 and 7. If we do not know α, about all that can be said is that values of β less than 1 are quite unlikely.

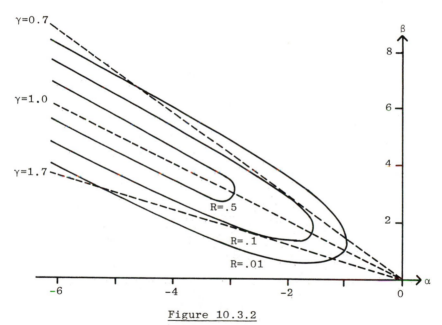

Figure 10.3.2

Contours of constant relative likelihood for
parameters of the logistic model (Example 10.3.2)

Let γ denote the ED50 as in Example 10.3.1. The three lines $\alpha + \beta\gamma = 0$ ($\gamma = 1.7, 1.0, 0.7$) are shown in Figure 10.3.2. Values of γ near 1 are quite plausible, since the corresponding lines cut through the region of highest relative likelihood. Values of γ less than 0.7 or greater than 1.7 are rather implausible. Although little can be said about either α or β individually, the

ED50 can be estimated fairly precisely.

Problems for Section 10.3.

1. (a) Show that the cubic equation (10.3.7) always has a positive
         real root when $0 < a < 3$ and $0 < b < 3$.
    (b) Show that (10.3.7) has no positive real root when $a = 0$ and
         $0 \le b \le 2$.
2. Find the maximum of (10.3.3) in the case $y = z = n$.
    Hint. Define $C = AB^{-1}$, and consider pairs (A,B) with C fixed.
3. Suppose that p_i, the probability of death when a drug is admini-
    stered in dose d_i, is given by

    $$\log[p_i/(1 - p_i)] = \alpha + \beta d_i.$$

    In n_i subjects receiving dose d_i, there are x_i deaths
    $(i = 1,2,\ldots,k)$, Show that the maximum likelihood equations for α
    and β are

    $$\sum(x_i - n_i p_i) = 0, \qquad \sum(x_i - n_i p_i)d_i = 0$$

    and the second derivatives of the log likelihood are

    $$-\sum v_i, \qquad -\sum v_i d_i, \qquad -\sum v_i d_i^2$$

    where $v_i = n_i p_i(1 - p_i)$.

†4. The probability of a normal specimen after radiation dose d is
    assumed to be $p = e^{\alpha + \beta d}$ where α and β are constants. The
    following table gives the number of normal specimens and the total
    number tested at each of five doses:

	0	1	2	3	4
d = Radiation dose					
x = Number of normals	4357	3741	3373	2554	1914
n = Number tested	4358	3852	3605	2813	2206

    (a) Plot $\log(x/n)$ against d to check whether the model seems
         reasonable, and obtain rough estimates of α and β from
         the graph.
    (b) Find the maximum likelihood equations and solve numerically
         for $\hat{\alpha}$ and $\hat{\beta}$ using the Newton-Raphson method or otherwise.
         Plot contours of constant relative likelihood, and obtain in-
         tervals for β and e^{α}.

*10.4   An Example from Learning Theory

In their book *Stochastic Models for Learning* (Wiley 1955),
R.R. Bush and F. Mosteller develop general probabilistic learning
models, and apply them to a variety of learning experiments. One of
the most interesting applications is to the Solomon-Wynne experiment
(R.L. Solomon and L.C. Wynne, Traumatic Avoidance Learning: Acquisi-
tion in Normal Dogs, *Psych. Monog.* 67(1953), No. 4). We shall first
describe this experiment, then develop the model, and finally use likelihood methods to estimate the two parameters of the model.

In the Solomon-Wynne experiment, 30 dogs learned to avoid
an intense electric shock by jumping a barrier. The lights were turned out in the dog's compartment and the barrier was raised. Ten
seconds later, an intense shock was applied through the floor of the
compartment to the dog's feet, and was left on until the dog escaped
over the barrier. The dog could avoid the shock only by jumping the
barrier during the ten-second interval after the lights were turned
out and before the shock was administered. Each trial could thus be
classified as a shock trial, or as an avoidance trial. The experimental record of 30 dogs, each of which had 25 trials, is shown in
Table 10.4.1, with 0 denoting a shock trial and 1 an avoidance
trial. (The dogs are numbered 13, 16, etc. for identification purposes, and no use is made of these numbers in the analysis.) Initially,
all of the dogs received shocks on almost every trial, but by trial 20,
all except dog number 32 had learned to avoid the shock by jumping
the barrier.

The Model

Consider the sequence of trials for one dog. Let S_n denote
the event that the nth trial is a shock trial, and let A_n denote
the event that the nth trial is an avoidance trial $(n = 0,1,\dots,24)$.
Since all dogs initially receive a shock, we shall assume that
$P(S_0) = 1$.

For $n > 0$, the probability that the dog receives a shock
at trial n will depend upon his past history in trials 0 to $n-1$.
As a result of the learning which takes place at trial $n-1$, the
probability of a shock should decrease from trial $n-1$ to trial n.
The amount of the decrease will depend upon whether there was shock or
avoidance at the $(n-1)$st trial.

---

* This section may be omitted on first reading.

Table 10.4.1

Data from 25 Trials with 30 Dogs in the Solomon-Wynne Experiment

0 = shock trial   1 = avoidance trial

Trial numbers

	0-4	5-9	10-14	15-19	20-24
Dog 13	0 0 1 0 1	0 1 1 1 1	1 1 1 1 1	1 1 1 1 1	1 1 1 1 1
16	0 0 0 0 0	0 0 1 0 0	0 0 0 0 1	1 1 1 1 1	1 1 1 1 1
17	0 0 0 0 0	1 1 0 1 1	0 0 1 1 0	1 0 1 1 1	1 1 1 1 1
18	0 1 1 0 0	1 1 1 1 0	1 0 1 0 1	1 1 1 1 1	1 1 1 1 1
21	0 0 0 0 0	0 0 0 1 1	1 1 1 1 1	1 1 1 1 1	1 1 1 1 1
27	0 0 0 0 0	0 1 1 1 1	0 0 1 0 1	1 1 1 1 1	1 1 1 1 1
29	0 0 0 0 0	1 0 0 0 0	0 0 1 1 1	1 1 1 1 1	1 1 1 1 1
30	0 0 0 0 0	0 0 1 1 0	0 1 1 1 1	1 1 1 1 1	1 1 1 1 1
32	0 0 0 0 0	1 0 1 0 1	1 0 1 0 0	0 1 1 1 1	1 0 1 1 0
33	0 0 0 0 1	0 0 1 1 0	1 0 1 1 1	1 1 1 1 1	1 1 1 1 1
34	0 0 0 0 0	0 0 0 0 0	1 1 1 1 1	1 0 1 1 1	1 1 1 1 1
36	0 0 0 0 0	1 1 1 1 1	0 0 1 1 1	1 1 1 1 1	1 1 1 1 1
37	0 0 0 1 1	0 1 0 0 1	1 1 1 1 1	1 1 1 1 1	1 1 1 1 1
41	0 0 0 0 1	0 1 1 0 1	1 1 1 1 1	1 1 1 1 1	1 1 1 1 1
42	0 0 0 1 0	1 1 0 1 1	1 1 1 1 1	1 1 1 1 1	1 1 1 1 1
43	0 0 0 0 0	0 0 1 1 1	1 1 1 1 1	1 1 1 1 1	1 1 1 1 1
45	0 1 0 1 0	0 0 1 0 1	1 1 1 0 1	1 1 1 1 1	1 1 1 1 1
47	0 0 0 0 1	0 1 0 1 1	1 1 1 1 1	1 1 1 1 1	1 1 1 1 1
48	0 1 0 0 0	0 1 0 0 0	1 1 1 1 1	1 1 1 1 1	1 1 1 1 1
46	0 0 0 0 1	1 0 1 0 1	1 0 1 0 1	1 1 1 1 1	1 1 1 1 1
49	0 0 0 1 1	1 1 1 0 1	1 1 1 1 1	1 1 1 1 1	1 1 1 1 1
50	0 0 1 0 1	0 1 1 1 1	1 1 1 1 1	1 0 0 1 1	1 1 1 1 1
52	0 0 0 0 0	0 0 1 1 1	1 1 1 1 1	1 1 1 1 1	1 1 1 1 1
54	0 0 0 0 0	0 0 0 1 1	1 0 1 0 0	0 1 1 0 1	1 1 1 1 1
57	0 0 0 0 0	0 1 0 1 1	1 1 0 1 0	1 1 1 1 1	1 1 1 1 1
59	0 0 1 0 1	1 1 0 1 1	0 1 1 1 1	1 1 1 1 1	1 1 1 1 1
67	0 0 0 0 1	0 1 1 1 1	1 1 1 1 1	1 1 1 1 1	1 1 1 1 1
66	0 0 0 1 0	1 0 1 1 1	0 1 0 1 1	1 1 1 1 1	1 1 1 1 1
69	0 0 0 0 1	1 0 0 1 1	1 0 1 0 1	0 1 0 1 1	1 1 1 1 1
71	0 0 0 0 1	1 1 1 1 1	0 1 0 1 1	1 1 1 1 1	1 1 1 1 1

Let $P(S_n|...)$ denote the probability of a shock at trial n given the dog's past history "...". We assume that, if there was shock at trial $n-1$, the probability of a shock at trial n is decreased by a factor α_0:

$$P(S_n|S_{n-1}...) = \alpha_0 P(S_{n-1}|...). \qquad (10.4.1)$$

Also, if there was avoidance at trial $n-1$, the probability of a shock at trial n is decreased by a factor α_1:

$$P(S_n | A_{n-1} \cdots) = \alpha_1 P(S_{n-1} | \cdots). \qquad (10.4.2)$$

We assume that the <u>shock parameter</u> α_0 and the <u>avoidance parameter</u> α_1 are constant over all trials.

If α_0 is small, then the effect of a shock trial is to greatly reduce the chance of a future shock. If $\alpha_0 = 1$, then nothing is learned from a shock trial. If $\alpha_0 = \alpha_1 = 1$, no learning takes place, and successive trials are independent.

The Likelihood Function

Given the above assumptions, we can find the probability of any sequence of shock and avoidance trials. For instance, the observed sequence for dog 13 was

$$S_0 S_1 A_2 S_3 A_4 S_5 A_6 A_7 \cdots A_{24}.$$

The probability of this sequence can be found as a product of twenty-five factors:

$$P(S_0)P(S_1|S_0)P(A_2|S_1S_0)P(S_3|A_2S_1S_0)\cdots P(A_{24}|A_{23}A_{22}\cdots S_0).$$

We have assumed that $P(S_0) = 1$, and each of the other factors can be determined using (10.4.1) and (10.4.2). From (10.4.1), we have

$$P(S_1|S_0) = \alpha_0 P(S_0) = \alpha_0;$$
$$P(S_2|S_1S_0) = \alpha_0 P(S_1|S_0) = \alpha_0^2$$

and hence

$$P(A_2|S_1S_0) = 1 - P(S_2|S_1S_0) = 1 - \alpha_0^2.$$

Now from (10.4.2) we obtain

$$P(S_3|A_2S_1S_0) = \alpha_1 P(S_2|S_1S_0) = \alpha_0^2 \alpha_1.$$

Continuing in this fashion, we find that

$$P(S_4|S_3A_2S_1S_0) = \alpha_0 P(S_3|A_2S_1S_0) = \alpha_0^3 \alpha_1$$
$$P(A_4|S_3A_2S_1S_0) = 1 - \alpha_0^3 \alpha_1$$
$$P(S_5|A_4S_3A_2S_1S_0) = \alpha_1 P(S_4|S_3A_2S_1S_0) = \alpha_0^3 \alpha_1^2$$

$$P(S_6|S_5A_4S_3A_2S_1S_0) = \alpha_0(\alpha_0{}^3\alpha_1{}^2) = \alpha_0{}^4\alpha_1{}^2$$

$$P(A_6|S_5A_4S_3A_2S_1S_0) = 1 - \alpha_0{}^4\alpha_1{}^2$$

and so on.

Note that, under this model, the probability of a shock at trial n depends only on the numbers of previous shock trials and avoidance trials. If there were k previous shock trials and $n-k$ avoidance trials, then

$$P(S_n|\dots) = \alpha_0{}^k\alpha_1{}^{n-k} = q_{n,k}; \qquad (10.4.3)$$

$$P(A_n|\dots) = 1 - P(S_n|\dots) = 1 - \alpha_0{}^k\alpha_1{}^{n-k} = p_{n,k}. \qquad (10.4.4)$$

The probability of any sequence can be built up as a product of p's and q's. For instance, the probability of the observed sequence for dog 13 is

$$q_{11}p_{22}q_{32}p_{43}q_{53}p_{64}p_{74}\cdots p_{24,4}$$

$$= \alpha_0(1 - \alpha_0{}^2)(\alpha_0{}^2\alpha_1)(1 - \alpha_0{}^3\alpha_1)(\alpha_0{}^3\alpha_1{}^2) \prod_{n=6}^{24} (1 - \alpha_0{}^4\alpha_1{}^{n-4}).$$

If we make the further assumptions that dogs are independent of one another and have the same parameter values (α_0, α_1), the likelihood function $L(\alpha_0, \alpha_1)$ can be found as the product of 30 such factors, one for each dog.

Let $y_{in} = 0$ if the ith dog receives a shock at the nth trial, and $y_{in} = 1$ if it avoids the shock. Then Table 10.4.1 gives the matrix (y_{in}). Let k_{in} be the number of shocks received by the ith dog prior to the nth trial, so that

$$k_{in} = n - \sum_{j=0}^{n-1} y_{ij}.$$

Then the likelihood function is

$$L(\alpha_0, \alpha_1) = \prod_{i=1}^{30} \prod_{n=1}^{24} (p_{nk_{in}})^{y_{ij}}(q_{nk_{in}})^{1-y_{in}}, \qquad (10.4.5)$$

where p and q are defined by (10.4.3) and (10.4.4).

89

Numerical Results

The likelihood function (10.4.5) is a product of 720 factors. It can be simplified somewhat by combining the q-terms and collecting like p-terms. However, it will still involve a product of more than 160 polynomial expressions in α_0 and α_1, and is not amenable to hand computation. Before high speed computers were available, the analysis of data such as these was very difficult, and often depended upon approximations whose accuracy could not be checked. Such approximations are now unnecessary. The computer is easily programmed to evaluate the log likelihood function for any given pair of values (α_0, α_1). Then, by a double application of the method of repeated bisection, we find that

$$\hat{\alpha}_0 = 0.924; \quad \hat{\alpha}_1 = 0.786$$

$$\ell(\hat{\alpha}_0, \hat{\alpha}_1) = \log L(\hat{\alpha}_0, \hat{\alpha}_1) = -273.9848.$$

We can now tabulate the log relative likelihood function

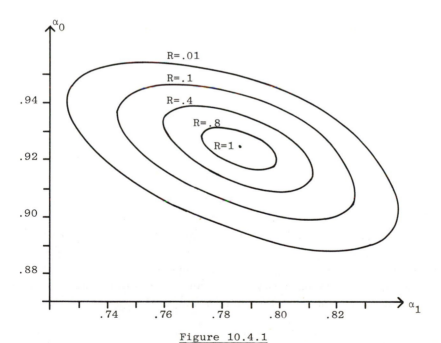

Figure 10.4.1
Contours of constant relative likelihood
for the avoidance and shock parameters

$$r(\alpha_0, \alpha_1) = \log L(\alpha_0, \alpha_1) + 273.9848,$$

and prepare the contour map in Figure 10.4.1. All of these calculations require only a minute or two on a fast computer.

The contour map gives a concise pictorial summary of the information provided by the data concerning α_0 and α_1. In particular, we see that the avoidance parameter α_1 is smaller than the shock parameter α_0. This means that an avoidance trial is more effective than a shock trial in teaching the dog to avoid shock on subsequent trials. In fact, since $\hat{\alpha}_1 \approx (\hat{\alpha}_0)^3$, we see that three shock trials are required to produce roughly the same effect as a single avoidance trial.

Each contour includes a larger range of values of α_1 than of α_0. For instance, the .1-contour extends from $\alpha_1 = 0.744$ to $\alpha_1 = 0.826$, and from $\alpha_0 = 0.899$ to $\alpha_0 = 0.946$. The experiment determines the value of α_0 more precisely than it does the value of α_1.

The contours also illustrate how knowledge of the value of one of the parameters would affect statements about the other. For instance, if α_1 were known to be 0.744, the most likely value of α_0 would be 0.935 rather than 0.924. The maximum of $R(\alpha_0, \alpha_1)$ with $\alpha_1 = 0.744$ is 0.1, and the relative likelihood function for α_0 alone would be

$$R(\alpha_0) = \frac{R(\alpha_0, .744)}{0.1}.$$

Since $R(.915, .744) = R(.954, .744) = 0.01$, the 10% likelihood interval for α_0 would now be $[0.915, 0.954]$.

*10.5   Elimination of Nuisance Parameters

It often happens that, although the probability model involves two parameters α and β, only one of them (β, say) is of real interest. The other parameter α is called a nuisance parameter. It is not of direct interest itself, but its presence complicates the problem of making statements about β. We wish to eliminate α from the likelihood function in order that we may make statements about β alone.

---

* This section may be omitted on first reading.

I.   Factorization of the Likelihood Function

By carefully designing the experiment and setting up the
model, we may be able to arrange that the likelihood function will
factor:

$$L(\alpha,\beta) = g(\alpha)h(\beta) \quad \text{for all} \quad \alpha,\beta.$$

Then the likelihood ratio for (α,β_1) versus (α,β_2) is

$$\frac{L(\alpha,\beta_1)}{L(\alpha,\beta_2)} = \frac{h(\beta_1)}{h(\beta_2)}$$

which does not depend upon α. We may then take

$$L(\beta) = h(\beta),$$

and proceed to make statements about β as in the one-parameter case
(Chapter 9).

Example 10.5.1.   Suppose that a radioactive particle A can decay in
three different ways to give a particle $B,C,$ or D. Particle B is
charged, while particles C and D are not. It is observed that, of
n A-particles, x decay to B, y to C, and z to D, where
$x + y + z = n$. Let α be the probability that the decay results in a
charged particle. Let β be the conditional probability of C, given
that an uncharged particle is produced. The probabilities of particle
types $B,C,$ and D are α, $(1-\alpha)\beta$, and $(1-\alpha)(1-\beta)$, respectively. The joint relative likelihood function for α and β is

$$L(\alpha,\beta) = \alpha^x[(1-\alpha)\beta]^y[(1-\alpha)(1-\beta)]^z = \alpha^x(1-\alpha)^{y+z}\beta^y(1-\beta)^z.$$

Because of the natural parametrization used, the joint likelihood
function factors into a function of α times a function of β. Statements about the individual parameters may now be based on the one-dimensional likelihood functions

$$L_1(\alpha) = \alpha^x(1-\alpha)^{y+z}; \quad L_2(\beta) = \beta^y(1-\beta)^z.$$

Alternatively, we might have taken the parameters to be α and γ
where γ is the unconditional probability of C. Then

$$L(\alpha,\gamma) = \alpha^x\gamma^y(1-\alpha-\gamma)^z,$$

which does not factor, and some other method of parameter elimination would be needed.

II. Maximum Relative Likelihood

One method of eliminating α from the joint relative likelihood function $R(\alpha,\beta)$ is to maximize over it. This yields a function of β only,

$$R_{max}(\beta) = \max_{\alpha} R(\alpha,\beta),$$

which is called the maximum relative likelihood function of β. Suppose that we fix the value of β, and let $\hat{\alpha}(\beta)$ denote the value of α at which $L(\alpha,\beta)$ is maximized. Then

$$R_{max}(\beta) = R(\hat{\alpha}(\beta),\beta).$$

The maximum relative likelihood function is similar to an ordinary RLF in that

$$0 \le R_{max}(\beta) \le 1; \quad R_{max}(\hat{\beta}) = 1.$$

If $R_{max}(\beta_1)$ is small, then $R(\alpha,\beta_1)$ is small for all values of α. There does not exist a value of α for which the pair (α,β_1) is plausible, and in this sense β_1 is an implausible value of β. On the other hand, if $R_{max}(\beta_1)$ is near 1, there exists at least one value of α for which $R(\alpha,\beta_1)$ is large. Without some information about the value of α, additional to that obtained from the experiment, we cannot claim that β_1 is an unlikely value.

We may think of the joint RLF $R(\alpha,\beta)$ as a "mountain" sitting in the (α,β) plane, and imagine that we are viewing the mountain from a distant point on the α-axis. Then the maximum RLF is the profile or silhouette of the mountain.

Sometimes a function of α and β, $\gamma = c(\alpha,\beta)$, is of interest. We may then change parameters from (α,β) to (γ,δ), where δ is chosen to make the transformation one-to-one. Then, using the invariance property of likelihood (Section 9.7), we may find the joint RLF of (γ,δ) and maximize over δ to obtain the maximum RLF of γ. Equivalently,

$$R_{max}(\gamma) = \max R(\alpha,\beta),$$

where the maximum is taken over all pairs (α,β) such that $c(\alpha,\beta) = \gamma$.

Example 10.5.2. Suppose that n measurements $x_1, x_2, \ldots, x_n$ are assumed to be independent values from a normal distribution $N(\mu, \sigma^2)$, as in Example 10.1.1. Then

$$L(\mu, \sigma) = \sigma^{-n} \exp\{-\textstyle\sum(x_i - \mu)^2 / 2\sigma^2\};$$

$$\ell(\mu, \sigma) = -n \log \sigma - \textstyle\sum(x_i - \mu)^2 / 2\sigma^2;$$

$$\frac{\partial \ell}{\partial \sigma} = \sigma^{-3}[\textstyle\sum(x_i - \mu)^2 - n\sigma^2];$$

$$\frac{\partial \ell}{\partial \mu} = \sigma^{-2}[\textstyle\sum x_i - n\mu].$$

Let $\hat{\sigma}(\mu)$ denote the MLE of σ for a fixed value of μ. The equation $\frac{\partial \ell}{\partial \sigma} = 0$ gives

$$[\hat{\sigma}(\mu)]^2 = \frac{1}{n} \textstyle\sum(x_i - \mu)^2,$$

and hence

$$L(\mu, \hat{\sigma}(\mu)) = [\hat{\sigma}(\mu)]^{-n} \exp\{-\tfrac{n}{2}\}.$$

The maximum relative likelihood function of μ is then

$$R_{max}(\mu) = \frac{L(\mu, \hat{\sigma}(\mu))}{L(\hat{\mu}, \hat{\sigma})} = \left[\frac{\hat{\sigma}(\mu)}{\hat{\sigma}}\right]^{-n} = \left[\frac{\sum(x_i - \mu)^2}{n\hat{\sigma}^2}\right]^{-n/2}$$

by (10.1.3). Now (10.1.2) gives

$$R_{max}(\mu) = \left[1 + \left(\frac{\hat{\mu} - \mu}{\hat{\sigma}}\right)^2\right]^{-n/2} \quad \text{for} \quad -\infty < \mu < \infty.$$

The range of plausible μ-values can be determined from a graph of this function or its logarithm.

If we fix the value of σ, the MLE of μ is obtained by solving the equation $\frac{\partial \ell}{\partial \mu} = 0$. We find that $\hat{\mu}(\sigma) = \hat{\mu} = \bar{x}$; that is, the MLE of μ is the same all values of σ. Substitution in (10.1.4) now gives

$$R_{max}(\sigma) = \left(\frac{\hat{\sigma}}{\sigma}\right)^n \exp\{\tfrac{n}{2}[1 - (\tfrac{\hat{\sigma}}{\sigma})^2]\} \quad \text{for} \quad \sigma > 0.$$

Other methods for making inferences about the normal distribution parameters are considered in Section 14.1. The quantity used to make inferences about σ with μ unknown is a constant times $(\hat{\sigma}/\sigma)^2$; the quantity used for inferences about μ with σ unknown is a constant times $(\hat{\mu} - \mu)/\hat{\sigma}$.

Example 10.5.3. In Section 10.2 we considered the likelihood analysis of data under the assumption of a Weibull distribution model. By (10.2.4), the MLE of λ when β is fixed is

$$\hat{\lambda}(\beta) = m/[\textstyle\sum x_i^{\beta} + T_j^{\beta}],$$

and hence, by (10.2.6),

$$R_{max}(\beta) = R(\beta, \hat{\lambda}(\beta)) = [\frac{\beta\hat{\lambda}(\beta)}{\hat{\beta}\hat{\lambda}}]^m (\Pi x_i)^{\beta - \hat{\beta}}.$$

The logarithm of this function is plotted in Figure 10.5.1 for the data of Example 10.2.1 (no censoring), and for the situation considered in Example 10.2.2 (censoring at 75 million).

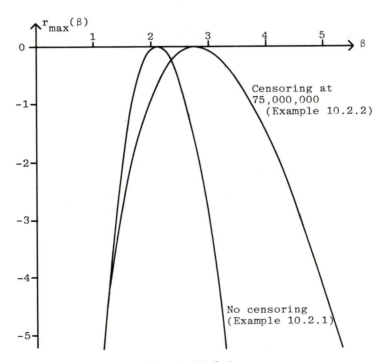

Figure 10.5.1

Log maximum relative likelihood function of the
Weibull shape parameter in Examples 10.2.1 and 10.2.2

The 10% likelihood intervals corresponding to $r_{max}(\beta) \geq -2.30$ in Figure 10.5.1 are [1.45,2.85] for no censoring and [1.60,4.35] for censoring at 75,000,000 revolutions. These are the vertical extremes of the 10% contours in Figures 10.2.1 and 10.2.2. In both cases, $r_{max}(1) < -5$, indicating that $\beta = 1$ (which corresponds to an exponential distribution) is extremely implausible.

The maximum relative likelihood function of θ (or of any other quantile) can also be found. However, numerical methods are required to determine $\hat{\beta}(\theta)$, the MLE of β when the value of θ is fixed.

Example 10.5.4. In Section 10.3 we considered the likelihood analysis of data from a three point assay assuming a logistic model. We shall now obtain the maximum relative likelihood function of the slope parameter, $\beta = \log B$, and of the ED50, $\gamma = -\alpha/\beta$.

If the value of B is fixed, we may find $\hat{A}(B)$, the MLE of A, by solving equation (10.3.4). This simplifies to a cubic equation in A,

$$A^3 B(3-d) + A^2(B^2 + B + 1)(2-d) + A(B^2 + B + 1)(1-d) - dB = 0,$$

where $d = (x + y + z)/n$. For $B > 0$ and $0 < d < 3$, the expression on the left hand side is negative at $A = 0$ and as $A \to -\infty$, and is positive at $A = -1$ and as $A \to +\infty$. The equation therefore has three real roots, one positive and two negative. We can find the positive root by the numerical methods of Section 9.2, and hence obtain

$$R_{max}(B) = R(\hat{A}(B), B).$$

The maximum RLF of β is then found by substituting $B = e^\beta$.

Figure 10.5.2 shows the log maximum relative likelihood function of β based on the data of Example 10.3.2 ($x = y = 0$, $z = 5$, $n = 10$). The joint RLF of α and β is shown in Figure 10.3.2. For large β, the most likely value of α is approximately $-\beta$, and the maximum relative likelihood of β approaches 1 as $\beta \to \infty$. From Figure 10.5.2 we see that values of β less than 0.45 are quite implausible ($R_{max}(\beta) < .01$), and values of β greater than 2.60 are quite plausible ($R_{max}(\beta) \geq 0.5$). Without some additional information concerning α, it is impossible to put an upper limit on the value of β.

In Example 10.3.1 we showed that if γ is the ED50, then $AB^\gamma = 1$. Hence $A = B^{-\gamma}$, and the joint RLF of B and γ is obtained by substituting $A = B^{-\gamma}$ in (10.3.9). For any given value of γ, we can maximize this function over B by the method of repeated bisection

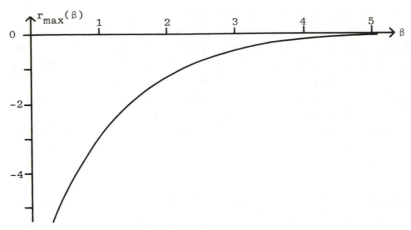

Figure 10.5.2

Log maximum RLF of the slope parameter in the
logistic model, based on the data of Example 10.3.2

(Section 9.2), and hence obtain $R_{max}(\gamma)$. Figure 10.5.3 shows the log
maximum RLF of γ based on the data of Example 10.3.2. Values of γ
between 0.85 and 1.18 have maximum relative likelihood greater than
0.5, while values of γ outside the interval [0.50,3.95] have maximum relative likelihood less than 0.01. Although the experiment was

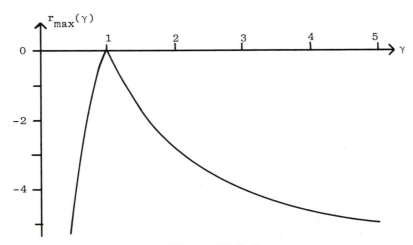

Figure 10.5.3

Log maximum RLF of the ED50 in the logistic model,
based on the data of Example 10.3.2

not very successful in pinning down the value of either α or β, the ratio $\gamma = -\alpha/\beta$ can be estimated fairly precisely.

III. Other Methods of Parameter Elimination

It is sometimes possible to eliminate α from the joint likelihood function by integrating it out using a suitable weight function w. The result is a function of β alone,

$$L_{int}(\beta) = \int_{-\infty}^{\infty} L(\alpha,\beta)w(\alpha)d\alpha,$$

and is called an integrated likelihood function of β. The weight function w would usually be the probability density function of a Bayesian prior distribution for α (Section 16.2), or of the fiducial distribution of α (Section 16.1). It is usually difficult to define a suitable weight function, and hence this method of parameter elimination is not often applicable.

It is sometimes possible to identify part of the data which may be said to contain all of the information about β, and whose marginal distribution depends only on β. For instance, in the case of a sample $x_1, x_2, \ldots, x_n$ from a normal distribution $N(\mu,\sigma^2)$, it can be argued that the variate $V \equiv \sum(X_i - \bar{X})^2$ contains all of the information about σ when μ is unknown. Furthermore, the marginal distribution of V does not depend upon μ. The marginal likelihood function of σ is then given by

$$L_{max}(\sigma) = kf(v),$$

where f is the marginal p.d.f. of V and v is the observed value.

Similarly, one can sometimes argue that all of the information about β is contained in a conditional distribution which does not depend upon α. The conditional likelihood function of β is then proportional to the p.d.f. or probability function of this conditional distribution.

The above three methods are not always applicable because they depend upon extra information about the nuisance parameter, or special properties of the experiment and probability model. However, when one of them can be used, it is preferable to maximum relative likelihood. Difficulties can arise with maximum relative likelihood if one attempts to maximize over a large number of nuisance parameters. All of the methods described give quite similar results when the number of observations is large in comparison with the number of nuisance parameters to be eliminated.

Problems for Section 10.5

†1. Find the maximum RLF for λ and the maximum RLF for c in Problem 10.1.8.

2. Derive the maximum relative likelihood function for μ_2 in Problem 10.1.6.

3. Find the maximum relative likelihood function for θ in Problem 10.1.9. Show that this is the same as the relative likelihood function for θ based on the conditional distribution of X_2, X_3 and X_4 given X_1.

†4. Find the maximum relative likelihood function for λ in Problem 10.1.10.

5. Suppose that the likelihood function factors,

$$L(\alpha, \beta) = g(\alpha)h(\beta) \quad \text{for all} \quad \alpha, \beta.$$

(a) Show that the maximum relative likelihood function of α is proportional to $g(\alpha)$.

(b) Suppose that we make a one-to-one parameter transformation from (α, β) to (α, γ). Show that the maximum relative likelihood function of α which is obtained by maximizing over γ will be proportional to $g(\alpha)$.

CHAPTER 11. CHECKING THE MODEL

Suppose that we have observations from n independent repetitions of an experiment. We wish to determine how well these data agree with a probability model for the experiment. We do this by comparing observed values from the experiment with corresponding theoretical values derived from the model.

In Section 1,2 and 3, we consider the comparison of a set of observed frequencies with expected frequencies under the model. Even if the model were exactly correct, one would anticipate some discrepancy between the observed and expected frequencies owing to random variation. A goodness of fit test may be used to determine whether the observed discrepancy is too great to be accounted for by chance.

When the number of possible outcomes of the experiment is large, it is necessary to combine these to form a small number of classes before constructing a frequency table; see Example 1.2.1. The choice of classes is usually rather arbitrary, and grouping the data may lead to a serious loss of information.

These problems may be partially overcome through use of the graphical procedures discussed in Sections 4 and 5. These are based on a comparison of the theoretical cumulative distribution function with an analogous function, the empirical c.d.f., which is computed from the sample.

11.1 Goodness of Fit Tests

One method for assessing how well an assumed model agrees with the data is to prepare a table of observed frequencies f_j and expected frequencies e_j under the model.

Event or class	A_1	A_2	A_3	$\cdots$	A_k	Total
Observed frequency	f_1	f_2	f_3	$\cdots$	f_k	n
Expected frequency	e_1	e_2	e_3	$\cdots$	e_k	n

The sample space S for a single repetition of the experiment is partitioned into k mutually exclusive classes or events, $S = A_1 \cup A_2 \cup \ldots \cup A_k$. The observed frequency f_j is the number of times that event A_j occurs in n independent repetitions of the experiment. The probability p_j that A_j occurs in one repetition may be found from the model assumed, and then the expected frequency of event A_j is $e_j = np_j$. Exactly one of the events A_j must occur at each repetition of the experiment, so

$$\textstyle\sum p_j = 1; \quad \sum e_j = \sum f_j = n.$$

In most applications, the probability model will involve one or more unknown parameters. These may be estimated by the method of maximum likelihood as in Section 9.1, and then the estimated parameter values are used in computing the expected frequencies.

Even if the proposed model were exactly correct, one would anticipate that there would be some discrepancy between the f_j's and e_j's owing to random variation in $f_1, f_2, \ldots, f_k$. A test of significance may be used to determine whether the observed discrepancy is too great to be accounted for by this chance variation.

To set up a test of significance, we require a measure D of the distance or discrepancy between a set of observed frequencies $(f_1, f_2, \ldots, f_k)$ and the set of expected frequencies $(e_1, e_2, \ldots, e_k)$. The distance measure D could be defined in many ways, depending upon what type of departure from the model we wish to detect; see Chapter 12. For the present, we consider only a general purpose measure, the goodness of fit criterion, or goodness of fit statistic,

$$D = \sum_{j=1}^{k} \frac{(f_j - e_j)^2}{e_j} = \sum \frac{(\text{obs. freq.} - \text{exp. freq.})^2}{\text{exp. freq.}}. \tag{11.1.1}$$

A large value of D indicates a bad fit; that is, a poor overall agreement between the observed frequencies f_j and the expected frequencies e_j.

The usual measure of distance from point $(f_1, f_2, \ldots, f_k)$ to point $(e_1, e_2, \ldots, e_k)$ is the square root of $\sum (f_j - e_j)^2$. We have not chosen this as the discrepancy measure because it treats all classes equally, whereas one would expect to get greater deviations $(f_j - e_j)^2$ in larger classes. If $e_j = 1000$, we will not be surprised or dismayed if we find that $(f_j - e_j)^2 = 100$; however the same deviation in a class with $e_j = 10$ would likely cause us to abandon the model. Division by the expected frequency e_j in (11.1.1) compensates for the fact that larger deviations $(f_j - e_j)^2$ are more probable in large classes.

Significance Level

Now let d be the observed value of D. The significance

<u>level</u> of the data is the probability of obtaining as large a discre-
pancy D as was observed; that is,

$$SL = P(D \geq d).$$

If SL is large (e.g. 20%), a discrepancy as great as that observed
would frequently occur as a result of chance variation if the model
were correct, and hence the data are consistent with the model. If
SL is small (e.g. 1%), such a large discrepancy would rarely occur by
chance if the model were correct, and hence there is evidence against
the model.

From Section 4.6, the joint distribution of the frequencies
$f_1, f_2, \ldots, f_k$ is multinomial:

$$P(f_1, f_2, \ldots, f_k) = \binom{n}{f_1\ f_2\ \ldots\ f_k} p_1^{f_1} p_2^{f_2} \ldots p_k^{f_k}. \qquad (11.1.2)$$

If all of the p_j's are known, the exact significance level may be
obtained (at least in theory) as a sum of multinomial probabilities.
The sum will extend over all sets of frequencies $(f_1, f_2, \ldots, f_k)$ with
$\sum f_j = n$ for which the value of D is at least as great as the observed value d; see Section 11.3.

The situation is more complicated when the model involves
one or more unknown parameters θ, since in this case probabilities
computed by summing (11.1.2) will generally be functions of the unknown θ. See Section 12.2 for discussion.

Calculation of the exact significance level will involve a
large amount of arithmetic when the expected frequencies are large.
In this case, we make use of the following result, which enables us to
obtain an approximate significance level from tables of the chi-square
distribution (Section 6.9).

<u>Theorem 11.1.1</u>. Suppose that the assumed probability model is correct,
and that the expected frequencies e_j are all large. Then

$$SL \approx P\{\chi^2_{(k-r-1)} \geq d\} \qquad (11.1.3)$$

where   d = observed value of the goodness of fit criterion;
        k = number of classes used in computing d;
        r = number of parameters estimated from the data.

<u>Pooling</u>. The accuracy of the approximation (11.1.3) is adversely af-

fected by large deviations in classes with small expected frequencies.
The usual rule of thumb is that the approximation can be applied safely
whenever all of the expected frequencies are at least 5. Often it
will be necessary to combine or pool classes in the frequency table to
get expected frequencies large enough that (11.1.3) can be used. The
value of k in (11.1.3) will then be the number of classes which remain after pooling.

Degrees of Freedom. If no parameters require estimation from the data
($r = 0$), the degrees of freedom in the χ^2 approximation is $k - 1$.
There are k frequencies $f_1, f_2, \ldots, f_k$ in which deviations can occur,
but their total is fixed: $\sum f_j = n$. Thus only $k - 1$ of them can vary
freely, the kth then being determined by subtraction from the total
n.

Now suppose that the p_j's are functions of an unknown parameter θ. From Section 9.1, the log-likelihood function of θ is

$$\ell(\theta) = \sum f_j \log p_j .$$

To estimate θ by the method of maximum likelihood, we set the derivative of $\ell(\theta)$ with respect to θ equal to zero. This gives

$$\sum a_j f_j = 0, \quad \text{where} \quad a_j = \frac{\partial}{\partial \theta} \log p_j.$$

Estimation of a parameter is thus equivalent to placing a further linear restriction on the f_j's. Now only $k - 2$ of them can vary freely,
the other two then being determined by the linear restrictions $\sum f_j = n$;
$\sum a_j f_j = 0$. Similarly, if r parameters are estimated, there will be
$r + 1$ linear restrictions, and only $k - (r + 1)$ degrees of freedom.

The degrees of freedom formula, $k - r - 1$, assumes that the
r parameters are estimated from the k frequencies used in computing
(11.1.1). In particular, the formula assumes that any pooling of classes to avoid small expected frequencies will be done before parameter
estimates are computed. In practice, we usually compute parameter estimates first, then find expected frequencies, and finally pool classes
if necessary; see Examples 11.2.2, 11.2.3, and 11.2.4. When this procedure is followed, the appropriate degrees of freedom will be greater
than $k - r - 1$, but less than $k - 1$. However, unless a substantial a-
mount of grouping is involved, the standard formula, $k - r - 1$, should
be close enough for most practical purposes.

Justification of χ^2 approximation

A general proof of Theorem 11.1.1 is difficult and beyond the scope of this book. However, we shall give a justification of the χ^2 approximation in the special case where there are no unknown parameters to estimate ($r = 0$).

Let $Y_1, Y_2, \ldots, Y_k$ be independent Poisson variates with means $\mu_1, \mu_2, \ldots, \mu_k$ and define $N \equiv \sum Y_j$. From Problem 4.6.5, the conditional distribution of the Y_j's given that $N = n$ is multinomial with parameters $(n, p_1, p_2, \ldots, p_k)$, where $p_j = \mu_j / \sum \mu_j$. We work out the approximate distribution of the goodness of fit criterion for independent Poisson variates, and then condition on N to get the result for multinomial frequencies.

From Section 6.8, a Poisson distribution with mean μ can be approximated by a normal distribution $N(\mu, \mu)$ if μ is large. Hence the standardized variates $U_j \equiv (Y_j - \mu_j)/\sqrt{\mu_j}$ for $j = 1, 2, \ldots, k$ are independent and approximately $N(0,1)$ if the μ_j's are all large. Now define $Z_1 \equiv \sum a_j U_j$, where the a_j's are constants with $\sum a_i^2 = 1$, and let $V \equiv \sum U_j^2 - Z_1^2$. Then, by Theorem 7.3.2, Z_1 and V are distributed independently, with $Z_1 \sim N(0,1)$ and $V \sim \chi^2_{(k-1)}$. Since Z_1 and V are independent (for $\mu_1, \mu_2, \ldots, \mu_k$ large), the conditional distribution of V given Z_1 is also approximately $\chi^2_{(k-1)}$.

Now take $a_j = \sqrt{p_j}$, so that $\sum a_j^2 = \sum p_j = 1$ as required. Then Z_1 is the standard form of the variate N:

$$Z_1 \equiv \sum \sqrt{p_j}\, U_j \equiv (N - \sum \mu_j) \div \sqrt{\sum \mu_j}.$$

Conditioning on Z_1 is thus equivalent to conditioning on N; the conditional distribution of V given N is approximately $\chi^2_{(k-1)}$.

Now let D be the usual goodness of fit criterion. After a little algebra we find that

$$D \equiv \sum_{j=1}^{k} \frac{(Y_j - np_j)^2}{np_j} \equiv \frac{\sum \mu_j}{n}[\sum_{j=1}^{k} U_j^2 - Z_1^2] \equiv \frac{\sum \mu_j}{n} \cdot V.$$

Hence, given that $N = n$, we have

$$\frac{n}{\sum \mu_j} \cdot D \approx \chi^2_{(k-1)}.$$

Finally, since $N \equiv \sum Y_j$ has mean and variance $\sum \mu_j$, the ratio

$N \div \sum \mu_j$ has mean 1 and variance $1/\sum \mu_j$. Since the μ_j's are assumed to be large, the variance of the ratio is very small, and Chebyshev's Inequality (5.2.7) implies that the ratio will be close to 1 with very high probability. Thus we have $D \approx \chi^2_{(k-1)}$ as required.

11.2 Examples: Approximate Goodness of Fit Tests

In this section we give several examples in which the goodness of fit criterion (11.1.1) and the approximation (11.1.3) are used to compare sets of observed and expected frequencies.

Example 11.2.1. Table 11.2.1 shows the observed frequencies in 100 rolls of a die, taken from Example 1.4.1. The expected frequencies are computed on the assumption that the six faces are equally probable.

Table 11.2.1

Frequency Table for 100 Rolls of a Die

Face j	1	2	3	4	5	6	Total
Observed frequency f_j	16	15	14	20	22	13	100
Expected frequency e_j	16.67	16.67	16.67	16.67	16.67	16.67	100.02
$(f_j - e_j)^2/e_j$	0.03	0.17	0.43	0.67	1.71	0.81	3.82

The last row of the table shows the calculation of the observed value, $d = 3.82$, for the goodness of fit criterion (11.1.1). Since all of the expected frequencies are fairly large, the approximation (11.1.3) may be used. There are $k = 6$ classes, and no parameters were estimated from the data $(r = 0)$. Hence the degrees of freedom is $k - r - 1 = 5$, and Table B4 gives

$$SL \approx P(\chi^2_{(5)} \geq 3.82) \approx 0.6.$$

If the model were correct, so that each face of the die had a probability of exactly $\frac{1}{6}$, a discrepancy of 3.82 or more would occur about 60% of the time. The observed discrepancy is certainly not unusually large, and hence the observed frequencies are consistent with the model.

Example 11.2.2. Table 11.2.2 gives f_x, the number of time intervals out of 2608 in which x α-particles were recorded by a counter, for x = 0,1,2,... . The table also gives the corresponding expected frequencies e_x under the assumption of a Poisson distribution. The mean of the Poisson distribution was estimated from the data to be 3.870; see Example 4.4.2. The observed value of the goodness of fit criterion is d = 12.90. Since all of the expected frequencies e_x are fairly large, the approximation (11.1.3) may be used. There are k = 11 classes and one parameter was estimated from the data (r = 1). Hence the degrees of freedom is k − r − 1 = 9, and Table B4 gives

$$SL \approx P(\chi^2_{(9)} \geq 12.90) \approx 0.17.$$

Table 11.2.2

Observed and Expected Frequencies with which

x α-particles were Recorded

x	f_x	e_x	$(f_x - e_x)^2/e_x$	x	f_x	e_x	$(f_x - e_x)^2/e_x$
0	57	54.4	0.12	6	273	253.8	1.45
1	203	210.5	0.27	7	139	140.3	0.01
2	383	407.4	1.46	8	45	67.9	7.72
3	525	525.5	0.00	9	27	29.2	0.17
4	532	508.4	1.10	≥10	16	17.1	0.07
5	408	393.5	0.53	Total	2608	2608.0	12.90

If the Poisson model were correct, the chance of obtaining such a large discrepancy between the observed and expected frequencies would be about 17%. The observed value 12.90 is therefore not unusually large, and the Poisson distribution gives a reasonably good overall fit to the data.

Example 11.2.3. A similar analysis may be applied to the observed and expected frequencies of flying bomb hits in Table 4.4.2. In order to be able to apply the approximation (11.1.3), we wish to have all of the expected frequencies fairly large. Hence we pool the data for x ≥ 4 to obtain a single class with observed frequency 8 and expected frequency 8.71; see Table 11.2.3. There are k = 5 classes, and one parameter was estimated (r = 1), giving k − r − 1 = 3. Hence

$$SL \approx P\{\chi^2_{(3)} \geq 1.02\} \approx 0.8,$$

and the Poisson distribution gives a very good fit to the data.

Table 11.2.3
Observed and Expected Frequencies with which
x   Flying-Bomb Hits were Recorded

No. of hits x	0	1	2	3	≥4	Total
Observed frequency f_x	229	211	93	35	8	576
Expected frequency e_x	226.74	211.39	98.54	30.62	8.71	576.00
$(f_x - e_x)^2/e_x$	0.02	0.00	0.31	0.63	0.06	1.02

Example 11.2.4. Consider the 109 waiting times between accidents
which we previously discussed in Sections 1.2 and 1.4. Table 1.4.1
gives the observed number of waiting times in each of twelve classes.
The expected frequencies were computed under the assumption that the
waiting time has an exponential distribution. The mean of the exponential distribution was estimated from the data to be 241 days.

Table 11.2.4
Observed and Expected Frequencies for Waiting
Times between Mining Accidents

Class	f_j	e_j	$(f_j - e_j)^2/e_j$
[0,50)	25	20.42	1.03
[50,100)	19	16.60	0.35
[100,150)	11	13.49	0.46
[150,200)	8	10.96	0.80
[200,250)	9	8.91	0.00
[250,300)	7	7.24	0.01
[300,350)	11	5.88	4.46
[350,400)	6	4.78	0.31
[400,600)	5	11.69	3.83
[600,∞)	8	9.04	0.12
Total	109	109.01	11.37

The last two classes in Table 1.4.1 have quite small expected frequencies. In order that we may use the approximation (11.1.3),
we combine these with the third class prior to computing d; see Table
11.2.4. There are now k = 10 classes, and r = 1 parameter was estimated, so that

$$SL \approx P\{\chi^2_{(8)} \geq 11.37\} \approx 0.2.$$

If the exponential model were correct, the chance of obtaining such a

large discrepancy between the observed and expected frequencies would
be about 20%. Therefore, this test does not provide evidence against
the exponential model. Nevertheless, an exponential distribution with
a constant mean θ does not provide a satisfactory model for the accident data. As was pointed out in Section 1.4, the mean time between
accidents is not constant. In fact, it is apparent from Table 1.4.1
that there are serious difficulties in the right hand tail of the distribution. These difficulties are not reflected in the significance
level because the last three classes were pooled.

This example illustrates an important general point: a large
significance level cannot be interpreted as proof that the probability
model is correct. There may still be substantial inconsistencies between the model and the data, but of a type to which this particular
test of significance is not sensitive.

Note. In the preceding three examples, we estimated the unknown parameter before pooling the data rather than after. As we noted in Section 11.1, the appropriate degrees of freedom will then be slightly
larger than $k - r - 1$, but less than $k - 1$. Since $r = 1$ in all three
examples, the appropriate degrees of freedom would lie between $k - 2$
and $k - 1$, whereas we used $k - 2$. Increasing the degrees of freedom
would slightly increase the significance level in each case, but the
general conclusions would not be affected.

Example 11.2.5. A sociologist wished to determine whether the propen-
sity of an individual to become an alcoholic depends upon the size of
the family from which he comes. Accordingly, he determined the number
of siblings (brothers and sisters) of each of 242 alcoholics, with
the following results:

No. of siblings	0	1	2	3	4	5	6	7	8	9	10	11	Total
Observed frequency	21	32	40	47	29	23	20	11	10	3	3	3	242

From census data, he determined the number a_j of families with j
children in the entire population:

j	a_j	j	a_j	j	a_j
1	207,756	5	32,080	9	2,859
2	156,111	6	18,128	10	1,353
3	95,779	7	10,511	11	575
4	56,275	8	5,621	12	326

The total number of children in the entire population is

$$N = \sum_j ja_j = 1,469,626.$$

Of these, the number belonging to families having j children is ja_j, and each such child has $j-1$ siblings. Hence, if a child is randomly chosen from the population, the probability that he has exactly $j-1$ siblings is

$$p_{j-1} = \frac{ja_j}{N}, \qquad j = 1,2,\ldots,12.$$

If 242 children are randomly selected, the expected number having exactly $j-1$ siblings is

$$e_{j-1} = 242\, p_{j-1} = 242\, ja_j/N; \qquad j = 1,2,\ldots,12.$$

Table 11.2.5 now shows the observed frequencies, and the frequencies that would be expected if the 242 alcoholics were a random sample from the whole population. Clearly, there are more alcoholics from large families and many fewer from small families than would be expected in a random sample from the population represented by the census.

Table 11.2.5
Observed and Expected Numbers of Alcoholics
having Exactly j Brothers and Sisters

j	f_j	e_j	$(f_j - e_j)^2/e_j$	j	f_j	e_j	$(f_j - e_j)^2/e_j$
0	21	34.21	5.10	6	20	12.12	5.12
1	32	51.41	7.33	7	11	7.40	1.75
2	40	47.32	1.13	8	10	4.24	
3	47	37.07	2.66	9	3	2.23	14.44
4	29	26.41	0.25	10	3	1.04	
5	23	17.91	1.45	11	3	0.64	
				Total 242		242.00	39.23

To see whether the observed departures could reasonably have arisen by chance, we compute the significance level of the data using the goodness of fit criterion (11.1.1) and the approximation (11.1.3). After pooling the last four classes we obtain $d = 39.23$, based on $k = 9$ classes. The f_j's were not used to estimate any parameters ($r = 0$), and hence

$$SL \approx P\{\chi^2_{(8)} \geq 39.23\} < 0.001.$$

The observed discrepancy is much too great to be attributable to chance. There is strong evidence that the distribution of family size for alcoholics is different from that in the census.

There is some doubt here about the appropriateness of the census data used. It is not clear how "family size" should be defined, what part of the census data is appropriate, or even which year's census should be used. Because of these difficulties, one would be reluctant to conclude that children from large families are more likely to become alcoholics.

It is important to remember that the goodness of fit criterion (11.1.1) and the approximation (11.1.3) are appropriate for comparing observed frequencies f_j with their expected values e_j. For instance, it would be incorrect to use them to compare the fractions $f_j/(j+1)$ with their expected values $e_j/(j+1)$. For further discussion, see D.A. Sprott, Use of Chi-square, *Journal of Abnormal and Social Psychology* 69 (1964), pages 101–3.

Problems for Section 11.2

1. In a biological experiment, a square millimeter of yeast culture was subdivided into 400 equal-sized squares, and the number of yeast cells in each small square was recorded. The results are summarized in the following frequency table:

Number of cells	0	1	2	3	4	5	6	≥7
Frequency observed	129	137	83	38	10	2	1	0

    If yeast cells are randomly distributed over the area examined, the number of yeast cells per square should have a Poisson distribution. Test whether a Poisson model is consistent with the data.

†2. (a) A city police department kept track of the number of traffic accidents involving personal injury on sixty week-day mornings. The results were as follows:

Number of accidents	0	1	2	3	4	5	≥6
Frequency observed	17	17	16	7	2	1	0

    Is a Poisson distribution model consistent with these data?

   (b) The police department also records the number of persons injured in traffic accidents for the same sixty mornings, with the following results:

Number injured	0	1	2	3	4	5	6	7	≥8
Frequency observed	17	8	9	8	10	4	2	2	0

    If injuries were randomly distributed over time, the number of

injuries per morning would have a Poisson distribution. Show
that this model is contradicted by the data, and indicate
which of the assumptions for a Poisson process is violated.

3. Of the 83 accidents recorded in 2(a), 22 occurred on Mondays,
13 on Tuesdays, 11 on Wednesdays, 12 on Thursdays, and 25
on Fridays. Are these results consistent with the hypothesis that
accidents are equally likely to occur on any day of the week?

4. Six coins are tossed, and the number of heads is recorded. The
following table gives the observed frequencies in 210 repetitions
of this experiment:

Number of heads	0	1	2	3	4	5	6	Total
Frequency observed	3	19	49	58	57	22	2	210

Test whether these results are consistent with the hypothesis that
trials are independent and the coins are balanced.

5. Twelve dice were rolled 26306 times. Each time, the number of
dice showing 5 or 6 uppermost was recorded. The results are
summarized in the following table:

No. of 5's and 6's	0	1	2	3	4	5	6
Frequency observed	185	1149	3265	5475	6114	5194	3067

No. of 5's and 6's	7	8	9	10	11	12	Total
Frequency observed	1331	403	105	14	4	0	26306

Compute expected frequencies under the assumption that trials are
independent and the dice are balanced. Test for consistency, and
give a possible explanation for the poor agreement.

†6. In an experiment on human behaviour, a sociologist asks four men
and four women to seat themselves at a rectangular table. There
are three chairs at each side of the table, and one chair at each
end. The two end seats are special in that they give the people
sitting there a dominant position at the table.

(a) If people choose their seats at random, what is the probability
that the end seats will be occupied by two men?

(b) The experiment is performed 28 times using new subjects each
time. It is observed that the end seats are occupied by two
men on 10 occasions, by two women on 4 occasions, and by a
man and a woman on the other 14 occasions. Test whether
these results are consistent with the assumption that people
choose their seats at random.

7. Carry out goodness of fit tests in the following problems:
9.1.4(b), 9.1.5(b), 9.1.9, 9.1.10(b), 9.7.3, 10.1.7(a), 10.1.12(c).

11.3 Examples: Exact Goodness of Fit Tests

When the model does not involve unknown parameters, the exact significance level in a goodness of fit test is obtained as a sum of the multinomial probabilities (11.1.2). The sum is taken over all sets of frequencies $(f_1, f_2, \ldots, f_k)$ for which the value of the goodness of fit criterion (11.1.1) is at least as great as its observed value d. In this section, we illustrate this computational procedure in the cases k = 2 and k = 3, and compare the exact results with what would be obtained from the approximation (11.1.3).

Binomial case (k = 2)

When k = 2, there are just two classes, which we may call success and failure. Let X be the number of successes and n - X the number of failures in n independent repetitions. Their expected values are np and n(1 - p), where the probability p of success is determined by the probability model. The frequency table now appears as follows:

Class	Success	Failure	Total
Observed frequency	X	n - X	n
Expected frequency	np	n(1 - p)	n

The goodness of fit criterion (11.1.1) now becomes

$$D \equiv \frac{[X - np]^2}{np} + \frac{[(n - X) - n(1 - p)]^2}{n(1 - p)} \equiv \frac{[X - np]^2}{np(1 - p)} . \qquad (11.3.1)$$

The multinomial distribution (11.1.2) simplifies to a binomial distribution:

$$P(x) = \binom{n}{x} p^x (1 - p)^{n-x} \quad \text{for} \quad x = 0, 1, \ldots, n.$$

In an exact test, we compute d, the observed value of D, and then sum P(x) over all x such that $D \geq d$. This amounts to adding up the probabilities of all values of X which are at least as far from the mean np as is the observed value of X.

If np and n(1 - p) are large, the binomial distribution can be approximated by a normal distribution N(np, np(1 - p)); see Section 6.8. It follows that

$$\frac{X - np}{\sqrt{np(1 - p)}} \approx N(0, 1). \qquad (11.3.2)$$

But, from Example 6.9.1, the square of a $N(0,1)$ variate has a χ^2 distribution with one degree of freedom, and hence

$$D \equiv \frac{(X - np)^2}{np(1 - p)} \approx \chi^2_{(1)}, \qquad (11.3.3)$$

provided that the expected frequencies are large. Now we have

$$SL = P(D \geq d) \approx P(\chi^2_{(1)} \geq d),$$

which is the special case $k = 2$, $r = 0$ of (11.1.3). In this case, the χ^2 approximation to the distribution of the goodness of fit statistic D is equivalent to the normal approximation to the binomial distribution.

Example 11.3.1. In a particular Ontario county, a very large number of people are eligible for jury duty, and half of these are women. The judge is supposed to prepare a jury list by randomly selecting individuals from all those eligible. In an important 1974 murder trial, the jury list of 82 people contained 58 men and 24 women. Could such an extreme imbalance in the sexes reasonably have occurred by chance?

Solution. Suppose that the jury list is drawn at random. Successive draws will be practically independent because of the large number eligible, and the probability of drawing a male is 0.5. Under this model, the expected frequencies of males and females will be equal:

Class	Male	Female	Total
Observed frequency	58	24	82
Expected frequency	41	41	82

By (11.3.1), we have

$$D \equiv \frac{(X - 41)^2}{20.5} \; ; \qquad d = \frac{(58 - 41)^2}{20.5} = \frac{17^2}{20.5} = 14.10.$$

The significance level is then

$$SL = P(D \geq d) = P\left\{\frac{(X - 41)^2}{20.5} \geq \frac{17^2}{20.5}\right\} = P\{|X - 41| \geq 17\}$$

$$= P(X \geq 58) + P(X \leq 24)$$
$$= \sum_{x \geq 58} P(x) + \sum_{x \leq 24} P(x)$$

where $P(x) = \binom{82}{x}(\frac{1}{2})^{82}$. We first compute $P(58)$, and then determine $P(59), P(60), \ldots$ recursively as in Section 2.5. Each value greater than 64 has probability less than 10^{-7}.

x	P(x)	x	P(x)
58	.0000674	62	.0000013
59	.0000274	63	.0000004
60	.0000105	64	.0000001
61	.0000038		
		Total	.0001109

By symmetry, $P(X \leq 24) = P(X \geq 58)$, and hence the exact significance level is

$$SL = 2(0.00011) = 0.00022.$$

The chance of obtaining such an extreme imbalance in the sexes is extremely small, and it is almost certain that the jury was not selected at random. This result was used by the defence in arguing for a retrial.

An approximate significance level can be obtained using the χ^2 approximation (11.3.3), or equivalently, the normal approximation (11.3.3):

$$SL \approx P\{\chi^2_{(1)} \geq 14.10\} = P\{|Z| \geq 3.75\} < 0.001$$

from Table B4 or Table B2. If more extensive tables are consulted, one finds the approximate significance level to be 0.00017. If a correction for continuity is used (see Section 6.8), we obtain

$$SL \approx P\left\{\chi^2_{(1)} \geq \frac{16.5^2}{20.5}\right\} = P\{|Z| \geq \frac{16.5}{\sqrt{20.5}}\} = 0.00029.$$

The exact value lies about midway between the two approximations.

Trinomial case (k = 3)

When $k = 3$ and there are no unknown parameters in the model, the exact significance level will be obtained as a sum of trinomial probabilities. Now (11.1.3) and (6.9.1) give

$$SL \approx P(\chi^2_{(2)} \geq d) = \int_{d}^{\infty} \frac{1}{2} e^{-x/2}\, dx = e^{-d/2}, \tag{11.3.4}$$

where d is the observed value of the goodness of fit criterion.

Example 11.3.2. Two coins are tossed, and the total number of heads is recorded. The model assumed is that the coins are balanced and tossed independently, so that the probabilities of 0,1,2 heads are

$$p_0 = 0.25, \quad p_1 = 0.50, \quad p_2 = 0.25.$$

In $n = 8$ independent repetitions the expected frequencies will be $e_1 = 2$, $e_2 = 4$, and $e_3 = 2$. The goodness of fit criterion (11.1.1) becomes

$$D(f_0, f_1, f_2) = \frac{1}{2}(f_0 - 2)^2 + \frac{1}{4}(f_1 - 4)^2 + \frac{1}{2}(f_2 - 2)^2,$$

and the probability function of (f_0, f_1, f_2) is

$$P(f_0, f_1, f_2) = \binom{8}{f_0 \ f_1 \ f_2}(\frac{1}{4})^{f_0}(\frac{1}{2})^{f_1}(\frac{1}{4})^{f_2} = \frac{8! \, 2^{f_1}}{f_0! f_1! f_2!}(\frac{1}{2})^{16}.$$

Table 11.3.1 gives the probability and value of D for each of the 45 possible outcomes (f_0, f_1, f_2) with $\sum f_j = 8$.

Table 11.3.1
Probabilities and Values of D
for a Goodness of Fit Example with $k = 3$ Classes

f_0	f_1	f_2	$2^{16}P(f_0,f_1,f_2)$	D	f_0	f_1	f_2	$2^{16}P(f_0,f_1,f_2)$	D
0	0	8	1	24.00	2	6	0	1792	3.00
0	1	7	16	16.75	3	0	5	56	9.00
0	2	6	112	11.00	3	1	4	560	4.75
0	3	5	448	6.75	3	2	3	2240	2.00
0	4	4	1120	4.00	3	3	2	4480	0.75
0	5	3	1792	2.75	3	4	1	4480	1.00
0	6	2	1792	3.00	3	5	0	1792	2.75
0	7	1	1024	4.75	4	0	4	70	8.00
0	8	0	256	8.00	4	1	3	560	4.75
1	0	7	8	17.00	4	2	2	1680	3.00
1	1	6	112	10.75	4	3	1	2240	2.75
1	2	5	672	6.00	4	4	0	1120	4.00
1	3	4	2240	2.75	5	0	3	56	9.00
1	4	3	4480	1.00	5	1	2	336	6.75
1	5	2	5376	0.75	5	2	1	672	6.00
1	6	1	3584	2.00	5	3	0	448	6.75
1	7	0	1024	4.75	6	0	2	28	12.00
2	0	6	28	12.00	6	1	1	112	10.75
2	1	5	336	6.75	6	2	0	112	11.00
2	2	4	1680	3.00	7	0	1	8	17.00
2	3	3	4480	0.75	7	1	0	16	16.75
2	4	2	6720	0.00	8	0	0	1	24.00
2	5	1	5376	0.75					

Suppose that we observe the frequencies (1,2,5). The observed value of D is then 6.00. The exact significance level is obtained by summing the probabilities of all outcomes (f_0, f_1, f_2) for which D ≥ 6.00. There are 22 such outcomes, and their total probability is

$$SL = 3904 \div 2^{16} = 0.060.$$

If the model were correct, there would be only a 6% chance of obtaining such a large discrepancy. Hence the observed frequencies (1,2,5) provide some (rather weak) evidence against the model.

By (11.3.4), the approximate significance level when d = 6.00 is

$$SL \approx P(\chi^2_{(2)} \geq 6) = e^{-3} = 0.050.$$

The exact and approximate significance levels may be computed in this way for each possible value of D, and these values are given in Table 11.3.2. The agreement between the exact and approximate significance levels is suprisingly good in view of the rather small expected frequencies 2,4,2.

Table 11.3.2

Exact and Approximate Significance Levels

in a Goodness of Fit Example with k = 3 Classes

d	$P(D \geq d)$	$P(\chi^2_{(2)} \geq d)$	d	$P(D \geq d)$	$P(\chi^2_{(2)} \geq d)$
0.00	1.000	1.000	6.75	0.0391	0.0342
0.75	0.897	0.687	8.00	0.0151	0.0183
1.00	0.597	0.607	9.00	0.0102	0.0111
2.00	0.460	0.368	10.75	0.0085	0.0046
2.75	0.371	0.253	11.00	0.0050	0.0041
3.00	0.248	0.223	12.00	0.0016	0.0025
4.00	0.142	0.135	16.75	0.0008	0.0002
4.75	0.108	0.093	17.00	0.0003	0.0002
6.00	0.060	0.050	24.00	0.0000	0.0000

Problems for Section 11.3

1. A radio station claims that its weather forecast is correct 75% of the time. A listener decides to test this claim by taking note of the weather on the next 8 Saturdays. He finds that, 5 times out of 8, the forecast is incorrect. Are these results compatible with the station's claims?

†2. In research on drugs to counteract the intoxicating effect of alco-

hol, twenty subjects were used to compare the relative merits of benzedrine and caffeine. Each subject received the drugs in a random order on two different occasions far enough apart to eliminate carry-over effects. Benzedrine brought about the more rapid recovery in 14 subjects, while caffeine was judged better in the other 6 cases. Are these results consistent with the hypothesis that the two drugs are equally good?

3. In January, party A won 53% of a very large number of votes in an election. Six months later, a poll of 200 randomly selected votors showed that only 48% would vote for party A if another election were called. Could these results reasonably be due to chance, or is there evidence of a real change in the support for party A?

4. Each of 25 individuals is given a sample of butter and a sample of margarine in random order. 15 of them state a preference for the butter, while 10 prefer the margarine. Are these observations consistent with the hypothesis that there is no detectable difference between butter and margarine?

†5. A seed merchant states that 80% of the seeds of a certain plant will germinate. Each of 4 customers buys one packet and sows 100 seeds from it. The numbers of seedlings appearing are 73, 76, 74 and 77.

(a) Discuss whether any customer, on the basis of his observation only, has adequate cause to claim that the stated germination rate is erroneous.

(b) If the 4 packets are from a homogeneous stock of seed, is the total germination record consistent with the stated rate?

6. Determine whether each of the following is consistent with the hypothesis that the sex ratio (probability of a boy ÷ probability of a girl) is equal to one.

(a) 293 girls and 299 boys in 592 births;

(b) 2930 girls and 2990 boys in 5920 births;

(c) 29300 girls and 29900 boys in 59200 births.

7. (a) Suppose that 12 pairs of identical twins of whom one smokes and one does not are observed until one twin in each pair has died. It is observed that, of the twelve who have died, eight are smokers and four are non-smokers. Are these results consistent with the hypothesis that smokers and non-smokers have equal chances of dying first?

(b) Suppose that 60 pairs are observed. The smoker dies first in 40 cases and the non-smoker in 20 cases. Are these results consistent with the hypothesis in (a)?

†8. Suppose that the experiment described in Problem 11.2.6 was per-
     formed only ten times. The two end seats were occupied by men 4
     times and once by women. Using an exact test, determine whether
     these results are consistent with the hypothesis that seats are
     chosen at random.

11.4 Empirical Cumulative Distribution Function

         Suppose that we have n independent repetitions of an expe-
riment, yielding n observed values $x_1, x_2, \ldots, x_n$ of a variable X.
The probability model specifies that X is a variate with some particular cumulative distribution function F, say. The problem is to determine whether there is reasonable agreement between the theoretical
distribution and the sample values $x_1, x_2, \ldots, x_n$.

         A possibility which we have already considered is to compare
sets of observed and expected frequencies in a goodness of fit test.
Typically, X will have many possible values, and these are grouped
into a small number of classes in forming the frequency table; see
Example 1.2.1. This condensation of the data may conceal interesting
patterns, particularly in the tails of the distribution. Furthermore,
the choice of classes may be quite arbitrary, and a different grouping
of the data may lead to different conclusions about the fit of the
model.

         We now consider some informal graphical procedures which are
based on the comparison of cumulative frequencies. These methods do
not require the data to be grouped, and hence some of the difficulties
with goodness of fit tests are avoided.

Empirical c.d.f.

         Let x be any real value. Then K(x), the cumulative fre-
quency at point x, is defined to be the number of sample values which
do not exceed x. According to the model, the probability that any
particular sample value is at most x is

$$p = P(X \leq x) = F(x)$$

where F is the cumulative distribution function of X. Since trials
are independent, K(x) has a binomial (n,p) distribution. From
Section 6.3, the mean and variance of K(x) are

$$E\{K(x)\} = np; \quad var\{K(x)\} = np(1 - p).$$

Dividing $K(x)$ by the sample size n gives the fraction (relative frequency) of sample values which do not exceed x. The function S_n defined by

$$S_n(x) \equiv \frac{1}{n} K(x) \quad \text{for} \quad -\infty < x < \infty$$

is called the _empirical c.d.f._, or sample c.d.f. By (5.1.5) and (5.2.4), the mean and variance of $S_n(x)$ are

$$E\{S_n(x)\} = p; \quad \text{var}\{S_n(x)\} = \frac{p(1-p)}{n}$$

where $p = F(x)$.

Note that $E\{S_n(x)\} = F(x)$ for all x. Furthermore, since $\text{var}\{S_n(x)\} \to 0$ as $n \to \infty$, Chebyshev's Inequality (5.2.7) implies that, with probability 1, $S_n(x) \to F(x)$ as $n \to \infty$ for all values of x. Substantial differences between $S_n(x)$ and $F(x)$ may thus be taken as evidence that the theoretical distribution F is not appropriate. The graphical methods to be described here involve the comparison of $S_n(x)$ with $F(x)$.

The empirical c.d.f. for an observed sample $x_1, x_2, \ldots, x_n$ is a step function, with a step of size $\frac{1}{n}$ corresponding to each sample value. For instance, Figure 11.4.1 shows the empirical c.d.f. for the following sample of size $n = 10$:

$$1.2 \quad 0.7 \quad 0.6 \quad 1.4 \quad 1.5 \quad 0.9 \quad 0.7 \quad 2.1 \quad 1.4 \quad 0.7$$

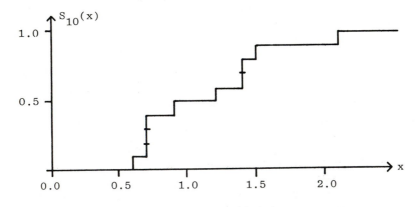

Figure 11.4.1

Empirical c.d.f. for a Sample of Size 10

Sample value 0.7 occurs three times, and hence the corresponding step size in $\frac{3}{10}$. Similarly, the step size for 1.4 is $\frac{2}{10}$.

From the e.c.d.f. one can recover the entire sample arranged in order of magnitude. For example, from Figure 11.4.1 we see that the ordered sample of size 10 is

0.6 0.7 0.7 0.7 0.9 1.2 1.4 1.4 1.5 2.1

However we cannot recover the order in which these values occur in the original sample.

In general, from the empirical c.d.f. of a sample $x_1, x_2, \ldots, x_n$ we can recover the ordered sample, or order statistics,

$$x_{(1)} \leq x_{(2)} \leq x_{(3)} \leq \ldots \leq x_{(n)}$$

where $x_{(i)}$ denotes the ith smallest sample value. However, we cannot recover the order in which the values occur in the original sample. If we wished to question the independence of successive X-values, then we would need to know the order in which they occur. However, if the independence of successive values is to be assumed, the order in which the sample values occur is irrelevant, and there will be no loss of information in replacing the original sample by the order statistics or empirical c.d.f.

Comparison of empirical and theoretical c.d.f.'s

One way to compare the sample with the model is to plot S_n, the empirical c.d.f., on the same graph as the theoretical c.d.f. F; see Figure 11.4.2. If the difference between the two curves is "small", the model is judged to be satisfactory. As in comparing sets of frequencies, the question which arises is whether the observed difference is sufficiently small to have occurred by chance, or whether there is evidence that the model is defective. It would be desirable to have an objective test to determine whether the discrepancy is large or small, so that it is not necessary to rely completely on subjective impressions.

In setting up a test of significance to compare S_n with F, it is necessary to define a measure of the "distance" between S_n and F, and this can be done in many different ways. A particular measure which has been discussed extensively in the statistical literature is the Kolmogorov-Smirnov statistic,

$$D \equiv \sup_x |S_n(x) - F(x)|.$$

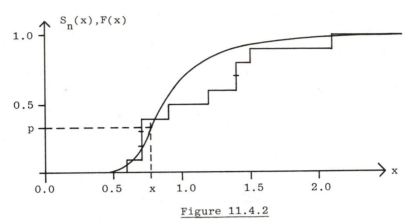

Figure 11.4.2

Comparison of Empirical and Theoretical C.d.f.'s

When $S_n(x)$ and $F(x)$ are both plotted against x as in Figure 11.4.2, D represents the maximum vertical separation between the two curves. An unusually large value of D would cast suspicion on the model. The significance level is the probability of obtaining a value of D as large as the observed value, assuming the model to be correct. This probability may be obtained from special tables of the distribution of D.

There is, however, a serious difficulty with the Kolmogorov-Smirnov statistic. The variance of S_n is largest for p near 0.5, and decreases to zero as $p \to 0$ or $p \to 1$. This means that we can expect fairly large differences between $S_n(x)$ and $F(x)$ near the centre of the distribution, but the differences in the tails of the distribution should be much smaller. Whether or not an observed difference d is large enough to cause concern will thus depend upon whether it occurs near the centre of the distribution or in the tails. However, the Kolmogorov-Smirnov statistic involves only maximum separation of the curves, without regard to where it occurs. A test based on this statistic may well fail to detect substantial departures from the model if they occur in the tails of the distribution, while exaggerating the importance of departures in the middle of the distribution. This is similar to what would happen if we took the goodness of fit statistic to be $\sum(f_j - e_j)^2$, and ignored the fact that large deviations are more probable in larger classes. Because of this difficulty, tests based on the Kolmogorov-Smirnov statistic are not recommended.

Other distance measures could be considered. For instance one could divide the Kolmogorov-Smirnov measure by the standard deviation of $S_n(x)$, or one might look at the maximum horizontal separation between S_n and F. However, we shall leave this problem now, and turn to another graphical procedure based on the empirical c.d.f.

11.5 Quantile Plots

In Figure 11.4.2, the empirical c.d.f. $S_n(x)$ and the theoretical c.d.f. F(x) are both plotted against x. We now consider two additional graphical methods for comparing S_n with F.

In the first method, we plot $S_n(x)$ versus F(x) for selected values of x. One point in this graph is obtained by choosing a particular x in Figure 11.4.2, then taking the height of the theoretical c.d.f. as abscissa and the height of the empirical c.d.f. as ordinate in the new graph. The resulting graph is called a probability plot, or PP-plot, because the scale on each axis is probability.

In the second method, we plot $S_n^{-1}(p)$ versus $F^{-1}(p)$ for selected values of p. One point in this graph is obtained by choosing a particular probability p in Figure 11.4.2. The corresponding x-value from the theoretical distribution is taken as abscissa, and that from the empirical distribution is taken as ordinate. From Section 6.1, the value x such that F(x) = p is called the pth quantile (100pth percentile) of the distribution. Because of this, the graph is called a quantile plot, or QQ-plot.

In Section 11.4 we showed that $E\{S_n(x)\} = F(x)$, and that $S_n(x) \rightarrow F(x)$ as $n \rightarrow \infty$. Thus, if the theoretical model is correct, we should have $S_n(x) \approx F(x)$. Thus the ordinates and abscissae should be approximately equal in a PP-plot. If the model is correct, the points in this graph should be scattered about a straight line through the origin with unit slope. Similarly, we should have $S_n^{-1}(p) \approx F^{-1}(p)$ for all p, so points in a quantile plot should also be scattered about a straight line through the origin with unit slope. Systematic deviations from this line would indicate poor agreement between the empirical and theoretical c.d.f.'s. The type of deviation indicates how the actual distribution differs from the theoretical one. Hence a PP-plot or quantile plot may help us to select a more suitable model.

Preparation of quantile plots

The remainder of this section deals mainly with quantile plots because these are the more widely used, but similar comments also apply to probability plots.

By convention, the p-values used in generating the quantile plot are taken to be

$$p = \frac{i-.5}{n} \quad \text{for} \quad i = 1, 2, \ldots, n.$$

At the ith smallest sample value $x_{(i)}$, S_n increases from $\frac{i-1}{n}$ to $\frac{i}{n}$. Hence the x-value corresponding to $p = \frac{i-.5}{n}$ will be $x_{(i)}$. The ordinates in the quantile plot are thus the ordered sample values $x_{(1)}, x_{(2)}, \ldots, x_{(n)}$. The abscissae are the p-quantiles of F where

$$p = \frac{.5}{n}, \frac{1.5}{n}, \frac{2.5}{n}, \ldots, \frac{n-.5}{n}.$$

Example 11.5.1. The following sample of 50 values was obtained by computer using a random number generator. If the random number generator is working properly, these should look like 50 observations from the continuous uniform distribution on (0,1). The agreement may be checked by means of a quantile plot.

0.533	0.713	0.821	0.844	0.427
0.197	0.962	0.463	0.498	0.045
0.250	0.441	0.108	0.371	0.156
0.828	0.775	0.281	0.854	0.892
0.711	0.977	0.573	0.951	0.284
0.061	0.741	0.797	0.664	0.110
0.338	0.824	0.309	0.594	0.592
0.483	0.822	0.398	0.149	0.394
0.043	0.426	0.817	0.016	0.566
0.820	0.416	0.999	0.487	0.510

To get the ordinates for the quantile plot, we arrange the sample values in non-decreasing order:

$$0.016 \quad 0.043 \quad 0.045 \quad 0.061 \quad \ldots \quad 0.977 \quad 0.999.$$

By (6.2.1), the c.d.f. of U(0,1) is F(x) = x. The p-quantile of U(0,1) satisfies the equation $F(Q_p) = p$, and it follows that $Q_p = p$. Now for n = 50 and $p = \frac{.5}{n}, \frac{1.5}{n}, \ldots, \frac{n-.5}{n}$, we have

$$Q_p = 0.01, 0.03, 0.05, 0.07, \ldots, 0.97, 0.99$$

and these are the abscissae for the quantile plot. The 50 points

$$(.01, .016), (.03, .043), (.05, .045), \ldots, (.99, .999)$$

are plotted in Figure 11.5.1. As expected, these points are scattered about a straight line through the origin with unit slope, and there is no indication of serious departures from a uniform distribution. (The

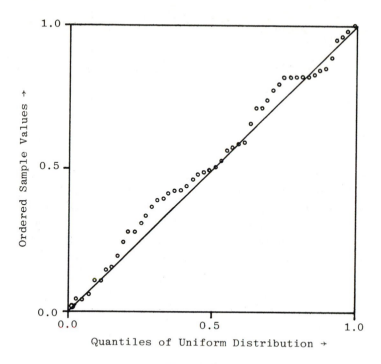

Figure 11.5.1

Quantile Plot for a Sample of Size 50 from
the Uniform Distribution U(0,1)

agreement with a uniform distribution could also be checked by grouping
the data to form a frequency table, and then performing a goodness of
fit test.)

Standardization

Usually the probability model will involve unknown parameters.
These may be estimated by the method of maximum likelihood, and their
estimated values may then be used in computing quantiles. Alternatively, one can sometimes avoid the necessity to estimate parameters by
suitably standardizing the theoretical distribution before computing
quantiles.

Suppose that a,b are constants with a > 0, and define
X' ≡ (X − b)/a. Let Q_p and Q'_p be the p-quantiles of X and X',
so that p = P(X ≤ Q_p) = P(X' ≤ Q'_p). Then

$$p = P(\frac{X - b}{a} \le Q'_p) = P(X \le aQ'_p + b) = P(X \le Q_p)$$

from which it follows that $Q_p = aQ'_p + b$. Thus, if the model is correct, we have

$$x_{(i)} \approx Q_p = aQ'_p + b; \quad p = \frac{i-.5}{n} .$$

If we plot $x_{(i)}$ versus Q'_p, we should get points scattered about a straight line with slope a and vertical intercept b.

If X has a normal distribution $N(\mu, \sigma^2)$, it is convenient to take $a = \sigma$ and $b = \mu$, so that $X' \equiv (X - \mu)/\sigma$ has a standardized normal distribution. The necessary quantiles of $N(0,1)$ can be obtained from tables. When we plot the ordered sample values $x_{(i)}$ against the $\frac{i-.5}{n}$ -quantiles of $N(0,1)$, we should obtain n points scattered about a straight line with slope σ and vertical intercept μ. Systematic departures from a straight line would indicate that the normal distribution does not provide an adequate model for the data.

If X has an exponential distribution with mean θ, then by (6.2.4), the p-quantile of X is

$$Q_p = -\theta \log(1 - p) = \theta Q'_p$$

where $Q'_p = -\log(1 - p)$ is the p-quantile of the unit exponential distribution $(\theta = 1)$. We have

$$x_{(i)} \approx Q_p = \theta Q'_p \quad \text{where} \quad p = \frac{i-.5}{n} .$$

Plotting $x_{(i)}$ versus Q'_p should give points scattered about a straight line through the origin with slope θ. This corresponds to taking $a = \theta$, $b = 0$ in the argument above.

Example 11.5.2. Quantile plots may be used to compare the mining accident data in Section 1.2 with an exponential distribution model. We noted previously that the mean time between accidents is increasing. Hence we shall prepare a quantile plot for only the first 20 waiting times, where the assumption of a constant mean is not too unrealistic. The ordinates in the quantile plot are the 20 ordered sample values:

4	11	15	15	31	36	50	55	59	72
93	96	120	124	137	176	203	215	315	378

The abscissae are the unit exponential quantiles $Q'_p = -\log(1 - p)$ where $p = \frac{i-.5}{20}$; $i = 1, 2, \ldots, 20$. These are found to be

```
0.03  0.08  0.13  0.19  0.25  0.32  0.39  0.47  0.55  0.64
0.74  0.86  0.98  1.12  1.29  1.49  1.74  2.08  2.59  3.69
```

The quantile plot in Figure 11.5.2 can now be prepared. The points are
scattered about a straight line through the origin, and hence there is
no indication of departures from an exponential distribution.

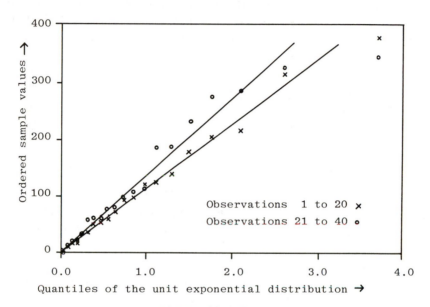

Figure 11.5.2

Quantile plot for mining accident data

Figure 2 also shows a quantile plot for the 21st to 40th
observations in the same example. Once again, the points are scattered
about a straight line, but the slope is greater, indicating a slightly
larger value of θ. The exponential distribution appears to give a
satisfactory model for these accident data, but as previously noted,
the average interval between accidents is increasing with time.

Example 11.5.3. Cuckoos lay their eggs in the nests of other birds.
The following are the lengths in millimeters of 24 cuckoos' eggs
found in nests of reed-warblers and wrens. (The data are from an article by O.H. Latter in *Biometrika* [1902].)

Lengths of eggs found in reed-warblers' nests			Lengths of eggs found in wrens' nests				
21.2	21.6	21.9	19.8	20.0	20.3	20.8	20.9
22.0	22.0	22.2	20.9	21.0	21.0	21.0	21.2
22.8	22.9	23.2	21.5	22.0	22.0	22.1	22.3

Because of the Central Limit Theorem (Section 6.7), many biological measurements have distributions that are nearly normal. We shall use quantile plots to check the agreement of these two data sets with normal distribution models.

For the 9 values in the first sample, the abscissae will be the p-quantiles of $N(0,1)$, where $p = \frac{.5}{9}, \frac{1.5}{9}, \ldots, \frac{8.5}{9}$. Table B2 gives $P(Z \le 1.59) = \frac{17}{18}$, and (6.6.4) then gives $P(Z \le -1.59) = \frac{1}{18}$. In this way, the abscissae are found to be

$$0, \pm 0.28, \pm 0.59, \pm 0.97, \pm 1.59.$$

A scatter plot of the 9 points

$$(-1.59, 21.2), \ (-0.97, 21.6), \ldots, (1.59, 23.2)$$

can now be prepared as in Figure 11.5.3. A plot of the 15 values in the second sample against the $\frac{i-.5}{15}$ -quantiles of $N(0,1)$ is shown in the same diagram.

In each case, the points are scattered about a straight line, indicating reasonable agreement with a normal distribution model. The lines have nearly equal slopes, indicating that the standard deviations are approximately the same. However, the plot for the first sample lies above that for the second, showing the greater average length of eggs laid in reed-warblers' nests. It appears reasonable to assume that lengths of eggs laid in reed-warblers' nests are distributed as $N(\mu_1, \sigma^2)$, while lengths of eggs laid in wrens' nests are $N(\mu_2, \sigma^2)$. It appears from the data that $\mu_1 > \mu_2$. It would be of some interest to know whether the observations give conclusive proof that $\mu_1 \ne \mu_2$, or whether as great a difference in average length could be attributed to chance. We shall consider this problem in Section 14.4.

Interpretation of Quantile Plots

If the points in a quantile plot deviate markedly from a straight line, the theoretical distribution F is in poor agreement with the data. The configuration of points in the plot gives an indication of the ways in which the actual distribution differs from the model assumed. For instance, in Figure 11.5.4(i), observed values in the lower tail of the distribution are larger than expected, and observed values in the right tail are also larger than expected. This means

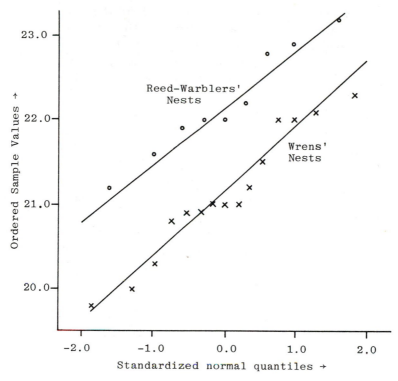

Figure 11.5.3

Normal probability plots for lengths of cuckoo's eggs

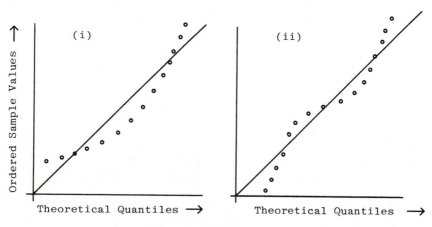

Figure 11.5.4

Interpretation of Quantile Plots

that the actual distribution has a shorter tail on the left and a longer tail on the right than has the theoretical distribution. In Figure 11.5.4(ii), observed values in the lower tail of the distribution are smaller than expected, and observed values in the upper tail are larger than expected. The extreme tails of the actual distribution contain more probability than do those of the theoretical distribution.

Even if the model were correct, the points in a quantile plot would not lie exactly in a straight line, but would involve some random scatter. We encounter the same question which arose in comparing frequency tables in Section 11.1 and distribution functions in Section 11.4: are the observed departures too great to be accounted for by chance alone? Some formal quantitative methods, analogous to goodness of fit tests and Kolmogorov-Smirnov tests, are discussed in the statistical literature. However, statisticians usually evaluate quantile plots informally, using personal judgement based on experience with similar plots. Such experience can be built up by using the computer to generate samples from a theoretical distribution (see Section 6.3) and prepare quantile plots. Modern computer facilities permit many such plots to be generated in a short time. By examining these, one can get a good indication of the frequencies with which various types of departures would occur by chance if the model were correct.

Probability Paper

An advantage of quantile plots is the ease with which they can be programmed for the computer. However, if they are to be prepared manually, the amount of labour involved can sometimes be reduced through the use of special graph paper. For instance, normal probability paper has selected percentiles of the standardized normal distribution marked in along one axis. From these, the necessary $\frac{i-.5}{n}$-quantiles may be located, and the use of tables is eliminated. Quantile plots for the exponential distribution may be prepared on semi-log paper. Probability paper is also available commercially for some other distributions.

Problems for Section 11.5

1. Use a goodness of fit test and a quantile plot to assess the adequacy of an exponential distribution model for the following data:

```
1.9   8.3  10.1  12.4  19.9  11.7   1.2   3.3  16.8  18.6
14.1  21.1   3.2   4.6   6.4   0.4   4.2   8.7  16.8  10.6
10.7   5.1  12.6   3.8   0.7   8.7  14.2  16.4   0.8   3.2
```

†2. The following are the examination marks of 36 students in one section of a statistics course:

```
70  74  83  59  31  85  75  86  67  61  40  62
78  64  76  88  89  58  71  57  79  86  76  74
88  61  80  80  73  75  90  42  70  65  77  85
```

Prepare a quantile plot to check the agreement of the data with a normal distribution, and discuss your findings.

3. The following are the measured resistances of 50 eight-ohm resistors produced by a small electronics firm.

```
8.02  7.89  7.92  8.08  8.14  7.86  7.96  8.06  7.88  7.90
8.20  8.10  8.04  8.06  8.00  7.94  7.98  8.12  7.92  8.02
8.00  8.00  8.04  8.02  7.98  8.18  7.96  7.94  8.16  8.04
8.14  7.90  7.84  7.82  7.96  8.02  8.04  7.98  8.02  8.02
7.92  8.06  7.94  8.04  8.02  7.92  8.00  7.98  8.00  8.02
```

Prepare a quantile plot to check their agreement with a normal distribution model.

4. Prepare a quantile plot for checking whether the observations

$$2.53 \quad 2.70 \quad 2.20 \quad 1.98 \quad 1.50$$

could reasonably have come from the continuous distribution with probability density function

$$f(x) = \frac{1}{x} \quad \text{for} \quad 1 < x < e.$$

Comment on the interpretation of the plot.

†5. An experiment involves measuring 5 angles in radians. It is hypothesized that the angles are independent and identically distributed, with probability density function

$$f(x) = \frac{1}{2} \sin x \quad \text{for} \quad 0 < x < \pi.$$

The measurements of the five angles were as follows:

$$1.7 \quad 0.6 \quad 2.6 \quad 1.1 \quad 1.4$$

Prepare a quantile plot, and comment on whether or not there is good agreement with the model.

6. Suppose that we have two samples $x_1, \ldots, x_n$ and $y_1, \ldots, y_m$, and denote their empirical c.d.f.'s by S_n and S_m. The two samples are combined to give a single sample of size $n + m$. Show that the e.c.d.f. for the combined sample is

$$S_{n+m}(x) = \frac{n}{n+m} S_n(x) + \frac{m}{n+m} S_m(x).$$

7. The probability density function for an exponential distribution with guarantee time $c > 0$ is

$$f(x) = \frac{1}{\theta} e^{-(x-c)/\theta} \quad \text{for} \quad c \leq x < \infty.$$

What would one expect to see if an ordered sample from this distribution were plotted against quantiles of the unit exponential distribution?

†8. Use a quantile plot to check the agreement of the model with the data in Problem 9.6.4.

*9. (a) Derive an expression for the p-quantile of the Weibull distribution (10.2.2). What would one expect to see if $x_{(i)}$ were plotted against $[-\log(1 - \frac{i-.5}{n})]^{1/\beta}$? if $\log x_{(i)}$ were plotted against $\log[-\log(1 - \frac{i-.5}{n})]$?

(b) Use the two graphical procedures in (a) to check the Weibull distribution model in Example 10.2.1.

†*10. (a) Check the agreement of the data in Problem 10.2.1 with a Weibull distribution model.

(b) As an alternate model, it is suggested that the log failure times are normally distributed. Check the agreement of the data in Problem 10.2.1 with this lognormal model.

CHAPTER 12. TESTS OF SIGNIFICANCE

Consider an experiment for which a probability model has been postulated. The model will generally be based upon a number of simplifications and assumptions whose appropriateness should be checked against the data. This is done by comparing observed values from the experiment with values to be expected according to the model. Some differences are to be anticipated owing to random variation. The question which arises is whether the observed differences are too great to be accounted for purely by chance. Tests of significance are used to answer such questions.

The goodness of fit tests discussed in Chapter 11 are examples of tests of significance. They are applicable in the situation where a set of observed (multinomial) frequencies is to be compared with a set of expected frequencies, and no particular type of departure is anticipated. In this chapter we give some further discussion of tests of significance, and derive tests suitable for some other situations.

We begin with a general discussion of tests of significance in Section 1. In Section 2 we distinguish between simple and composite hypotheses, and develop exact conditional tests for some composite hypotheses. Tests in contingency tables are discussed in Sections 3 and 4. Likelihood ratio tests are defined and illustrated in Sections 5 and 6, and tests defined in terms of tail areas are considered in Section 7. In Section 8 we define ancillary statistics and illustrate their use in defining conditional significance tests. Throughout this chapter, extensive use is made of the material on sufficient statistics in Section 9.8.

The power of a test statistic against an alternative hypothesis is defined in Section 9. This concept is sometimes helpful in a theoretical comparison of two or more possible test statistics.

12.1   Formulation and Interpretation of Significance Tests

The goodness of fit tests considered in Section 11.1 are
examples of significance tests. A test of significance is a procedure
for measuring the strength of the evidence provided by the experimental
data against an hypothesis. It involves the comparison of an observed
event with certain other outcomes which might have occurred. These outcomes are ranked according to how well they agree with the hypothesis.
Assuming the hypothesis to be true, one computes the total probability
of outcomes whose agreement with the hypothesis is no better than that
of the observed outcome. This probability is called the significance
level of the data in relation to the hypothesis. A significance level
near zero indicates that, if the hypothesis were true, a result as extreme as that observed would rarely occur, and hence there is evidence
that the hypothesis is false. A significance level near one indicates
that the observed event is not an extreme one, and hence there is no
evidence against the hypothesis.

Ingredients of a significance test

A test of significance has two ingredients. The first ingredient is the reference set, which consists of an observed event and
certain other events with which it is to be compared in the test. The
significance level is computed from a probability distribution defined
on the reference set under the assumption that the hypothesis is true.
The second ingredient is a ranking of the events in the reference set
according to how well they agree with the hypothesis. This ranking is
usually defined by means of a discrepancy measure D, also called the
test criterion or test statistic. D measures the "distance" of possible outcomes from what one would expect if the hypothesis were true.
The larger the value of D, the poorer the agreement between the outcome and the hypothesis.

When the reference set and ranking have been specified, the
significance level is computed by summing the probabilities of all e-
vents in the reference set whose agreement with the hypothesis is no
better than that of the observed event. If a discrepancy measure D
is used to define the ranking, the significance level is $P(D \geq d)$
where d is the observed value of D.

Example 12.1.1. Consider an experiment to investigate whether the
probability p that a child is male depends upon the age of the mother.
The sex of the child is recorded at n births to women more than forty
years old. It is known that, for the population as a whole, 52% of

single births produce males. The problem is to determine whether the experiment gives evidence that $p \neq 0.52$.

We assume that births are independent and that p is the same at each. Then X, the number of male births, is a sufficient statistic for p; see Example 9.8.1. In testing $H: p = 0.52$, we ignore the order in which male and female births occur. The reference set for the test is the range of possible values of X. When H is true, X has a binomial distribution with $p = 0.52$:

$$f(x) = \binom{n}{x}(.52)^{x}(.48)^{n-x} \quad \text{for} \quad x = 0,1,\dots,n,$$

and this distribution is used in computing the significance level.

The second ingredient of the test is a ranking of the possible values of X according to how well they agree with $H: p = 0.52$. One possibility is to rank them according to their distance from the mean $np = 0.52n$; that is, we take $D \equiv |X - 0.52n|$. Some other test criteria are considered later.

(a) Suppose that $n = 10$, so that $D \equiv |X - 5.2|$. An outcome x is considered to be in good agreement with $H: p = 0.52$ if $|x - 5.2|$ is small, and in poor agreement if $|x - 5.2|$ is large. The outcomes are thus ranked as follows:

0 (worst agreement), 10,1,9,2,8,3,7,4,6,5 (best agreement).

The significance level is the sum of the probabilities of all those values of X which give no better agreement than the observed value.

If we observed $X = 2$, the observed value of D would be $d = |2 - 5.2| = 3.2$, and the significance level would be

$$\begin{aligned} SL &= P\{|X - 5.2| \geq 3.2\} \\ &= f(0) + f(10) + f(1) + f(9) + f(2) = 0.0568. \end{aligned}$$

If $H: p = 0.52$ were true, the chance of getting such poor agreement would be only 0.0568, and hence there is some indication that $p \neq 0.52$. However, the evidence is rather weak, and it would be desirable to collect additional data to settle the question.

If we observed $X = 6$, the significance level would be

$$\begin{aligned} SL &= P\{|X - 5.2| \geq 0.8\} \\ &= f(0) + f(10) + f(1) + \dots + f(4) + f(6) \\ &= 1 - f(5) = 0.756. \end{aligned}$$

If $H: p = 0.52$ were true, the chance of getting such poor agreement would be 0.756. The large significance level indicates good agreement

between the hypothesis and the data. There is no evidence that the
hypothesis is false. Neither is there evidence that H:p = 0.52 is
true, since there will be many other values of p which are also in
good agreement with the data.

      (b) Suppose that n = 50, so that $D \equiv |X - 26|$. Outcomes
are now ranked as follows:

    0 (worst);1;2 & 50 (equal); 3 & 49;...;25 & 27;26 (best).

If we observed X = 16, the significance level would be

$$SL = P\{|X - 26| \geq 10\} = P\{X \leq 16\} + P\{X \geq 36\}$$
$$= f(0) + f(1) + \ldots + f(16) + f(36) + f(37) + \ldots + f(50).$$

These binomial probabilities may be computed recursively and summed to
give SL = 0.0066. Alternatively, SL may be found approximately as
illustrated in Example 11.3.1. If H:p = 0.52 were true, an outcome
as extreme as X = 16 would very rarely occur. Hence the observed result X = 16 gives strong evidence against H:p = 0.52.

Choice of the reference set

      In choosing the reference set for a test of significance, we
are focusing attention on the part of the data which carries information relevant to the hypothesis in question. In the preceding example,
we assumed independent trials with a constant value of p, and the
hypothesis concerned the value of p. The test was therefore based on
the distribution of X, the total number of males observed, which
carries all of the information about p.

      In general, if the hypothesis concerns the value of an unknown
parameter θ, and if T is a sufficient statistic for θ, then the
reference set is the range of possible values of T. The significance
level is computed from the distribution of T when the hypothesis is
true.

      Choice of an appropriate reference set and probability dis-
tribution may present difficulties when the hypothesis to be tested is
composite. Frequently a conditional reference set and distribution
would seem to be appropriate; see Section 12.2. Conditional tests are
also called for when ancillary statistics are present; see Section 12.8.

Rankings and discrepancy measures

      The choice of D (and hence the ranking for the test) will
depend upon what sort of departures from the hypothesis we wish to de-

tect. The goodness of fit statistic is a "general purpose" measure for
use with multinomial data when no particular type of departure is anticipated. It would not be appropriate if, before doing the experiment,
we had reason to expect that there might be a large deviation in one
particular cell of the table. Nor would it be appropriate for the
mining accident data in Example 11.2.4 if we wished to detect a possible change over time in the mean of the exponential distribution.

It is the ranking produced by the discrepancy measure D
which is important, not the actual magnitude of D. If two discrepancy
measures give the same ranking of events in the reference set, they
will give equal significance levels for the same set of data, and hence
they define equivalent tests. In Example 12.1.1, we took $D \equiv |X - np|$;
three other possible discrepancy measures are

$$D_1 \equiv |\hat{p} - p|; \quad D_2 \equiv (X - np)^2; \quad D_3 \equiv \frac{1}{np(1-p)} (X - np)^2$$

where $\hat{p} \equiv X/n$. The last of these is the goodness of fit criterion;
see Section 11.3. These three test statistics also rank values of X
according to their distance from the mean np. The significance level
will be the same whether we use D, D_1, D_2, or D_3 as the discrepancy
measure.

Ranking by distance from the mean seems reasonable when the
probability distribution is fairly symmetrical. If not, it is better
to use the <u>probability ranking</u>, in which outcomes having the smallest
probabilities under the hypothesis are defined to be those in poorest
agreement with the hypothesis. This is equivalent to taking D to be
some strictly decreasing function of $f(x)$, for instance $D(x) = 1/f(x)$,
where $f(x)$ is the probability of outcome x under the hypothesis.
For instance, when $n = 10$ and $p = 0.52$, the distribution of X in
Example 12.1.1 is as follows:

x	0	1	2	3	4	5	6	7	8	9	10
f(x)	.0006	.0070	.0343	.0991	.1878	.2441	.2204	.1364	.0554	.0133	.0014

The values of X in order of increasing probability are

0 (worst),10,1,9,2,8,3,7,4,6,5 (best).

In this case, the probability ranking is identical with that defined
by $D \equiv |X - np|$, and will thus lead to the same significance level.
However, different rankings would be obtained if we were testing a
value of p near 0 or 1, since then the binomial distribution is
highly skewed.

The use of the likelihood function to rank outcomes for a test of significance will be considered in Section 12.5. Rankings based on "tail areas" will be discussed in Section 12.7.

Interpretation

The test of significance in statistics is similar to the method of proof by contradiction in mathematics. In each case one assumes an hypothesis to be true, and then checks to see whether this assumption leads to an inconsistency. If a contradiction is found, the hypothesis is disproved. If no contradiction is found, the method of proof fails, and the hypothesis could be either true or false.

In statistical applications, there will rarely be a logical inconsistency between the hypothesis and the data. If we got 100 heads in 100 tosses of a coin, we could not prove mathematically that the coin was biassed, since this result could arise from 100 tosses of a balanced coin. Nevertheless, we would be quite sure that the coin was biassed, since a balanced coin would almost never produce such an extreme result.

The significance level is the probability of a result as extreme as that observed. A significance level close to zero means that such an extreme result would almost never occur if the hypothesis were true. The smaller the significance level, the greater the inconsistency, and hence the stronger the evidence that the hypothesis is false.

A large significance level, such as SL = 0.9, does not mean that the hypothesis is "probably true". The probability statement applies to the data, not the hypothesis. A large significance level merely indicates that no inconsistency has been detected by the test used. There will be many other hypotheses which, if tested in a similar way, would also show reasonable agreement with the data. Furthermore, there may be substantial departures from the hypothesis, but of a type to which the test is not sensitive; see Example 11.2.4.

In practice, a significance level of 5% or less is usually considered necessary before one can claim to have evidence against the hypothesis. Of course, this convention is arbitrary, and should not be taken too seriously. The interpretation of 4.9% and 5.1% significance levels will be much the same, even though these values lie on opposite sides of the magic 5% level.

Detection versus estimation

In Example 12.1.1, the significance level measures the strength of the evidence that p differs from 0.52, but it does not indicate how large the difference is. The following two numerical examples were considered earlier:

(a) n = 10, x = 2, SL = 0.0568;
(b) n = 50, x = 16, SL = 0.0066.

In (b), there is strong evidence that $p \neq 0.52$, whereas the evidence in (a) is weak. However, the difference $|\hat{p} - 0.52|$ is larger in (a) where $\hat{p} = 0.2$ than in (b) where $\hat{p} = 0.32$.

A small significance level shows only that there is a "real" departure from the hypothesis; that is, a departure which cannot readily be explained by chance. A small significance level does not mean that the departure is large or important, since with a large amount of data, very small departures can be detected. On the other hand, with a small sample, a large and important difference may go undetected. To investigate the magnitude of the difference, we can examine the relative likelihood function as in Section 9.4, or compute a confidence interval as in Section 13.1.

Tests suggested by the data

Examination of a data set will often reveal an interesting pattern which had not been anticipated before the experiment was performed. There is then a temptation to design a test of significance which will "prove" that this pattern could not have arisen by chance. This is a misuse of a significance test. For a valid test, one must specify what sort of discrepancy is being sought before the data are examined. For instance, there are 2^{100} possible sequences of outcomes from 100 coin tosses. With a little imagination, one could find an unusual and improbable pattern in any one of them, and then design a test guaranteed to give a small significance level! It would be a mistake to take such a pattern too seriously, unless it had been predicted in advance, or could be reproduced in subsequent experiments.

12.2 Tests of Composite Hypotheses

In testing problems, there will be a set of basic assumptions which are not being questioned, at least for the time being. We shall refer to this set of assumptions as the basic model. The hypothesis H is an additional statement about the model. Taken together, the basic model and hypothesis make up the hypothesized model. If the hypothesized model does not involve any unknown parameters, H is called a simple hypothesis; otherwise, H is called a composite hypothesis.

In Example 12.1.1, the basic model is that trials are independent, with a constant probability p of a male birth at each trial. The hypothesis $H:p = 0.52$ is a simple hypothesis, because it specifies a numerical value for the only unknown parameter. The hypothesis $H:p \leq 0.52$ is composite; it is composed of simple hypotheses $H:p = p_0$ where p_0 is a real number between 0 and 0.52.

More generally, in goodness of fit tests (Section 11.1), the basic assumption is that we have n independent repetitions of an experiment with k possible outcomes; that is, trials are independent, and the probability p_j of the jth outcome is the same at all n trials $(j = 1,2,...,k)$. The hypothesis H is a statement about the values of the p_j's. If H gives numerical values of $p_1, p_2, ..., p_k$, it is a simple hypothesis; see Examples 11.2.1, 11.2.5, 11.3.1, and 11.3.2. If H specifies the p_j's only as functions of one or more unknown parameters, it is a composite hypothesis; see Examples 11.2.1, 11.2.3, and 11.2.4.

If H is composite, the probability of the data cannot be found numerically, but only as a function of the unknown parameter(s) θ. If there is a large amount of data, it may be reasonable to replace θ by its maximum likelihood estimate $\hat{\theta}$, and then proceed as though θ were known. This is not a satisfactory procedure in small samples because then the estimate $\hat{\theta}$ may be quite imprecise. A possibility in this case would be to determine the range of plausible values for θ from the relative likelihood function, and then carry out a series of tests using several different values of θ.

In this section, we consider several examples in which there exists a sufficient statistic T for the unknown parameter θ. By (9.8.3), the conditional distribution of outcomes given the value of T does not depend upon θ. This conditional distribution, together with a suitably chosen discrepancy measure D, will give a test of significance for the composite hypothesis.

Comparison of Binomial Proportions

Consider a different experiment to determine whether the probability of a male child depends upon the age of the mother. The sex of the child is recorded at n births to mothers less than twenty years old, and at m births to mothers aged more than forty years. The basic assumptions are that births are independent, and that there is a constant probability p_1 of a male child for women under twenty, and p_2 for women over forty. We wish to determine whether there is evidence that $p_1 \neq p_2$, and we do this by testing the hypothesis $H:p_1 = p_2 = p$, say. The value of p is not given, and hence this is a composite hypothesis.

Under the basic model, the likelihood function of p_1 and p_2 is

$$L(p_1, p_2) = p_1^X(1 - p_1)^{n-x}p_2^Y(1 - p_2)^{m-y},$$

where X and Y are the numbers of males born to younger and older mothers, respectively. Under the hypothesis, there is just one unknown parameter p, with likelihood function

$$L(p) = p^{x+y}(1 - p)^{n+m-x-y} = p^t(1 - p)^{n+m-t}$$

where $T \equiv X + Y$. T is a sufficient statistic for p, and the test will be based on the conditional distribution of X and Y given the observed value of T.

Under the hypothesis, X and Y are independent binomial (n,p) and (m,p) variates, and T is a binomial $(n+m,p)$ variate. The conditional probability function of X and Y given that $T = t$ is thus

$$f(x,y|t) = \frac{\binom{n}{x}p^X(1 - p)^{n-x}\binom{m}{y}p^Y(1 - p)^{m-y}}{\binom{n+m}{t}p^t(1 - p)^{n+m-t}} = \frac{\binom{n}{x}\binom{m}{y}}{\binom{n+m}{t}}, \qquad (12.2.1)$$

where $x + y = t$. Since only pairs (x,y) with $x + y = t$ are considered, variable y is superfluous, and we may write (12.2.1) as a distribution of X only:

$$g(x) = \binom{n}{x}\binom{m}{t-x}/\binom{n+m}{t} \quad \text{for} \quad x = 0,1,\ldots,t. \qquad (12.2.2)$$

This is a hypergeometric distribution. It does not depend upon p, and will form the basis for a test of the composite hypothesis H. The reference set for the test is the set of pairs $(x,t-x)$ with probability distribution (12.2.2).

The second ingredient of the test is a ranking of outcomes $(x,t-x)$ according to how well they agree with $H:p_1 = p_2$. One possibility would be to rank outcomes according to their conditional probabilities (12.2.2). Alternatively, the mean of the hypergeometric distribution (12.2.2) is $\frac{nt}{n+m}$, and we could take

$$D \equiv |X - \frac{nt}{n+m}|. \qquad (12.2.3)$$

Yet another possibility is to take

$$D_1 \equiv |\hat{p}_1 - \hat{p}_2| \equiv |\frac{X}{n} - \frac{Y}{m}|.$$

Since only outcomes (x,y) with $x + y = t$ are compared in the test, we may substitute $Y \equiv t - X$ to give

$$D_1 \equiv |\frac{X}{n} - \frac{t-X}{m}| \equiv \frac{n+m}{nm}|X - \frac{nt}{n+m}| \equiv \frac{n+m}{nm} D.$$

Hence D and D_1 produce identical rankings of the reference set, and they will lead to the same (conditional) significance level.

Example 12.2.1. Suppose that births to 10 younger and 10 older mothers were observed, with the following results:

	Male birth	Female birth	Total
Mothers under 20	x = 7	3	10 = n
Mothers over 40	y = 2	8	10 = m
Total	t = 9	11	20

Are these results consistent with the hypothesis that the probability of a male birth is the same for both age groups?

Solution. In this case, (12.2.2) and (12.2.3) become

$$g(x) = \binom{10}{x}\binom{10}{9-x}/\binom{20}{9} \quad \text{for} \quad x = 0,1,\ldots,9;$$

$$D \equiv |X - 4.5|.$$

The observed discrepancy is $d = |7 - 4.5| = 2.5$, and hence

$$SL = P\{|X - 4.5| \geq 2.5 | T = 9\}$$
$$= g(0) + g(1) + g(2) + g(7) + g(8) + g(9)$$
$$= 0.070.$$

Such poor agreement would occur reasonably often (7 times in 100) if p_1 and p_2 were equal, and hence we cannot claim to have proof that $p_1 \neq p_2$.

Comparison of Poisson Means

Suppose that bacteria counts are made for $n + m$ water samples, each of volume v. These are taken from two different locations in a lake. At the first location there are n samples, with counts $X_1, X_2, \ldots, X_n$. At the second location there are m samples with counts $Y_1, Y_2, \ldots, Y_m$. We wish to determine whether there is evidence of different concentrations of bacteria at the two locations.

The basic assumptions are that bacteria are randomly distributed throughout the lake water, with λ_1 per unit volume at the first location and λ_2 per unit volume at the second location. The hypothesis to be tested is $H: \lambda_1 = \lambda_2 = \lambda$, say. Since the value of λ is not given, H is a composite hypothesis.

Under the basic model, the X_i's and Y_j's are independent Poisson variates with means $\lambda_1 v$ and $\lambda_2 v$, respectively. The joint distribution of the X_i's and Y_j's is thus

$$f(x_1, \ldots, x_n, y_1, \ldots, y_m) = \prod_{i=1}^{n} (\lambda_1 v)^{x_i} e^{-\lambda_1 v} / x_i! \cdot \prod_{j=1}^{m} (\lambda_2 v)^{y_j} e^{-\lambda_2 v} / y_j!,$$

and the likelihood function of λ_1 and λ_2 is

$$L(\lambda_1, \lambda_2) = \lambda_1^{\sum x_i} \lambda_2^{\sum y_j} e^{-v(n\lambda_1 + m\lambda_2)}.$$

The totals $T_1 \equiv \sum X_i$ and $T_2 \equiv \sum Y_j$ are sufficient statistics for λ_1 and λ_2. Under the hypothesis, there is only one unknown parameter λ, with likelihood function

$$L(\lambda) = \lambda^{\sum x_i + \sum y_j} e^{-v\lambda(n+m)}.$$

The grand total $T \equiv T_1 + T_2$ is a sufficient statistic for λ. The significance test will be based on the conditional distribution of T_1 and T_2 given the observed value of T.

By the corollary to Example 4.5.5, a sum of independent Poisson variates has a Poisson distribution. Hence, under the hypothesis, T_1 and T_2 are independent Poisson variates with means $n\lambda v$ and $m\lambda v$, respectively. Similarly, T has a Poisson distribution with mean $(n+m)\lambda v$. The required conditional distribution is thus

$$f(t_1, t_2 | t) = \frac{(n\lambda v)^{t_1} e^{-n\lambda v}}{t_1!} \frac{(m\lambda v)^{t_2} e^{-m\lambda v}}{t_2!} \Big/ \frac{[(n+m)\lambda v]^{t} e^{-(n+m)\lambda v}}{t!}$$

$$= \frac{t!}{t_1! t_2!} \frac{n^{t_1} m^{t_2}}{(n+m)^{t}} \quad \text{for} \quad t_1 + t_2 = t. \qquad (12.2.4)$$

The reference set for a test of the composite hypothesis $H: \lambda_1 = \lambda_2 = \lambda$ is the set of pairs (t_1, t_2) with $t_1 + t_2 = t$, and the significance level is computed from the conditional distribution (12.2.4).

Upon substituting $t_2 = t - t_1$ and $p = \frac{n}{n+m}$, we see that (12.2.4) is just a binomial distribution:

$$g(t_1) = \binom{t}{t_1} p^{t_1} (1 - p)^{t - t_1} \quad \text{for} \quad t_1 = 0, 1, \ldots, t.$$

The problem of testing the composite hypothesis $H: \lambda_1 = \lambda_2$ has thus been reduced to that of testing the simple hypothesis $H: p = \frac{n}{n+m}$ in a binomial distribution. The test can now be completed as in Example 12.1.1.

Example 12.2.2. Suppose that the bacteria counts were as follows:

Location 1: 0,2,0,1,1,2,2,0,2,0,0,1 $(n = 12, \ t_1 = 11)$
Location 2: 3,1,2,1,3,2,3,3,1,2,2,1,3,3,1 $(m = 15, \ t_2 = 31)$.

Here we have $p = \frac{12}{12 + 15} = \frac{4}{9}$, and the observed grand total is $t = 42$. A test of the hypothesis $\lambda_1 = \lambda_2$ will thus be based on the binomial distribution with parameters $(42, \frac{4}{9})$. The mean of this distribution is $tp = 18.67$, and we choose the discrepancy measure to be

$$D \equiv |T_1 - 18.67|.$$

The observed discrepancy is $|11 - 18.67| = 7.67$, and hence

$$SL = P\{|T_1 - 18.67| \geq 7.67 | T = 42\}.$$

The significance level can now be obtained exactly by summing binomial probabilities as in Example 12.1.1, or approximately using the normal approximation to the binomial distribution. We find that $SL \approx 0.02$, and hence there is reasonably strong evidence that the concentration of bacteria is different in the two locations.

Another natural choice for D would be $|\hat{\lambda}_1 - \hat{\lambda}_2|$. It can be shown that, since the value of T is held constant in the test, this will lead to the same significance level as $D \equiv |T_1 - tp|$.

The Hardy-Weinberg Law

In some simple cases, the inheritance of a characteristic is governed by a single gene which occurs in two forms, R and W say.

Each individual has a pair of these genes, one obtained from each parent, so that there are three possible genotypes: RR, RW, and WW. Suppose that the gene forms R and W occur with relative frequencies θ and $1 - \theta$, respectively, in both the male and female populations. Then it can be shown that, if mating is at random with respect to this gene, the proportions of individuals having genotypes RR, RW, and WW in the next generation will be

$$p_1 = \theta^2, \quad p_2 = 2\theta(1 - \theta), \quad \text{and} \quad p_3 = (1 - \theta)^2. \qquad (12.2.5)$$

Furthermore, if random mating continues, these proportions will remain approximately constant for generation after generation. This famous result from genetics is called the Hardy-Weinberg Law.

Suppose that n individuals (e.g. pea plants) are selected at random and are classified according to genotype. There are f_1 with genotype RR (red flowers), f_2 with genotype RW (pink flowers), and f_3 with genotype WW (white flowers), where $\sum f_j = n$. We wish to determine whether these observed frequencies are consistent with the Hardy-Weinberg Law (12.2.5).

The joint distribution of f_1, f_2, f_3 is trinomial, and substituting for the p_i's from (12.2.5) gives

$$P(f_1, f_2, f_3) = \binom{n}{f_1 \ f_2 \ f_3} [\theta^2]^{f_1} [2\theta(1 - \theta)]^{f_2} [(1 - \theta)^2]^{f_3}$$

$$= \binom{n}{f_1 \ f_2 \ f_3} 2^{f_2} \theta^{2f_1 + f_2} (1 - \theta)^{f_2 + 2f_3}. \qquad (12.2.6)$$

Hence the likelihood function of θ is

$$L(\theta) = \theta^t (1 - \theta)^{2n - t}$$

where $T \equiv 2f_1 + f_2$.

The maximum likelihood estimate of θ is $\hat{\theta} = t/2n$, and using this value we may compute expected frequencies

$$n\hat{\theta}^2, \quad 2n\hat{\theta}(1 - \hat{\theta}), \quad \text{and} \quad n(1 - \hat{\theta})^2.$$

The observed and expected frequencies can now be compared in a goodness of fit test. If the three expected frequencies are large, the χ^2 approximation (11.1.3) may be used. There are $k = 3$ classes and $r = 1$ parameter to estimate, so the degrees of freedom is $k - r - 1 = 1$.

If one or more of the expected frequencies is small, the approximation (11.1.3) may be inaccurate, and an exact test is advisable. Since T is a sufficient statistic for θ, we base the exact test on the conditional distribution of the f_j's, given the observed value of T.

The probability of the observed T-value may be obtained by summing the trinomial probabilities (12.2.6):

$$P(T = t) = \sum P(f_1, f_2, f_3)$$

where the sum extends over all (f_1, f_2, f_3) such that $2f_1 + f_2 = t$, and of course $f_1 + f_2 + f_3 = n$. This is actually a one-dimensional sum, since under these conditions we have

$$f_2 = t - 2f_1; \qquad f_3 = n - t + f_1. \qquad (12.2.7)$$

As soon as f_1 is specified, the values of f_2 and f_3 are determined by (12.2.7). Only one of the frequencies is free to vary in a conditional test, and this is directly related to the single degree of freedom in the χ^2 approximation.

In this problem, there is no simple algebraic expression for $P(T = t)$. To determine this probability, we list all sets of frequencies satisfying (12.2.7) and sum their probabilities. We then divide each of the probabilities by the total to obtain the required conditional probabilities. Because we consider only sets of frequencies for which $2f_1 + f_2 = t$, θ will cancel out of the ratio, and the conditional probabilities will not depend upon θ. Then, using the goodness of fit criterion or the probability ranking, we may compute the exact significance level.

Example 12.2.3. Suppose that $n = 20$ individuals were observed, and the observed frequencies of the three genotypes were as follows:

Genotype	RR	RW	WW	Total
Frequency	5(2.8)	5(9.4)	10(7.8)	20

Here we have $t = 2f_1 + f_2 = 15$, so that $\hat{\theta} = t/2n = 0.375$. The expected frequencies can now be obtained, and these are shown in parentheses above. Since one of them is rather small, an exact test is advisable.

In Table 12.2.1 we have listed all possible outcomes (f_1, f_2, f_3) with $\sum f_j = 20$ and $2f_1 + f_2 = 15$, together with their probabilities. All of the probabilities are proportional to the same function of θ,

$$c(\theta) = 10^{10} \theta^{15} (1 - \theta)^{25}.$$

We sum these probabilities to obtain $P(T = 15)$, which is also proportional to $c(\theta)$. When we divide the probabilities by their total, $c(\theta)$ cancels out and we obtain the conditional probabilities shown in the second last column. The last column gives the values of the goodness of fit criterion D. The observed discrepancy is 4.36, and we sum the conditional probabilities of outcomes with $D \geq 4.36$ to obtain

$$SL = 0.0126 + 0.0370 + 0.0028 + 0.0001 = 0.0525.$$

The probability ranking would give the same result in this case. We conclude that the observed frequencies do give some (rather weak) evidence against the Hardy-Weinberg Law.

Table 12.2.1

Calculations for an Exact Test of the Hardy-Weinberg Law

f_1	f_2	f_3	$P(f_1, f_2, f_3)$	$P(f_1, f_2, f_3 \mid T = 15)$	D
0	15	5	0.0508c	0.0126	7.20
1	13	6	0.4445c	0.1105	2.99
2	11	7	1.2383c	0.3078	0.60
3	9	8	1.4189c	0.3527	0.03
4	7	9	0.7095c	0.1764	1.28
*5	5	10	0.1490c	0.0370	4.36
6	3	11	0.0113c	0.0028	9.25
7	1	12	0.0002c	0.0001	15.96
	Total		4.0226c	0.9999	–

Problems for Section 12.2

1. In a pilot study, a new deoderant was found to be effective for 20% of the ten men tested, and for 80% of the five women tested. Are these data consistent with the hypothesis that the deoderant is equally effective for men and women?

†2. Two manufacturing processes produce defective items with probabilities p_1 and p_2, respectively. It was decided to examine four items from the first process and sixteen items from the second. In each case, two defectives were found. Are these results consistent with the hypothesis $p_1 = p_2$?

†3. Two manufacturing processes produce defective items with probabilities p_1 and p_2, respectively. Items were examined from the first process until the rth defective had been obtained, by which time there had been x_1 good items. The second process gave x_2 good items before the rth defective.

(a) Write down the joint probability function of X_1 and X_2. Show that, if $p_1 = p_2 = p$, then $T \equiv X_1 + X_2$ is a sufficient statistic for p. Hence set up a test of significance for the hypothesis $p_1 = p_2$.

(b) For each process, items were examined until $r = 2$ defectives had been found. Process 1 gave 2 good items, and process 2 gave 14 good items. Test the hypothesis $p_1 = p_2$, and compare the significance level with that obtained in Problem 2.

4. Show that, in the comparison of two Poisson means, $|\hat{\lambda}_1 - \hat{\lambda}_2|$ and $|T_1 - tp|$ define the same conditional test of the hypothesis $\lambda_1 = \lambda_2$.

5. Suppose that X and Y are independent and have Poisson distributions with means μ and ν, respectively. Set up a conditional test of significance for the hypothesis $\mu = k\nu$, where k is a given constant.

6. Twelve pea plants were observed, and there were four of each of the genotypes RR, RW, and WW. Test whether these results are consistent with the Hardy-Weinberg Law.

7. Articles coming off a production line may be classified as acceptable, repairable or useless. If n items are examined let X_1, X_2 and X_3 be the number of acceptable, repairable and useless items found. Suppose that it is twice as probable that an item is acceptable as it is that it is repairable.

(a) Show that $X_1 + X_2$ is a sufficient statistic for p, the probability of a repairable item.

(b) Of six items examined, one is acceptable, four are repairable, and one is useless. Test the consistency of the model with these data.

†8. In a certain factory there are three work shifts: days (#1), evenings (#2), and nights (#3). Let X_i denote the number of accidents in the ith shift ($i = 1,2,3$). The X_i's are assumed to be independent Poisson variates with means μ_1, μ_2, and μ_3. There are only half as many workers on the night shift as on the other two. Hence, if the accident rate is constant over the three shifts, we should have $\mu_1 = \mu_2 = 2\mu_3$. Set up exact and approximate tests of significance for this hypothesis.

9. Suppose that n families each with three children are observed. Let X_i be the number of such families which contain i boys and $3 - i$ girls ($i = 0,1,2,3$). If births are independent, the probability that a family of 3 has i boys will be given by

$$p_i = \binom{3}{i}\theta^i(1-\theta)^{3-i} \quad \text{for} \quad i = 0,1,2,3$$

where θ is the probability of a male child.

(a) Show that $T \equiv X_1 + 2X_2 + 3X_3$ is a sufficient statistic for θ and has a binomial distribution with parameters $(3n,\theta)$.

(b) In 8 families there were 3 with three boys, 2 with one boy, and 3 with no boys. Use an exact test to investigate whether these results are consistent with the model.

10. A lethal drug is administered to n rats at each of k doses $y_1, y_2, \ldots, y_k$. Let the numbers of deaths be $X_1, X_2, \ldots, X_k$. The logistic model gives the probability of death at dose y_i to be

$$p(y_i) = AB^{y_i}/(1 + AB^{y_i}),$$

where A and B are unspecified parameters (see Section 11.3).

(a) Show that $S \equiv \sum X_i$ and $T \equiv \sum y_i X_i$ are sufficient statistics for A and B.

(b) Show that the conditional probability function of the X_i's given S and T is $c\binom{n}{x_1}\binom{n}{x_2}\ldots\binom{n}{x_k}$, where c is chosen so that the total conditional probability is 1.

(c) In an experiment with 10 rats at each of 3 doses $-1, 0, 1$, the numbers of deaths observed were $3, 0$, and 10, respectively. Are these frequencies consistent with the logistic model?

(d) In an experiment with 10 rats at each of the 4 doses $-3, -1, 1, 3$, the numbers of deaths observed were $1, 6, 4$, and 10, respectively. Are these frequencies consistent with the logistic model?

†11. In an experiment to detect linkage of genes, there are four possible types of offspring. According to theory, these four types have probabilities $\frac{p}{2}$, $\frac{1-p}{2}$, $\frac{1-p}{2}$, and $\frac{p}{2}$, where p is an unknown parameter called the recombination fraction. Let X_1, X_2, X_3, and X_4 be the frequencies of the four offspring types in n independent repetitions.

(a) Find a sufficient statistic for p.

(b) If the genes are not linked, they lie on different chromosomes, and $p = \frac{1}{2}$. Evidence against the hypothesis $p = \frac{1}{2}$ is thus evidence that the genes are linked. Set up an exact test for this hypothesis.

(c) Describe exact and approximate tests of the model when p is unknown.

12.3 Tests in 2 × 2 Contingency Tables

Many interesting statistical applications involve the analysis of cross-classified frequency data. For instance, in a study to evaluate three treatments for cancer, one might classify each of n patients according to the treatment received, and also according to whether or not the patient survived a five-year period. The results could be displayed in a 3 × 2 array, with one row for each treatment category and one column for each survival category. The body of the table would give the number of patients in each of the six classes. A cross-tabulation of frequency data such as this is called a contingency table.

In the example just described, we have a two-way or two-dimensional table. If we also classified patients by sex, we would have a three-way (3 × 2 × 2) contingency table containing 12 frequencies. We shall restrict the discussion here to two-way tables only. Many examples of higher dimensional contingency tables may be found in Bishop, Fienberg and Holland, *Discrete Multivariate Analysis,* MIT Press (1975).

A question of interest in the cancer study would be whether there is a connection or association between the column classification (survival) and the row classification (treatment). This can be investigated by testing the hypothesis that the row and column classifications are independent. If this hypothesis is contradicted by the data, then there is evidence of an association between the two classifications. Tests for independence in 2 × 2 tables are considered in this section, and tests for independence in larger two-way tables are discussed in Section 12.4.

Tests for independence in 2 × 2 tables

Consider n independent repetitions of an experiment, and suppose that the outcome at each trial is classified in two ways, according to the occurrence or non-occurrence of two events, A and B. There are $k = 4$ possible classes: AB, $A\bar{B}$, $\bar{A}B$, $\bar{A}\bar{B}$. We denote their probabilities by p_1, p_2, p_3, p_4 where $\sum p_i = 1$, and their frequencies by X_1, X_2, X_3, X_4 where $\sum X_i \equiv n$. These frequencies may be arranged in a 2 × 2 contingency table as follows:

	B	$\bar{B}$	Total
A	X_1	X_2	R
$\bar{A}$	X_3	X_4	n–R
Total	C	n–C	n

Here, R and C are variates denoting the first row and first column totals. Since we are assuming independent repetitions, the distribution of the table is multinomial with $k = 4$ classes:

$$f(x_1, x_2, x_3, x_4) = \binom{n}{x_1 \ x_2 \ x_3 \ x_4} p_1^{x_1} p_2^{x_2} p_3^{x_3} p_4^{x_4} . \qquad (12.3.1)$$

Conditions are appropriate for applying goodness of fit tests as in Section 11.1.

The hypothesis of independence states that A and B are independent events,

$$H : P(AB) = P(A) \cdot P(B) . \qquad (12.3.2)$$

An equivalent form of the independence hypothesis is

$$H : P(B|A) = P(B|\overline{A}) = P(B) . \qquad (12.3.3)$$

Under H, there are two unknown parameters, $\alpha = P(A)$ and $\beta = P(B)$, and the four class probabilities are

$$p_1 = \alpha\beta ; \quad p_2 = \alpha(1 - \beta) ; \quad p_3 = (1 - \alpha)\beta ; \quad p_4 = (1 - \alpha)(1 - \beta) .$$

The likelihood function for α and β is

$$L(\alpha, \beta) = p_1^{x_1} p_2^{x_2} p_3^{x_3} p_4^{x_4} = \alpha^r (1 - \alpha)^{n-r} \beta^c (1 - \beta)^{n-c} \qquad (12.3.4)$$

where $r = x_1 + x_2$ and $c = x_1 + x_3$. This function is maximized for $\hat{\alpha} = \frac{r}{n}$ and $\hat{\beta} = \frac{c}{n}$. The expected frequency for class AB is then

$$n\hat{p}_1 = n\hat{\alpha}\hat{\beta} = \frac{rc}{n} .$$

Similarly, the expected frequencies for the other classes are

$$n\hat{p}_2 = \frac{r(n - c)}{n} ; \quad n\hat{p}_3 = \frac{(n - r)c}{n} ; \quad n\hat{p}_4 = \frac{(n - r)(n - c)}{n} .$$

Under the independence hypothesis, the expected frequency for any class is obtained by multiplying the corresponding row and column totals, and then dividing by the grand total.

If the four expected frequencies are arranged in a 2×2

table, it will have the same row and column totals as the original table. Hence, as soon as one expected frequency has been computed, the other three may be obtained by subtraction from the marginal totals.

If all of the expected frequencies are large, an approximate goodness of fit test can be carried out using (11.1.3). There are $k = 4$ classes and $r = 2$ parameters require estimation, so there is just one degree of freedom for the test.

Example 12.3.1. (R.A. Fisher, *Smoking and the Cancer Controversy*, Oliver and Boyd, 1959). Seventy-one pairs of twins were examined with respect to their smoking habits. For each pair, it was determined whether they were identical twins (A) or fraternal twins ($\overline{A}$), and whether their smoking habits were alike (B), or unlike ($\overline{B}$). The data are as follows:

	Like habits	Unlike habits	Total
Identical twins	44 (39.56)	9 (13.44)	53
Fraternal twins	9 (13.44)	9 (4.56)	18
Total	53	18	71

Note that 83% of identical twin pairs have like habits, but only 50% of fraternal twin pairs have like habits. Could such a large difference reasonably have occurred by chance, or is the probability of like habits different for the two types of twins?

Solution. We wish to know whether the probability of B could be the same for both identical twins (A) and fraternal twins ($\overline{A}$); that is, we wish to evaluate the strength of the evidence against the hypothesis $P(B|A) = P(B|\overline{A})$. This is the hypothesis of independence. Under this hypothesis, the expected frequency in the first class is $rc/n = 39.56$. The remaining expected frequencies can be obtained by subtraction from the marginal totals, and are given in parentheses above. The observed value of the goodness of fit criterion is

$$d = \frac{(4.44)^2}{39.56} + \frac{(4.44)^2}{13.44} + \frac{(4.44)^2}{13.44} + \frac{(4.44)^2}{4.56} = 7.76$$

and (11.1.3) gives

$$SL \approx P(\chi^2_{(1)} \geq 7.76)$$

which may be looked up in Table B4. Alternatively, by the result derived in Example 6.9.1, we can take square roots and use tables of the standardized normal distribution:

$$SL \approx P\{|Z| \geq \sqrt{7.76}\} = 0.0053$$

from Table B2. It is not reasonable to attribute the observed discre-
pancies to chance, and hence there is strong evidence against the
hypothesis of independence. The probability of like smoking habits is
greater for identical twins than for fraternal twins.

Example 12.3.2. Twenty-seven of the pairs of identical twins considered in Example 12.3.1 had been separated at birth, whereas the other 26
pairs had been raised together. The frequencies of like and unlike
smoking habits for the two groups are as follows:

	Like habits	Unlike habits	Total
Separated	23 (22.42)	4 (4.58)	27
Not separated	21 (21.58)	5 (4.42)	26
Total	44	9	53

The figures in parentheses are the expected frequencies under the assumption that the two classifications are independent. We do not need
a formal test of significance to tell us that the agreement is extremely good. There is no evidence that the probability of like smoking
habits is different for the two groups.

The greater similarity between smoking habits of identical
twins (Example 12.3.1) could be accounted for in two ways. Firstly, it
could be due to the fact that identical twins have the same genotype,
whereas fraternal twins are no more alike genetically than ordinary
brothers and sisters. Secondly, it could be due to greater social
pressures on identical twins to conform in their habits. If the latter were the case, one would expect to find less similarity in the
smoking habits of identical twins who had been separated at birth.
Since this is not the case, it appears that genetic factors are primarily responsible for the similarity of smoking habits.

The possibility that genetic factors may influence smoking
habits has interesting implications for the smoking and cancer controversy, since these same genetic factors might also produce an increased
susceptibility to cancer. See Fisher's pamphlet for further discussion.

Exact Tests for independence

An exact (conditional) test for the hypothesis of independence
can be derived by the methods used in Section 12.2. From (12.3.4), the

totals R and C are sufficient statistics for the unknown parameters α and β. The conditional distribution of the X_i's given the observed values of R and C does not depend upon α and β, and this distribution will be used in computing the significance level.

Since R is the number of times event A occurs in n independent trials, the distribution of R is binomial (n,α). Similarly, C is the number of times that event B occurs in n independent trials, and its distribution is binomial (n,β). Under the hypothesis, A and B are independent events, and hence R and C will be independent variates. Their joint probability function is

$$f(r,c) = \binom{n}{r}\alpha^r(1-\alpha)^{n-r} \cdot \binom{n}{c}\beta^c(1-\beta)^{n-c}. \qquad (12.3.5)$$

This result could also be obtained by summing (12.3.1) over all sets of frequencies x_1,x_2,x_3,x_4 such that

$$x_1 + x_2 + x_3 + x_4 = n; \quad x_1 + x_2 = r; \quad x_1 + x_3 = c. \qquad (12.3.6)$$

Since the four x_i's satisfy three linear restrictions, the sum is one-dimensional, corresponding to the single degree of freedom in the χ^2 approximation.

To obtain the required conditional distribution, we substitute for the p_j's in (12.3.1) and divide it by (12.3.5), giving

$$f(x_1,x_2,x_3,x_4|r,c) = \binom{n}{x_1 \ x_2 \ x_3 \ x_4}/\binom{n}{r}\binom{n}{c}.$$

Conditions (12.3.6) may be used to express this as a function of x_1 only:

$$g(x_1) = \binom{r}{x_1}\binom{n-r}{c-x_1}/\binom{n}{c} \quad \text{for} \quad x_1 = 0,1,2,\dots . \qquad (12.3.7)$$

This hypergeometric distribution is used in computing the exact significance level.

To complete the test, we need a ranking of all outcomes satisfying (12.3.6) according to how well they agree with the independence hypothesis. We could rank outcomes according to their conditional probabilities, or we could use the goodness of fit criterion,

$$D \equiv \sum_{i=1}^{4} \frac{(X_i - n\hat{p}_i)^2}{n\hat{p}_i} \equiv \frac{n^3}{r(n-r)c(n-c)}(X_1 - \frac{rc}{n})^2. \qquad (12.3.8)$$

The latter expression is obtained by substituting for x_2, x_3, and x_4 from (12.3.6). It shows that the goodness of fit criterion ranks outcomes according to the distance of X_1 from the mean $\frac{rc}{n}$ of the hypergeometric distribution (12.3.7).

A normal approximation to the hypergeometric distribution was considered in Section 6.8, and the χ^2 approximation to the distribution of D can be deduced from this. The approximation can be improved slightly by means of a correction for continuity: each deviation $|x_i - n\hat{p}_i| = |x_1 - \frac{rc}{n}|$ is decreased by 0.5 before the observed discrepancy is computed from (12.3.8). Usually, the significance level is overestimated when the continuity correction is used, and is underestimated when the correction is not used.

Example 12.3.1 (continued). Since $r = c = 53$ and $n = 71$, the conditional distribution (12.3.7) becomes

$$g(x_1) = \binom{53}{x_1}\binom{18}{53-x_1}/\binom{71}{53} \quad \text{for} \quad x_1 = 0,1,2,\ldots \quad .$$

Values less than 35 or greater than 53 have conditional probability zero, since they could not occur with the given marginal totals. The remaining probabilities can be computed recursively using (2.3.3), and are given in Table 12.3.1. Values from 48 to 53 are not listed because their conditional probabilities are less than 10^{-6}.

Table 12.3.1

Conditional Distribution in an Exact Test for Independence

x_1	$g(x_1)$	x_1	$g(x_1)$	x_1	$g(x_1)$
35	0.00208	40	0.23200	45	0.00125
36	0.01872	41	0.15938	46	0.00016
37	0.07309	42	0.07806	47	0.00001
38	0.16413	43	0.02746		
39	0.23673	*44	0.00693	Total	1.00000

By (12.3.8), the goodness of fit criterion ranks the values of X_1 according to their distance from the mean $rc/n = 39.56$. Values further from the mean than 44 are $35,45,46,47,\ldots$. These are also the values which have smaller conditional probabilities than 44. Hence, for both the goodness of fit criterion and the probability ranking, the exact significance level is

$$SL = g(35) + g(44) + g(45) + g(46) + \ldots = 0.01043.$$

Note, however, that the ordering of values 38,39,40,41 is different, and if one of these values were observed, the two tests would produce different significance levels.

We saw previously that the χ^2 approximation without continuity correction gives SL $\approx$ 0.0053. To use a correction, we replace the deviation 4.44 by 3.94 before computing d. This gives

$$d = (3.94)^2(\frac{1}{39.56} + \frac{1}{13.44} + \frac{1}{13.44} + \frac{1}{4.56}) = 6.11,$$

and the significance level is approximately

$$P(\chi^2_{(1)} \geq 6.11) = P\{|Z| \geq 2.47\} = 0.0135$$

from Table B2. The exact significance level lies between the two approximations, and is somewhat closer to the value obtained when a continuity correction is used.

Fixing the Marginal Totals

We have assumed that the distribution of a 2×2 contingency table is multinomial with $k = 4$ classes. Both the row and column totals were treated as variates, and only the grand total n was assumed to be fixed by the experimental plan.

If $\alpha = P(A)$ is small, the experiment will probably produce very few occurrences of A, and the observed value of R will be close to zero. The experiment is then incapable of providing much information about the independence or lack of independence of the two classifications. To guard against this, the experimenter can sometimes fix the row totals r and n-r in advance. The appropriate model for the resulting 2×2 table would be a pair of binomial distributions, one for each row, rather than a single multinomial distribution. The table considered in Example 12.2.1 is of this type. Tests for independence will not be affected by this change in the experimental design, because all of the marginal totals are held fixed at their observed values in such tests.

For instance, suppose that we are planning a study similar to that described in Example 12.3.1. Twenty pairs of twins are to be examined, and these are to be selected from a registry of twins which lists four times as many fraternal twins as identical twins. If we chose 20 pairs at random from the list, we would expect to get only 4 pairs of identical twins, and might well get even fewer. It is unlikely that even a large difference between $P(B|A)$ and $P(B|\overline{A})$ could be detected. A better plan would be to randomly select 10 pairs from the identical twins listed, and 10 pairs from the fraternal twins listed. This will give a 2×2 table with fixed row totals $r = n - r = 10$. The test for independence will be the same as if we had selected the 20 pairs at random from the entire registry, and had by chance obtained $r = r - n = 10$.

Testing equality of marginal probabilities

Although one will often be interested in testing independence, one should not assume that every 2×2 contingency table calls for such a test. Contingency tables can arise in a variety of ways, and the question of interest will depend upon the situation. To illustrate this point, we consider an example where one is interested in comparing the marginal probabilities, $P(A)$ and $P(B)$, rather than in testing independence.

Example 12.3.3. Two drugs are compared to see which of them is less likely to produce unpleasant side effects. Each of 100 subjects is given the two drugs on different occasions, and is classified according to whether or not the drugs upset his stomach. The results can be summarized in a 2×2 contingency table as follows:

	Nausea with drug B	No nausea with B	Total
Nausea with drug A	38	2	40
No nausea with A	10	50	60
Total	48	52	100

Drug B produced nausea in 48% of subjects, but drug A produced nausea in only 40% of subjects. Could this discrepancy reasonably be ascribed to chance, or is there evidence of a real difference between the two drugs?

The hypothesis of independence is not of interest in this example. Indeed, one would expect a patient who experiences nausea with one drug to be more susceptible to nausea from the other drug, and hence the two classifications are almost certainly not independent. The question of interest is whether the probability of nausea is the same for both drugs, and hence we consider the hypothesis of marginal homogeneity,

$$H : P(A) = P(B). \tag{12.3.9}$$

Since $P(A) = p_1 + p_2$ and $P(B) = p_1 + p_3$, (12.3.9) is equivalent to the hypothesis $H: p_2 = p_3$. Under this hypothesis, there are two unknown parameters, which we may take to be p_1 and p_2. From (12.3.1), the likelihood function of p_1 and p_2 is

$$L(p_1, p_2) = p_1^{x_1} p_2^{x_2} p_3^{x_3} p_4^{x_4} = p_1^{x_1} p_2^{x_2 + x_3} (1 - p_1 - 2p_2)^{x_4}. \qquad (12.3.10)$$

This function is maximized for $\hat{p}_1 = x_1/n$ and $\hat{p}_2 = (x_2 + x_3)/2n$, giving expected frequencies

$$n\hat{p}_1 = x_1; \quad n\hat{p}_2 = n\hat{p}_3 = \tfrac{1}{2}(x_2 + x_3); \quad n\hat{p}_4 = x_4.$$

Note that there is perfect agreement between observed and expected frequencies in the first and last classes, so these two classes will contribute zero to the goodness of fit statistic.

In the numerical example, the expected frequencies are 38, 6,6,50, and the observed value of the goodness of fit statistic is

$$d = \frac{0^2}{38} + \frac{4^2}{6} + \frac{4^2}{6} + \frac{0^2}{50} = 5.33.$$

There are $k = 4$ classes and $r = 2$ parameters were estimated, so (11.3.1) gives

$$SL \approx P(\chi^2_{(1)} \geq 5.33) = P(|Z| \geq 2.31) = 0.021.$$

Hence there is evidence that $P(A) \neq P(B)$.

Since the expected frequencies are rather small, we might also like to carry out an exact test. From (12.3.10), X_1, $T \equiv X_2 + X_3$, and $X_4 \equiv n - X_1 - T$ are sufficient statistics for p_1 and p_2. Their distribution is trinomial:

$$f(x_1, t, x_4) = \binom{n}{x_1\ t\ x_4} p_1^{x_1} (2p_2)^t (1 - p_1 - 2p_2)^{x_4}. \qquad (12.3.11)$$

Dividing (12.3.1) by (12.3.11) gives

$$f(x_1, x_2, x_3, x_4 | x_1, t, x_4) = \frac{t!}{x_2! x_3!} 2^{-t} = \binom{t}{x_2}(\tfrac{1}{2})^t. \qquad (12.3.12)$$

An exact test of $H: P(A) = P(B)$ is based on this binomial $(t, \tfrac{1}{2})$ distribution.

In the numerical example, we have $t = 12$, and (12.3.12) becomes

$$g(x_2) = \binom{12}{x_2}(\frac{1}{2})^{12}; \quad x_2 = 0,1,\ldots,12. \tag{12.3.13}$$

This distribution has mean 6, so we take $D \equiv |X_2 - 6|$ with observed value $d = |2 - 6| = 4$. Then

$$SL = P\{|X_2 - 6| \geq 4\} = g(0) + g(1) + g(2) + g(10) + g(11) + g(12)$$

$$= 0.039.$$

There is evidence against the hypothesis $P(A) = P(B)$, but it is somewhat weaker than suggested by the approximate test.

In testing the hypothesis $P(A) = P(B)$, we ignore all cases in which both drugs had the same effect, and consider only the $t = 12$ cases in which the drugs produced different effects. If H is true, classes $A\overline{B}$ and $\overline{A}B$ will be equally probable, and the probability that x_2 out of the 12 fall in $A\overline{B}$ rather than $\overline{A}B$ is given by (12.3.13).

A different test would be appropriate if we were willing to assume independence of the two classifications. Then R and C, the first row and column totals, are sufficient statistics for $P(A)$ and $P(B)$, as we showed earlier. They have independent binomial distributions, and the problem becomes that of comparing two binomial proportions (Section 12.2). However, as we noted previously, it would not usually be reasonable to assume independence in the situation described.

Problems for Section 12.3

1. In December 1897 there was an outbreak of plague in a jail in Bombay. Of 127 persons who were uninoculated, 10 contracted the plague. Of 147 persons who had been inoculated, 3 contracted the disease. Use an approximate test to determine whether there is conclusive evidence of an association between inoculation and failure to get the plague.

†2. A study to investigate the effectiveness of vitamin C in preventing the common cold was reported in the Canadian Medical Association Journal, Sept. 1972, pp. 503-8. The 818 individuals participating in the study were divided at random into two groups. The 407 individuals of one group received daily doses of vitamin C. Of these, 105 reported no illness over the period of the study, while the remainder had at least one episode. The 411 individuals in the other group received a placebo (a similar pill having

no active ingredients), and only 76 of these were free of illness over the period of the study. Are these data consistent with the hypothesis that vitamin C has no effect on the incidence of illness?

†3. An investigator wishes to learn whether the tendency to crime is influenced by genetic factors. He argues that, if there is no genetic effect, the incidence of criminality among identical twins should be the same as that among fraternal twins. Accordingly, he examines the case histories of 30 criminals with twin brothers, of whom 13 are identical and 17 are fraternal. He finds that 12 of the twin brothers have also been convicted of crime, but only two of these are fraternal twins. Perform an exact test of the hypothesis of no genetic effect. For comparison, compute approximate significance levels with and without a continuity correction.

4. It was noticed that married undergraduates seemed to do better academically than single students. Accordingly, the following observations were made: of 1500 engineering students, 297 had failed their last set of examinations; 157 of them were married, of whom only 14 had failed. Are these observations consistent with the hypothesis of a common failure rate for single and married students? Under what conditions would the information that there were more married students in 3rd and 4th years than in 1st and 2nd years affect your conclusion?

5. In the study referred to in Problem 11.3.2, it happened that 12 subjects received benzedrine on the first occasion, while only 8 received caffeine first. Of the 14 successes recorded with benzedrine, 10 occurred when it had been administered first. Are these data consistent with the hypothesis that the success rate for caffeine over benzedrine is the same whether it is administered first or second?

6. Derive (12.3.5) by summing the multinomial distribution (12.3.1) over all x_1, x_2, x_3, x_4 such that $\sum x_i = n$, $x_1 + x_2 = r$, and $x_1 + x_3 = c$.

12.4 Testing for Independence in $a \times b$ Contingency Tables

The discussion of 2×2 contingency tables in the preceding section can easily be extended to larger tables.

Suppose that there are n independent trials, and that the outcome of each trial is classified in two ways: according to which of the events $A_1, A_2, \ldots, A_a$ occurs, and according to which of the events $B_1, B_2, \ldots, B_b$ occurs. We assume that the A_i's (and similarly the B_j's) are a set of mutually exclusive and exhaustive events, so that each outcome belongs to exactly one of them. There are $k = ab$ possible classes $A_i B_j$, with class frequencies X_{ij}, where $1 \leq i \leq a$, $1 \leq j \leq b$, and $\sum\sum X_{ij} \equiv n$. These frequencies can be arranged in an $a \times b$ contingency table as follows:

	B_1	B_2	$\ldots$	B_b	Total
A_1	X_{11}	X_{12}	$\ldots$	X_{1b}	R_1
A_2	X_{21}	X_{22}	$\ldots$	X_{2b}	R_2
$\cdots$	. . . .				. .
A_a	X_{a1}	X_{a2}	$\ldots$	X_{ab}	R_a
Total	C_1	C_2	$\ldots$	C_b	n

Given an $a \times b$ contingency table, there are many different questions that one might wish to ask. We shall consider only the problem of determining whether or not their is evidence of an association or connection between the two classifications. We consider tests for the hypothesis of independence,

$$H: P(B_j | A_i) = P(B_j) \quad \text{for} \quad 1 \leq i \leq a, \quad 1 \leq j \leq b. \tag{12.4.1}$$

This hypothesis states that all of the events $B_1, B_2, \ldots, B_b$ are independent of all of the events $A_1, A_2, \ldots, A_a$, so that

$$P(A_i B_j) = P(A_i) P(B_j) \quad \text{for} \quad 1 \leq i \leq a, \quad 1 \leq j \leq b.$$

If we find evidence against H, then we have evidence of an association between the row and column classifications.

A test for independence may be only the first step in the analysis of a large contingency table. If an association is detected by the test, there still remains the problem of measuring and explaining it.

Approximate Test for Independence

Under H, the following are unknown parameters:

$$P(A_1), P(A_2), \ldots, P(A_a); \quad P(B_1), P(B_2), \ldots, P(B_b).$$

However, since $\sum P(A_i) = 1$ and $\sum P(B_j) = 1$, there are actually only

$(a-1)+(b-1)=a+b-2$ parameters which require estimation. The degrees of freedom for the test will thus be

$$k-r-1 = ab - (a+b-2) - 1 = (a-1)(b-1).$$

As with 2×2 tables, it can be shown that the maximum likelihood estimates of $P(A_i)$ and $P(B_j)$ are r_i/n and c_j/n. Hence the expected frequency for class $A_i B_j$ is

$$n\hat{P}(A_i B_j) = n\hat{P}(A_i)\hat{P}(B_j) = r_i c_j /n.$$

Under the hypothesis of independence, the expected frequency of class $A_i B_j$ is obtained by multiplying the corresponding row and column totals and dividing by the grand total. If we compute all of the expected frequencies in the upper left hand $(a-1) \times (b-1)$ subtable, we can obtain the expected frequencies for the last row and last column of the table by subtraction from the marginal totals.

If all of the expected frequencies are reasonably large, an approximate goodness of fit test can now be done using (11.1.3). The approximate significance level will be obtained from tables of the χ^2 distribution with $(a-1)(b-1)$ degrees of freedom.

Example 12.4.1. In a study to determine whether laterality of hand is associated with laterality of eye (measured by astigmatism, acuity of vision, etc.), 413 subjects were classified with respect to these two characteristics. The results were as follows:

	Left-eyed	Ambiocular	Right-eyed	Total
Left-handed	34 (35.4)	62 (58.5)	28 (30.1)	124
Ambidextrous	27 (21.4)	28 (35.4)	20 (18.2)	75
Right-handed	57 (61.2)	105 (101.1)	52 (51.7)	214
Total	118	195	100	413

Is there any evidence that laterality of eye is related to laterality of hand?

Solution. Assuming that the two classifications are independent, the expected frequencies for the four classes in the upper left hand corner are

$$118 \times 124/413 = 35.4 \qquad 195 \times 124/413 = 58.5$$
$$118 \times 75/413 = 21.4 \qquad 195 \times 75/413 = 35.4.$$

The remaining expected frequencies are obtained by subtraction from the marginal totals, and are shown in parentheses above. The observed value of the goodness of fit criterion is

$$d = \frac{(1.4)^2}{35.4} + \frac{(3.5)^2}{58.5} + \ldots + \frac{(0.3)^2}{51.7} = 4.04.$$

Now (11.1.3) and Table B4 give

$$SL \approx P(\chi^2_{(4)} \geq 4.04) \approx 0.4.$$

The hypothesis that laterality of hand and eye are independent is consistent with the data, and there is no evidence of an association.

Example 12.4.2. Nine hundred and fifty school children were classified according to their nutritional habits and intelligence quotients, with the following results:

	Intelligence Quotient				
	<80	80-89	90-99	≥100	Total
Good nutrition	245 (252.5)	228 (233.3)	177 (173.8)	219 (209.4)	869
Poor nutrition	31 (23.5)	27 (21.7)	13 (16.2)	10 (19.6)	81
Total	276	255	190	229	950

Is there evidence of a relationship between nutrition and IQ?

Solution. If there is no relationship, the two classifications are independent, and the expected frequencies in the first three classes of the first row are

$$\frac{276 \times 869}{950}, \quad \frac{255 \times 869}{950}, \quad \frac{190 \times 869}{950}.$$

The remaining expected frequencies are obtained by subtraction from the marginal totals, and are given in parentheses above. The observed value of the goodness of fit criterion is

$$d = \frac{(7.5)^2}{252.5} + \frac{(5.3)^2}{233.3} + \ldots + \frac{(9.6)^2}{19.6} = 9.86.$$

Since all expected frequencies are reasonably large, we can apply (11.1.3) to obtain

$$SL \approx P(\chi^2_{(3)} \geq 9.86) \approx 0.02.$$

The observed frequencies provide reasonably strong evidence against the hypothesis of independence. There are substantially fewer children with poor nutrition and high IQ than one would have expected if there were no relationship. Poor nutrition and low IQ tend to occur together.

Although the data justify the conclusion that poor nutrition and low IQ go together, one should not be tempted to infer that poor nutrition causes a lowering of IQ. This inference would not be warranted on the data, because there is no indication that if nutrition were improved, children would become more intelligent. Two other explanations are equally well supported by the data. Firstly, it may be that children of lower intelligence do not know enough to eat properly, so that poor nutrition is caused by low intelligence. Secondly, there may be a third factor (such as poor home environment) which is responsible for both poor nutrition and low IQ. This type of data gives no justification for adopting one explanation in preference to the other two.

In general, the statement "A causes B" means that by manipulating A we can control B. If A is made to occur, the probability of the occurrence of B (within some reasonable time limit) is increased. On the other hand, the statement "A and B are associated" merely means that A and B tend to occur together. There is no guarantee that if we force A to occur, the chance of B's occurrence will subsequently be increased. Rigorous evidence of causation can be obtained only from a controlled experiment in which the cause A is manipulated at will by the experimenter.

Exact Test for Independence

An exact test for independence in a × b tables can be derived as in the 2 × 2 case. Under the hypothesis of independence, the marginal totals are sufficient statistics for the unknown parameters, and an exact test is based on the conditional distribution of the X_{ij}'s given the observed totals. The conditional probability of a table (x_{ij}) given totals (r_i) and (c_j) is

$$\binom{n}{x_{11}x_{12}\cdots x_{ab}} \Big/ \binom{n}{r_1 r_2 \cdots r_a}\binom{n}{c_1 c_2 \cdots c_a}$$

$$= \binom{r_1}{x_{11}\cdots x_{1b}}\binom{r_2}{x_{21}\cdots x_{2b}}\cdots\binom{r_a}{x_{a1}\cdots x_{ab}} \Big/ \binom{n}{c_1 \cdots c_b}$$

$$= \binom{c_1}{x_{11}\cdots x_{a1}}\binom{c_2}{x_{12}\cdots x_{a2}}\cdots\binom{c_b}{x_{1b}\cdots x_{ab}} \Big/ \binom{n}{r_1 \cdots r_a}.$$

A list of all possible tables with the observed marginal totals is prepared, and these tables are ranked according to their agreement with the hypothesis of independence. The probability ranking is the most convenient, but the goodness of fit criterion can also be used. The exact significance level is then obtained by summing the conditional probabilities of all tables which are in no better agreement than the observed table.

Note that, since the marginal totals are held fixed, tests for independence will not be affected if some or all of the marginal totals are fixed by the experimental plan.

Example 12.4.3. Is the following 2×3 contingency table consistent with the hypothesis of independent row and column classifications?

	B_1	B_2	B_3	Total
A_1	1 (1.8)	1 (3.0)	7 (4.2)	9
A_2	2 (1.2)	4 (2.0)	0 (2.8)	6
Total	3	5	7	15

Solution. The expected frequencies under the hypothesis of independence are shown in parentheses. Since these are small, an exact test is advisable.

The general form of tables with the observed marginal totals is

x	y	9−x−y	9
3−x	5−y	x+y−2	6
3	5	7	15

and the conditional probability is

$$f(x,y) = \binom{9}{x\ y\ 9-x-y}\binom{6}{3-x\ 5-y\ x+y-2}/\binom{15}{3\ 5\ 7}$$

$$= \binom{3}{x}\binom{5}{y}\binom{7}{9-x-y}/\binom{15}{9}.$$

There are 24 pairs (x,y) such that x ≤ 3 and y ≤ 5. Three of these would give a negative entry in the table (x + y < 2), and hence there are just 21 allowable pairs. These are listed in Table 12.4.1 together with their conditional probabilities.

Table 12.4.1
Exact Test for Independence in a 2 × 3 Contingency Table

x	y	$\binom{15}{9}f(x,y)$	x	y	$\binom{15}{9}f(x,y)$	x	y	$\binom{15}{9}f(x,y)$
0	2	10	1	4	525	2	5	63
0	3	70	1	5	105	3	0	7
0	4	105	2	0	3	3	1	105
0	5	35	2	1	105	3	2	350
*1	1	15	2	2	630	3	3	350
1	2	210	2	3	1050	3	4	105
1	3	630	2	4	525	3	5	7

The observed table $x = y = 1$ has conditional probability $15/\binom{15}{9}$. The significance level with respect to the probability ranking is then

$$SL = f(0,2) + f(1,1) + f(2,0) + f(3,0) + f(3,5) = 0.0084.$$

The observed table provides strong evidence against the hypothesis of independence.

Problems for Section 12.4

†1. Gregor Mendel grew 529 pea plants using seed from a single source, and classified them according to seed shape (round, round and wrinkled, wrinkled) and colour (yellow, yellow and green, green). He obtained the following data:

       38 round, yellow
       65 round, yellow and green
       60 round and wrinkled, yellow
      138 round and wrinkled, yellow and green
       28 wrinkled, yellow
       68 wrinkled, yellow and green
       35 round, green
       67 round and wrinkled, green
       30 wrinkled, green

(a) Test the hypothesis that the shape and colour classifications are independent.

(b) According to Mendel's theory, the frequencies of yellow, yellow and green, and green seeds should be in the ratio 1:2:1. Test whether this hypothesis is consistent with the data.

2. In a series of autopsies, indications of hypertension were found

in 37% of 200 heavy smokers, in 40% of 290 moderate smokers, in 45.3% of 150 light smokers, and in 51.3% of 160 non-smokers. Test the hypothesis that the probability of hypertension is independent of the smoking category.

3. In an experiment to detect a tendency of a certain species of insect to aggregate, 12 insects were released near two adjacent leaf areas, A and B, and after a certain period of time the number of insects that had settled on each was counted. The process was repeated 10 times, using the same two leaf areas. The observations are set out below.

Trial number	1	2	3	4	5	6	7	8	9	10
Number on A	7	3	3	9	0	0	5	5	7	4
Number on B	3	5	6	1	10	8	2	5	4	6

Do the observations suggest that insects tend to aggregate, or that they distribute themselves at random over the two areas?

4. A customer purchased 4 packets of sweet pea seeds, one packet of each of four colours. He planted 100 seeds from each packet, and observed the numbers germinating within one month:

Colour of Flower

	Red	White	Blue	Yellow	Total
Germination	75	66	81	74	296
No germination	25	34	19	26	104
Total	100	100	100	100	400

(a) Is there any evidence that the germination rate depends upon the colour?

(b) The observed germination rate for yellow flowers is 74%. Is this consistent with the seed dealer's guaranteed germination rate of 80%?

(c) The observed germination rate for all colours combined is 296 out of 400, or 74%. Is this consistent with the guaranteed rate of 80%?

†5. A study was undertaken to determine whether there is an association between the birth weights of infants and the smoking habits of their parents. Out of 50 infants of above average weight, 9 had parents who both smoked, 6 had mothers who smoked but fathers who did not, 12 had fathers who smoked but mothers who did not, and 23 had parents of whom neither smoked. The corresponding results for 50 infants of below average weight were 21,10,6, and 13, respectively.

(i) Test whether these results are consistent with the hypothesis that birth weight is independent of parental smoking habits.

(ii) Are these data consistent with the hypothesis that, given the smoking habits of the mother, the smoking habits of the father are not related to birth weight?

6. In the following table, 64 sets of triplets are classified according to the age of their mother at their birth and their sex distribution:

	3 boys	2 boys	2 girls	3 girls	Total
Mother under 30	5	8	9	7	29
Mother over 30	6	10	13	6	35
Total	11	18	22	13	64

(a) Is there any evidence of an association between the sex distribution and the age of the mother?

(b) Suppose that the probability of a male birth is 0.5, and that the sexes of triplets are determined independently. Find the probability that there are x boys in a set of triplets (x = 0,1,2,3), and test whether the column totals are consistent with this distribution.

7. n items were examined from the output of three similar machines in a factory. Ten percent were defective for the first machine, five percent for the second machine, and twelve percent for the third machine. A test of the hypothesis that the probability of a defective is the same for all three machines gave a significance level of 5%. How large was n?

†8. The following table records 292 litters of mice classified according to litter size and number of females in the litter.

		Number of females				
		0	1	2	3	4
Litter size	1	8	12			
	2	23	44	13		
	3	10	25	48	13	
	4	5	30	34	22	5

Suppose that the number of females in a litter of size i is binomially distributed with parameters (i, p_i).

(a) Test the hypothesis $p_1 = p_2 = p_3 = p_4$.

(b) Assuming the hypothesis in (a) to be true, test the significance of deviations from the binomial distribution model.

(c) How would the test in (b) be affected if equality of the p_i's was not assumed?

12.5 Likelihood Ratio Tests

Suppose that the probability model for an experiment involves an unknown parameter or vector of parameters θ, and that we wish to test an hypothesis concerning the value of θ. In many such problems, a likelihood ratio can be used to rank events for the significance test.

Let x denote a typical outcome of the experiment, with probability $P(x;\theta)$. From Section 9.1, the likelihood function of θ is proportional to this probability:

$$L(\theta;x) = k(x) \cdot P(x;\theta)$$

where $k(x)$ is positive and does not depend upon θ. The maximum likelihood estimate $\hat{\theta}(x)$ is the value of θ which maximizes $L(\theta;x)$.

First, consider a simple hypothesis $H: \theta = \theta_0$, where θ_0 is a given numerical value of θ. The likelihood ratio for $\theta = \theta_0$ versus the most likely value $\theta = \hat{\theta}$ is given by

$$R(\theta_0;x) = \frac{L(\theta_0;x)}{L(\hat{\theta};x)} = \frac{\text{Prob. of outcome x when } \theta = \theta_0}{\text{Maximum prob. of x for any } \theta}. \qquad (12.5.1)$$

This is also called the relative likelihood of $\theta = \theta_0$. In Section 9.4, we fixed x at its observed value and used $R(\theta;x)$ to rank possible values of θ according to their plausibilities. If, instead, we fix θ at its hypothesized value θ_0, we can use $R(\theta_0;x)$ to rank possible outcomes x. If $R(\theta_0;x)$ is near 1, then outcome x gives the hypothesized value a high relative likelihood, and is in good agreement with $H: \theta = \theta_0$. If $R(\theta_0;x)$ is near 0, outcome x makes the hypothesized value very unlikely, and hence x is in poor agreement with $H: \theta = \theta_0$.

If H is a composite hypothesis, the maximized likelihood ratio is often useful. Let $\hat{\theta}$ be the unrestricted MLE of θ as before, and let $\tilde{\theta}$ be the MLE of θ when the hypothesis is assumed to be true. The maximized likelihood ratio (maximum relative likelihood) is then

$$R(\tilde{\theta};x) = \frac{L(\tilde{\theta};x)}{L(\hat{\theta};x)} = \frac{\text{Max. prob. of x if H is true}}{\text{Max. prob. of x for any } \theta}. \qquad (12.5.2)$$

If H is simple, there is only one possible value of θ. Hence $\tilde{\theta} = \theta_0$ and (12.5.1) is just a special case of (12.5.2).

A test of significance in which the (maximized) likelihood ratio is used to rank outcomes is called a likelihood ratio test. Large values of R correspond to outcomes in good agreement with H, and small values of R indicate outcomes in poor agreement. Any strictly decreasing function of R can be selected as the discrepancy measure for a likelihood ratio test. A convenient choice is

$$D \equiv -2 \log R \qquad\qquad (12.5.3)$$

which we call the likelihood ratio statistic.

Likelihood ratio tests have several advantages. Because of the connection with the likelihood function, they seem less arbitrary than tests based on, say, distance from the mean or the goodness of fit statistic. Also, the ranking is meaningful in small samples, whereas the goodness of fit statistic relies on approximate normality in large samples. The likelihood ratio statistic is always a function of the appropriate set of sufficient statistics, and there is a convenient approximation to its distribution in large samples. The main disadvantage of likelihood ratio tests is that they may require substantially more computing than other test procedures.

We now give several examples of exact likelihood ratio tests.

Example 12.5.1. If X has a binomial (n,p) distribution, then $\hat{p} = x/n$ and the relative likelihood function (likelihood ratio) of p given x is

$$R(p;x) = \frac{L(p;x)}{L(\hat{p};x)} = \frac{p^x(1-p)^{n-x}}{\hat{p}^x(1-\hat{p})^{n-x}} = \left[\frac{np}{x}\right]^x\left[\frac{n(1-p)}{n-x}\right]^{n-x}.$$

The likelihood ratio statistic for a test of $H:p = p_0$ is

$$D(X) \equiv -2 \log R(p_0;X) \equiv 2X \log\left[\frac{X}{np_0}\right] + 2(n - X)\log\left[\frac{n-X}{n(1-p_0)}\right].$$

Taking $n = 10$ and $p_0 = 0.52$ as in Example 12.1.1(a) gives

$$D(X) \equiv 2X \log\left[\frac{X}{5.2}\right] + 2(10 - X)\log\left[\frac{10 - X}{4.8}\right].$$

The computed values of D are as follows:

x	0	1	2	3	4	5	6	7	8	9	10
D(x)	14.68	8.02	4.35	1.98	0.58	0.02	0.26	1.34	3.39	6.74	13.08

Values of X are now ranked in order of decreasing D-value:

0(worst agreement),10,1,9,2,8,3,7,4,6,5(best agreement).

This is the same ranking that we obtained in Section 12.1. In this example, the likelihood ratio test will give the same significance level as a test based on distance from the mean or the probability ranking.

Example 12.5.2. Suppose that $f_1, f_2, \ldots, f_k$ have a multinomial distribution with parameters $(n; p_1, p_2, \ldots, p_k)$ as in Section 11.1. The likelihood function of the p_j's is

$$L(p_1, p_2, \ldots, p_k) = p_1^{f_1} p_2^{f_2} \ldots p_k^{f_k} = \prod_{j=1}^{k} p_j^{f_j}$$

where $p_j \geq 0$ and $\sum p_j = 1$. This function is maximized for $\hat{p}_j = \frac{1}{n} f_j$ $(j = 1, 2, \ldots, k)$. The (maximized) likelihood ratio is

$$R(\tilde{p}_1, \tilde{p}_2, \ldots, \tilde{p}_k) = \left[\prod_{j=1}^{k} \tilde{p}_j^{f_j} \right] \div \left[\prod_{j=1}^{k} \hat{p}_j^{f_j} \right] = \prod_{j=1}^{k} \left[\frac{n \tilde{p}_j}{f_j} \right]^{f_j},$$

where $\tilde{p}_1, \tilde{p}_2, \ldots, \tilde{p}_k$ are the MLE's under the hypothesis. Hence the likelihood ratio statistic (12.5.3) becomes

$$D \equiv -2 \log R \equiv 2 \sum_{j=1}^{k} f_j \log(f_j/e_j) \qquad (12.5.4)$$

where $e_j = n \tilde{p}_j$ is the expected frequency for the jth class under the hypothesis. This statistic can be used instead of the goodness of fit criterion to rank outcomes for a test of significance.

In Example 12.2.3 we considered an exact test of the composite hypothesis

$$H: p_1 = \theta^2, \qquad p_2 = 2\theta(1 - \theta), \qquad p_3 = (1 - \theta)^2$$

in a trinomial distribution with $n = 20$. Under H, the MLE of θ is $\tilde{\theta} = 0.375$, giving expected frequencies $e_1 = 2.8$, $e_2 = 9.4$, $e_3 = 7.8$. The likelihood ratio criterion is

$$D \equiv 2[f_1 \log(f_1/2.8) + f_2 \log(f_2/9.4) + f_3 \log(f_3/7.8)],$$

and the D-values for the 8 outcomes listed in Table 12.2.1 are

9.57   3.22   0.60   0.04   1.30   4.97   9.86   18.69.

The likelihood ratio statistic implies the same ranking of these outcomes as does the goodness of fit criterion. In this example, the two tests will give the same exact (conditional) significance level.

In Example 11.3.2 we considered the simple hypothesis

$$H: p_0 = p_2 = \frac{1}{4}, \qquad p_1 = \frac{1}{2}$$

in a trinomial example with $n = 8$. The expected frequencies are $e_0 = 2$, $e_1 = 4$, $e_2 = 2$, so (12.5.4) becomes

$$D(f_0, f_1, f_2) = 2[f_0 \log(f_0/2) + f_1 \log(f_1/4) + f_2 \log(f_2/2)].$$

Values of D can be computed for the 45 outcomes listed in Table 11.3.1, and the significance level is then found by adding the probabilities of outcomes whose D-values are not less than the observed D-value. For instance, if we observe frequencies $(1,2,5)$, we find that $D(1,2,5) = 5.00$. There are 28 outcomes with $D \geq 5.00$, and their total probability is $SL = 0.142$. For comparison, an exact goodness of fit test gives $SL = 0.060$ in this case; see Example 11.3.2.

When the expected frequencies are large, the likelihood ratio test and goodness of fit test will give almost identical results. However, they can differ appreciably when some of the expected frequencies are small. In this case, it would seem preferable to use the likelihood ratio test.

Example 12.5.3. In Example 12.2.2 we considered an exact conditional test of $H: \lambda_1 = \lambda_2$, where λ_1 and λ_2 were the densities of bacteria in two locations. The maximized likelihood ratio is

$$R(\tilde{\lambda}_1, \tilde{\lambda}_2) = \frac{L(\tilde{\lambda}_1, \tilde{\lambda}_2)}{L(\hat{\lambda}_1, \hat{\lambda}_2)} = \frac{\tilde{\lambda}_1^{t_1} \tilde{\lambda}_2^{t_2} e^{-v(n\tilde{\lambda}_1 + m\tilde{\lambda}_2)}}{\hat{\lambda}_1^{t_1} \hat{\lambda}_2^{t_2} e^{-v(n\hat{\lambda}_1 + m\hat{\lambda}_2)}}.$$

The unrestricted MLE's of λ_1 and λ_2 are $\hat{\lambda}_1 = \frac{1}{nv} t_1$ and $\hat{\lambda}_2 = \frac{1}{mv} t_2$. Under $H: \lambda_1 = \lambda_2$, their common estimate is $\tilde{\lambda}_1 = \tilde{\lambda}_2 = \frac{1}{(n+m)v}(t_1 + t_2)$. Hence the likelihood ratio statistic is

$$D = -2 \log R = 2t_1 \log(\hat{\lambda}_1/\tilde{\lambda}_1) + 2t_2 \log(\hat{\lambda}_2/\tilde{\lambda}_2)$$

$$= 2t_1 \log \frac{t_1}{n} + 2t_2 \log \frac{t_2}{m} - 2(t_1 + t_2)\log \frac{t_1 + t_2}{n + m}.$$

The observed value of D is 5.98. From Section 12.2, the conditional reference set for the test is the set of pairs (t_1, t_2) with $t_1 + t_2 = 42$. With a little arithmetic, we find that $D \geq 5.98$ for $t_1 \leq 11$ and for $t_1 \geq 27$. The conditional distribution of T_1 given $T_1 + T_2 = 42$ is binomial $(42, \frac{4}{9})$, and the significance level is the sum of these binomial probabilities for $t_1 \leq 11$ and $t_1 \geq 27$. This gives the same result as the test based on distance from the conditional mean in Example 12.2.2.

Likelihood ratio test for normal mean

Let $X_1, X_2, \ldots, X_{n}$ be independent $N(\mu, \sigma^2)$ variates. We assume that the variance σ^2 is known, and derive a likelihood ratio test for $H: \mu = \mu_0$, where μ_0 is a specific numerical value of μ.

From Example 10.1.1 and equation (10.1.3), the likelihood function of μ and σ is

$$L(\mu, \sigma) = k\sigma^{-n}\exp\{- \frac{1}{2\sigma^2} \sum(x_i - \mu)^2\}$$

$$= k\sigma^{-n}\exp\{- \frac{1}{2\sigma^2} \sum(x_i - \bar{x})^2\} \cdot \exp\{- \frac{n}{2\sigma^2}(\bar{x} - \mu)^2\}.$$

We are assuming that σ is known. Only the last factor involves μ, and we may choose k so that

$$L(\mu) = \exp\{- \frac{n}{2\sigma^2} (\bar{x} - \mu)^2\} \quad \text{for} \quad -\infty < \mu < \infty.$$

This is maximized for $\hat{\mu} = \bar{x}$, and since $L(\hat{\mu}) = 1$, we have

$$R(\mu) = L(\mu) = \exp\{- \frac{n}{2\sigma^2} (\bar{x} - \mu)^2\}.$$

The likelihood ratio statistic is

$$D \equiv -2 \log R \equiv n(\frac{\bar{X} - \mu}{\sigma})^2 \equiv Z^2$$

where $Z \equiv \sqrt{n}(\bar{X} - \mu)/\sigma$. Note that D depends only on the sufficient statistic $\bar{X}$.

In this case, we can easily work out the probability distribution of D. We have $\bar{X} \sim N(\mu, \sigma^2/n)$ by (6.6.8), and thus $Z \sim N(0, 1)$ by (6.6.5). Since $D \equiv Z^2$, it follows from Example 6.9.1 that D has a χ^2 distribution with one degree of freedom.

For a likelihood ratio test of $H:\mu = \mu_0$, we compute the observed value of D:

$$d = z^2_{obs} \quad \text{where} \quad z_{obs} = \frac{\sqrt{n}(\bar{x} - \mu_0)}{\sigma} .$$

Since $D \equiv Z^2$, we have

$$SL = P(D \geq d) = P(|Z| \geq |z_{obs}|) \qquad (12.5.5)$$

which can be found from tables of $\chi^2_{(1)}$ or $N(0,1)$.

Most applications of this test are to situations where σ^2 is not known, but where n is large (40 or more). One may then replace σ^2 by the estimate

$$s^2 = \frac{1}{n-1} \sum (x_i - \bar{x})^2 = \frac{1}{n-1}[\sum x_i^2 - \frac{1}{n}(\sum x_i)^2]$$

and proceed as if σ^2 were known. This procedure will be justified when we consider tests of $H:\mu = \mu_0$ with σ^2 unknown; see Section 14.1.

Example 12.5.4. The relationship between the age of the parents and the incidence of mongolism in children was discussed in Section 7.5. It is known that the mean age of mothers at normal births is 31.25 years. The average age of mothers at $n = 50$ births of mongolian children was $\bar{x} = 37.25$ and the variance estimate was $s^2 = 49.35$. Are these observations consistent with the hypothesis $\mu = 31.25$?

Solution. We assume that mothers' ages at mongolian births are independent $N(\mu, \sigma^2)$. (The normality assumption could be checked via a quantile plot of the 50 ages.) Since n is large, we take $\sigma^2 = s^2 = 49.35$, and use the test derived above. We find that

$$z_{obs} = \frac{\sqrt{50}(37.25 - 31.25)}{\sqrt{49.35}} = 6.04;$$

$$SL = P\{|Z| \geq 6.04\} < 0.000001$$

from Table B1. There is overwhelming evidence against $H:\mu = 31.25$. The average age of mothers is significantly higher at mongolian births than at normal births.

Likelihood ratio test for exponential mean

Let $X_1, X_2, \ldots, X_n$ be independent variates having an exponential distribution with mean θ, and consider a test of the hypothesis $H: \theta = \theta_0$. From Example 9.5.1, the relative likelihood function of θ is

$$R(\theta) \equiv \frac{L(\theta)}{L(\hat{\theta})} \equiv \frac{\theta^{-n} e^{-\sum X_i/\theta}}{\hat{\theta}^{-n} e^{-\sum X_i/\hat{\theta}}} \equiv (\frac{T}{n\theta})^n \exp(n - \frac{T}{\theta}),$$

where $T \equiv \sum X_i$ and $\hat{\theta} \equiv T/n$. The likelihood ratio statistic is

$$D \equiv -2 \log R \equiv -2n \log (\frac{T}{n\theta}) - 2n + \frac{2T}{\theta}$$

which depends only on the sufficient statistic T.

In this case, D does not have a simple distribution, which can be looked up in standard tables. However, the variate $U \equiv 2T/\theta$ has a χ^2 distribution with $2n$ degrees of freedom (see Problem 6.9.5), and we shall use this result to compute the significance level. We can express D as a function of U:

$$D \equiv U - 2n [1 + \log \frac{U}{2n}]. \qquad (12.5.6)$$

Given d, the observed value of D, we determine the range of U-values for which $D \geq d$, and then consult tables of $\chi^2_{(2n)}$.

Example 12.5.5. In Example 9.5.1 we considered $n = 10$ lifetimes with observed total $T = 288$. Suppose that we wish to test $H: \theta = 20$. Under this hypothesis, we have observed values

$$u = \frac{2 \times 288}{20} = 28.8; \quad d = 28.8 - 20 [1 + \log \frac{28.8}{20}] = 1.507.$$

By evaluating D numerically at several values of U, we find that $D \geq 1.507$ for $U \geq 28.8$ and for $U \leq 13.206$. Hence

$$SL = P(D \geq 1.507) = P(U \geq 28.8) + P(U \leq 13.206).$$

Finally, since U has χ^2 distribution with 20 degrees of freedom, we consult Table B4 to get

$$SL = P(\chi^2_{(20)} \geq 28.8) + P(\chi^2_{(20)} \leq 13.206) \approx 0.09 + 0.12 = 0.21.$$

The hypothesis $\theta = 20$ is therefore consistent with the data.

Similarly, for a test of $H: \theta = 50$ we compute

$$u = \frac{2 \times 288}{50} = 11.52; \quad d = 11.52 - 20[1 + \log \frac{11.52}{20}] = 2.553.$$

Trial and error gives $D \geq 2.553$ for $U \leq 11.52$ and for $U \geq 31.88$, so that

$$SL = P(\chi^2_{(20)} \leq 11.52) + P(\chi^2_{(20)} \geq 31.88) \approx 0.07 + 0.04 = 0.11.$$

The hypothesis $\theta = 50$ is also consistent with the data.

Problems for Section 12.5

†1. The measurement errors associated with a set of scales are independent normal with known standard deviation $\sigma = 1.3$ grams. Ten weighings of an unknown mass μ give the following results (in grams):

$$\begin{array}{ccccc} 227.1 & 226.8 & 224.8 & 228.2 & 225.6 \\ 229.7 & 228.4 & 228.8 & 225.9 & 229.6 \end{array}$$

(a) Are these data consistent with the hypothesis $\mu = 226$?

(b) Are they consistent with $\mu = 229$?

(c) For which values of μ will the significance level be exactly 5%? at least 5%?

2. Survival times for patients treated for a certain disease may be assumed to be exponentially distributed. Under the standard treatment, the expected survival is 37.4 months. Ten patients receiving a new treatment survived for the following times (in months):

$$\begin{array}{ccccc} 99 & 8 & 30 & 6 & 53 \\ 60 & 44 & 12 & 105 & 17 \end{array}$$

(a) Are these data consistent with a mean survival time of 37.4 months?

(b) The doctor who developed the new treatment claims that it gives a 50% increase in mean survival time. Are the data consistent with this claim?

3. Let $X_1, X_2, \ldots, X_n$ be independent $N(\mu, \sigma^2)$ with known mean μ and unknown variance σ^2 . Find the likelihood ratio statistic for testing the hypothesis $\sigma = \sigma_0$.

4. Let $X_1, X_2, \ldots, X_n$ be independent normal variates having known var-

iances $\sigma_1{}^2, \sigma_2{}^2, \ldots, \sigma_n{}^2$, and the same unknown mean μ. Derive the likelihood ratio statistic for testing $H : \mu = \mu_0$, find its distribution, and show how the significance level may be found from tables of the standardized normal distribution.

†5. Let $X_1, X_2, \ldots, X_n$ be independent $N(\mu_1, \sigma^2)$ variates and let $Y_1, Y_2, \ldots, Y_m$ be independent $N(\mu_2, \sigma^2)$ variates. Assuming σ to be known, derive the likelihood ratio statistic for a test of the hypothesis $\mu_1 = \mu_2$.

6. Let $X_1, X_2, \ldots, X_n$ be independent exponential variates with mean θ_1, and let $Y_1, Y_2, \ldots, Y_m$ be independent exponential variates with mean θ_2. Show that the likelihood ratio statistic for testing $H : \theta_1 = \theta_2$ depends only on the ratio of the sample means, $\overline{X}/\overline{Y}$. (This ratio has an F distribution; see Problem 14.5.5.)

12.6 Approximate Likelihood Ratio Tests

Suppose that, under the basic model, the vector of unknown parameters θ can take any value in a parameter space Ω with dimension p. Further, suppose that the hypothesis H restricts θ to a subspace of Ω with dimension $r < p$; that is, there are r unknown parameters under the hypothesis. Let R be the (maximized) likelihood ratio for the hypothesis. Then, in large samples and under suitable regularity conditions, the distribution of the likelihood ratio statistic under the hypothesis is approximately chi-square with $p - r$ degrees of freedom:

$$- 2 \log R \approx \chi^2_{(p-r)}. \tag{12.6.1}$$

Using this result, one can obtain an approximation to the significance level in a likelihood ratio test from tables of the χ^2 distribution.

The precise regularity conditions needed to establish (12.6.1) will depend upon the situation. In general, one must rule out cases such as that in Example 9.8.3 where the range of the probability distribution depends upon the parameter. The method of proof for (12.6.1) usually involves expansion of the log likelihood function in a Taylor's series about the true value $\theta = \theta_0$. Consequently, θ_0 is assumed to be an interior point of the parameter space Ω, and the first three derivatives of the log likelihood with respect to θ are assumed to exist in the neighbourhood of θ_0. For further discussion of regularity conditions, see *Theoretical Statistics* by D.R. Cox and D.V. Hinkley.

Example 12.6.1. Tests for multinomial probabilities

In the multinomial case (Example 12.5.2), the k class probabilities sum to 1, and hence the parameter space Ω has dimension
$k-1$. If r parameters require estimation under the hypothesis, the
degrees of freedom for the test is $p-r=k-r-1$. The large-sample
distribution of the likelihood ratio statistic is thus the same as
that for the goodness of fit criterion. In fact, it can be shown that
these two statistics are equivalent for very large expected frequencies.

For the five examples in Section 11.2, the values of the
goodness of fit criterion were found to be

$$3.82 \quad 12.90 \quad 1.02 \quad 11.37 \quad 39.23$$

with 5,9,3,8, and 8 degrees of freedom. The observed values of
the likelihood ratio statistic (12.5.4) in these examples are

$$3.70 \quad 13.94 \quad 1.00 \quad 11.49 \quad 35.88$$

with the same degrees of freedom. The difference between the two
statistics is not great enough to affect the conclusions in any of
these examples.

Example 12.6.2. Test for normal mean

In the last section we considered the likelihood ratio test
of $H:\mu=\mu_0$ in a normal distribution with known variance. Since σ^2
is assumed to be known, there is just one unknown parameter $(p=1)$,
and its value is determined by the hypothesis. Hence $r=0$, and
(12.6.1) gives $-2\log R \approx \chi^2_{(1)}$. In fact, we showed in the last section that the likelihood ratio criterion has <u>exactly</u> a χ^2 distribution
in this case.

Example 12.6.3. Test for exponential mean

The likelihood ratio test of $H:\theta=\theta_0$ in an exponential distribution was also considered in Section 12.5. Again there is just one
unknown parameter, and its value is specified by the hypothesis. Thus
(12.6.1) gives $D \approx \chi^2_{(1)}$.

The observed value of D for a test of $H:\theta=20$ in Example
11.5.5 was $d=1.507$. The approximate significance level is

$$P\{\chi^2_{(1)} \geq 1.507\} = P\{|Z| \geq \sqrt{1.507}\} = 0.22$$

from table B2. Similarly, for a test of $H:\theta=50$ we get

$$SL \approx P\{\chi^2_{(1)} \geq 2.553\} = P\{|Z| \geq 1.60\} = 0.11.$$

The approximate values agree closely with the exact results found in Example 12.5.5.

12.7   Two-tail Tests

In preceding sections, we have considered several methods of ranking outcomes for a significance test. The most appealing and widely applicable of these is the ranking by likelihood ratio, although it can present computational problems. In this section we consider rankings based on tail areas. The main advantage of these would seem to be computational, in that they permit significance levels to be found easily from tables of cumulative probability.

Suppose that the test is to be based on the distribution of a single statistic T. For instance, if we were testing $H: \theta = \theta_0$ where θ is a one-dimensional parameter, T would be a sufficient statistic for θ. Let F denote the cumulative distribution function of T when H is true.

Usually, both large and small values of T (i.e. values in both tails of the distribution) would be considered as in poor agreement with the hypothesis, while values of T near the centre of the distribution are in good agreement. If $F(t)$ is near $\frac{1}{2}$, then t is in good agreement, while if $F(t)$ is near 0 or 1 then t is in poor agreement with the hypothesis. This suggests that the discrepancy measure be defined as follows:

$$D \equiv \left| F(T) - \frac{1}{2} \right|.$$

The test of significance is then called a two-tail test, because values in both the lower (left-hand) tail and upper (right-hand) tail give evidence against the hypothesis. The two-tail significance level is

$$SL = P(D \geq d) = P\{F(T) - \frac{1}{2} \geq d\} + P\{F(T) - \frac{1}{2} \leq -d\}$$
$$= P\{F(T) \geq \frac{1}{2} + d\} + P\{F(T) \leq \frac{1}{2} - d\}.$$

If T is continuous, then $F(T)$ has a uniform distribution between 0 and 1; see Section 6.3. Hence $P\{F(T) \leq u\} = u$, and we have

$$SL = 1 - (\frac{1}{2} + d) + (\frac{1}{2} - d) = 1 - 2d = 1 - 2 \cdot \left| F(t) - \frac{1}{2} \right|,$$

where t is the observed value of T. If t is in the upper tail (i.e. greater than the median), then $F(t) > \frac{1}{2}$, and

$$SL = 1 - 2[F(t) - \tfrac{1}{2}] = 2[1 - F(t)] = 2 \cdot P(T \geq t).$$

If t is in the lower tail (i.e. less than the median), then

$$SL = 1 - 2[\tfrac{1}{2} - F(t)] = 2F(t) = 2 \cdot P(T \leq t).$$

The observed value t divides the distribution into a smaller tail (with probability less than $\tfrac{1}{2}$) and a larger tail (with probability greater than $\tfrac{1}{2}$). When T is continuous, the two-tail significance level equals twice the probability in the smaller tail.

      If the distribution of T is symmetrical and the mean exists, a two-tail test will give the same significance level as a test based on the distance from the mean, $D \equiv |T - E(T)|$. The results will agree closely if T is reasonably symmetrical. If the distribution of T is badly skewed, a two-tail test or likelihood ratio test would be preferable to one based on distance from the mean.

Example 12.7.1. Two-tail test for a normal mean

      Suppose that $X_1, X_2, \ldots, X_n$ are independent $N(\mu, \sigma^2)$ with known variance σ^2. Then $\overline{X}$ is a sufficient statistic for μ; see Section 12.5. Under $H : \mu = \mu_0$, $\overline{X}$ has a $N(\mu_0, \sigma^2/n)$ distribution, and $Z \equiv \sqrt{n}(\overline{X} - \mu_0)/\sigma$ is $N(0,1)$. Let $\overline{x}$ and z_{obs} denote their observed values. If $\overline{x} > \mu_0$ $(z_{obs} > 0)$, there is less probability to the right of x to the left, and

$$SL = 2 \cdot P(\overline{X} \geq \overline{x}) = 2 \cdot P(Z \geq z_{obs}).$$

Similarly, if $\overline{x} < \mu_0$ $(z_{obs} < 0)$ we have

$$SL = 2 \cdot P(\overline{X} \leq \overline{x}) = 2 \cdot P(Z \leq z_{obs}).$$

But, since $N(0,1)$ is symmetrical about the origin, we have

$$P(Z \geq z) = P(Z \leq -z) = \tfrac{1}{2} P(|Z| \geq z)$$

for any $z \geq 0$, and hence in either case

$$SL = P(|Z| \geq |z_{obs}|).$$

Thus a two-tail test of $H : \mu = \mu_0$ gives the same significance level as the likelihood ratio test (Section 12.5).

Example 12.7.2. Two-tail test for an exponential mean

Let $X_1, X_2, \ldots, X_n$ be independent exponential variates with mean θ. Then $T \equiv \sum X_i$ is a sufficient statistic for θ; see Section 12.5. Under $H: \theta = \theta_0$, the variate $U \equiv 2T/\theta_0$ is distributed as $\chi^2_{(2n)}$. Let t and u denote the observed values of T and U, and let m be the median (50% point) of $\chi^2_{(2n)}$ from Table B4. The significance level in a two-tail test is given by

$$SL = \begin{cases} 2 \cdot P(\chi^2_{(2n)} \geq u) & \text{if } u > m \\ 2 \cdot P(\chi^2_{(2n)} \leq u) & \text{if } u \leq m \end{cases}$$

For instance, in Example 12.5.5, we had $n = 10$ and $t = 288$. Under $H: \theta = 20$, we have $u = 28.8$. From Table B4, the 50% point of $\chi^2_{(20)}$ is $m = 19.34$. Since $u > m$, we find that

$$SL = 2 \cdot P(\chi^2_{(20)} \geq 28.8) \approx 2(0.09) = 0.18$$

from Table B4 again. Under $H: \theta = 50$, we have $u = 11.52 < m$, and hence

$$SL = 2 \cdot P(\chi^2_{(20)} \leq 11.52) \approx 2(0.07) = 0.14.$$

Note that a two-tail test does not give the same significance level as a likelihood ratio test (Example 12.5.5) in this case. The significance level is found as a sum of two unequal tail areas in the likelihood ratio test, whereas equal tail areas are used in a two-tail test. The two-tail test requires less computation because it is not necessary to calculate the U-value in the opposite tail of the distribution.

Two-tail tests with discrete distributions

A slight complication arises when T is a discrete variate, for then it is usually impossible to take equal areas in the two tails of the distribution. The significance level in a two-tail test is then less than twice the probability in the smaller tail. For instance, consider a two-tail test of $H: p = 0.52$ in a binomial distribution with $n = 10$ (Example 12.1.1). The probability function of X when $p = 0.52$ is as follows:

x	0	1	2	3	4	5	6	7	8	9	10
f(x)	.0006	.0070	.0343	.0991	.1878	.2441	.2204	.1364	.0554	.0133	.0014

Suppose that we observe $X = 2$, which is in the lower tail. The lower tail probability is $f(0) + f(1) + f(2) = 0.0419$. To obtain the two-tail significance level, we add to this the largest possible upper tail probability which does not exceed 0.0419, giving

$$SL = 0.0419 + f(9) + f(10) = 0.0566.$$

If we observed $X = 8$, we would first compute the upper tail probability $f(8) + f(9) + f(10) = 0.0701$. To this we add the largest possible left tail probability which does not exceed 0.0701, giving

$$SL = 0.0701 + f(0) + f(1) + f(2) = 0.1120.$$

In this example, the two-tail significance levels agree with those obtained by ranking values of X according to their probabilities, or according to their distance from the mean.

One-tail tests

A two-tail test is appropriate when both large and small values of T give evidence against the hypothesis. It sometimes happens that only large values of T give evidence against the hypothesis, and small values of T are considered to be in good agreement. In this situation, a one-tail test is appropriate. T itself, or any increasing function of T may be selected as the discrepancy measure, and the significance level equals the probability in the upper tail of the distribution of T,

$$SL = P(T \geq t),$$

where t is the observed value. Similarly, a one-tail test is appropriate when small values of T give evidence against the hypothesis but large values do not. The significance level then equals the probability in the lower tail of the distribution,

$$SL = P(T \leq t).$$

Note. It is the logic of the situation, and not the observed value of T, which determines whether one should use the lower tail, the upper tail, or both tails of the distribution. One should be prepared to specify which of these is to be used before examining the data. If one decides to accept only large values of T as evidence against the hypothesis, then small values of T, however extreme, are by definition in very good agreement with H. In most of the examples to be considered, an extreme result in either tail would cast doubt on the hypothesis, and hence a two-tail test is appropriate.

Effects of changing variables

Suppose that T is a sufficient statistic for a parameter θ. Then from Section 9.8, so is any one-to-one function, $U \equiv h(T)$. The information content of U and T is the same, and a test of $H: \theta = \theta_0$ could equally well be based on the distribution of U as on the distribution of T. Thus, we might ask whether the various methods discussed for ranking outcomes are affected by a change of variables.

The "distance from the mean" discrepancy clearly will be affected by a nonlinear transformation from T to U, since in general $E\{h(T)\} \neq h\{E(T)\}$.

The probability ranking will not be affected if T is discrete, but if T is continuous, the Jacobian involved in a change of variables may alter the ranking quite substantially. The p.d.f. of U is

$$g(u) = f(t) \cdot \left| \frac{dt}{du} \right|,$$

and values of u for which $g(u)$ is large need not correspond to values of t for which $f(t)$ is large. For this reason, it does not seem appropriate to rank by probability density when T is continuous.

A continuous one-to-one transformation is monotonic, and will not alter tail areas (although a decreasing transformation will interchange the two tails). A change of variables will not affect the likelihood ratio either, because the Jacobian appears in both the numerator and the denominator in (12.5.1). The significance level in a two-tail test or a likelihood ratio test will not be altered by a change of variables, and this is an advantage of these procedures in the continuous case.

Problems for Section 12.7

1. Repeat Problem 12.5.2, using a two-tail test rather than a likelihood ratio test.

*12.8 Ancillary Statistics

Conditional tests for some composite hypotheses were considered in Section 12.2. Conditional procedures are also appropriate when ancillary statistics are present.

As we noted in Section 9.8, there will sometimes exist a single statistic T which carries all of the information concerning an unknown parameter θ. However, in many cases, a set of statistics will be needed.

Suppose we can find a pair of statistics (T,A) such that
(i) T and A together are minimally sufficient for θ;
(ii) the distribution of A does not depend upon θ.

Then A is called an <u>ancillary statistic</u> for θ.

Because T and A are jointly sufficient for θ, their joint distribution carries all of the information about θ. Since A is ancillary, we have

$$f(t,a;\theta) = f_1(t|a;\theta)f_2(a) \qquad\qquad (12.8.1)$$

where the second factor does not depend upon θ. All of the information about θ is carried by the first factor. Thus, tests of significance, confidence intervals (Section 13.1) and the fiducial argument (Chapter 16) will be based upon the conditional distribution of T given the observed value of the ancillary statistic A.

An ancillary statistic A gives no direct information about the magnitude of θ. Given only the observed value of A, we could learn nothing about the magnitude of θ, because the distribution of A does not depend upon θ. The primary information about θ is provided by T, with A giving supplementary or ancillary information.

We can usually interpret an ancillary statistic as a measure of the precision with which it is possible to estimate θ. The various possible outcomes of the experiment may differ greatly in the amount of information about θ which they are capable of yielding. If we are lucky, we may get an outcome which permits the value of θ to be determined quite precisely; if we are unlucky, we may obtain an uninformative outcome from which relatively little can be learned about θ.

In problems of inference, it is necessary to take into account the informativeness of the data actually obtained. The fact that one might obtain a more informative (or less informative) result if the experiment were repeated should be considered in designing future experiments, but it is irrelevant to the interpretation of the data at hand.

* This section may be omitted on first reading.

The observed value of the ancillary statistic indicates the informativeness of the data actually obtained. It is therefore appropriate to
base inferences about θ on the conditional distribution of T given
the observed value of A.

It is not necessary to worry about ancillary statistics when
likelihood methods (Chapters 9 and 10) or Bayesian methods (Chapter
16) are used. These depend only on the observed likelihood function.
Since the second factor in (12.8.1) does not depend upon θ, the likelihood function based on the conditional distribution of T given A
will proportional to that based on the original sample.

Example 12.8.1.   Random sample size

Consider the situation described in Example 9.8.1, where N
people are to be examined for tuberculosis. In the previous discussion, we assumed that N, the sample size, was fixed and known prior
to the experiment. However, it may be that N itself is subject to
random variation. For instance, the sample size might depend upon the
amount of money and laboratory space, and the number of personnel
available for the study, and perhaps none of these is under the strict
control of the experimenter. Or perhaps unforeseen circumstances un-
related to the incidence of tuberculosis cause the experiment to be
terminated after 150 people have been examined, although it was ori-
ginally planned to examine 200.

Suppose, then, that the sample size N is a random variable
with probability function g(n) not depending upon θ. The result of
the experiment will be a pair of values (x,n), where x is the number with tuberculosis out of the n examined. Given n, X has a binomial distribution with probability function

$$f_1(x|n) = \binom{n}{x}\theta^x(1 - \theta)^{n-x} \quad \text{for} \quad x = 0,1,\ldots,n.$$

The joint p.f. of X and N is thus

$$f(x,n) = f_1(x|n)g(n) = \binom{n}{x}\theta^x(1 - \theta)^{n-x}g(n),$$

and the likelihood function of θ is

$$L(\theta) = \theta^x(1 - \theta)^{n-x} \quad \text{for} \quad 0 < \theta < 1.$$

From this we see that (X,N) is a set of sufficient statistics for θ.
Since the distribution of N does not depend upon θ, N is an ancillary statistic for θ. In testing a value of θ or setting up a con-

fidence interval, we use the conditional (binomial) distribution of X
given the observed sample size   n.   Even though the sample size   N   is
random, we take it as fixed in the analysis.   The fact that we might
get a different sample size if we repeated the experiment is irrelevant
to the interpretation of the data actually obtained.

In the above discussion, we have assumed that the distribu-
tion of   N   does not depend upon θ.   Of course, if this distribution
depended upon θ,   we could not ignore it without losing information.
For instance, one might decide to keep examining people until three
with tuberculosis had been found and then stop.   Then the distribution
of the sample size would depend upon θ,   and it would not be appropriate to condition upon   N.

Example 12.8.2.   Suppose that there are two different techniques for
determining the concentration (in standard units) of a chemical in a
solution.   The first technique gives a reading $X \sim N(\mu, 1)$,   where μ
is the true concentration, while the second technique gives $X \sim N(\mu, 100)$.
Although the first technique is more precise, it is possible to use it
for only   50% of solutions.   Each solution is allocated at random
$(p = 1/2)$ to one or the other of the two techniques.   A specific solution is analysed, and we wish to obtain information about μ.

Let us define $Z = 0$ if the first technique is used, and
$Z = 1$ otherwise.   The result of the experiment will be a pair of
values (x, z).   Given z, X has standard deviation 10^Z,   and p.d.f.

$$f(x|z) = \frac{1}{\sqrt{2\pi}\ 10^Z} \exp\{-\tfrac{1}{2}(x-\mu)^2/10^{2z}\} \quad \text{for} \quad -\infty < x < \infty,$$

and the joint distribution of X and Z is given by

$$f(x,z) = f(x|z)f_2(z) = \tfrac{1}{2}\ f(x|z) \quad \text{for} \quad -\infty < x < \infty; \quad z = 0, 1.$$

Hence the likelihood function for μ is

$$L(\mu) = \exp\{-\tfrac{1}{2}(x-\mu)^2/10^{2z}\} \quad \text{for} \quad -\infty < \mu < \infty,$$

so that (X, Z) is a pair of sufficient statistics for μ.   Since the
distribution of Z does not depend upon μ, Z is an ancillary statistic, and inferences about μ will be based on the conditional distribution of X given Z.

For instance, if $z = 0$,   inferences about μ will be based
on the distribution $X \sim N(\mu, 1)$.   The fact that the measurement might

have been made by the second technique is irrelevant, and does not enter into the calculations. It is the precision associated with the
technique actually used which is important.

The same comment applies for other methods of allocating
solutions to techniques (e.g. at random with a fixed $p \neq \frac{1}{2}$). If the
method of allocation does not depend upon μ, it will be appropriate
to hold Z fixed at its observed value in making inferences about μ.

Example 12.8.3. A total of n clouds are to be observed in an experiment to determine the effectiveness of cloud seeding in producing rain.
For each cloud it is decided whether or not to seed by flipping a balanced coin. Hence Z, the number of clouds to be seeded, has a binomial $(n,\frac{1}{2})$ distribution.

Let X be the number of seeded clouds which produce rain,
and let Y be the number of unseeded clouds which produce rain. We
assume that clouds are independent, and that the probability of rain
is θ_1 for a seeded cloud and θ_2 for an unseeded cloud. Then,
given that z clouds are seeded, X has a binomial (z,θ_1) distribution, and Y has a binomial $(n-z,\theta_2)$ distribution independently
of X. We observe (x,y,z), and wish to make inferences about θ_1
and θ_2. In particular, we would wish to know whether there is evidence that $\theta_1 \neq \theta_2$.

The joint probability function of X,Y, and Z is

$$f(x,y,z) = f(x,y|z)f(z) = f(x|z)f(y|z)f(z)$$
$$= \binom{z}{x}\theta_1^{x}(1-\theta_1)^{z-x}\binom{n-z}{y}\theta_2^{y}(1-\theta_2)^{n-z-y}\binom{n}{z}(\frac{1}{2})^{n}.$$

The likelihood function of θ_1 and θ_2 is thus

$$L(\theta_1,\theta_2) = \theta_1^{x}(1-\theta_1)^{z-x}\theta_2^{y}(1-\theta_2)^{n-z-y},$$

so that X,Y, and Z are jointly sufficient for θ_1 and θ_2. Since
the distribution of Z does not depend upon θ_1 or θ_2, Z is an
ancillary statistic, and inferences about θ_1 and θ_2 should be based
on the conditional distribution of X,Y given the observed value of
Z (i.e. given the number of clouds actually seeded). In particular,
a test of the hypothesis $\theta_1 = \theta_2$ can be performed as in Section 12.2
(comparison of binomial proportions).

The relationship between the value of Z and precision is
easily seen in this example. If one should get $z = 0$ (improbable but
still possible), then one would not seed any clouds, and thus would

obtain no information about θ_1. Similarly, with $z = n$, one would obtain no information about θ_2. In both cases, the experiment would be incapable of giving evidence against the hypothesis $\theta_1 = \theta_2$. However, if one obtained $z \approx n/2$, the experiment would give a reasonable amount of information about θ_1 and θ_2, and hence would be capable of showing that they are different. The observed value of Z thus indicates the precision which is possible in inferences about θ_1 and θ_2. Although Z is a random variable, we regard Z as fixed at its observed value in the analysis.

In this case, the existence of the ancillary statistic Z shows up a defect in the design of the experiment. It would be better to set up the experiment so that the value of Z was fixed in advance near $n/2$. This could be done by drawing balls at random without replacement from an urn containing $n/2$ white balls and $n/2$ black balls, and seeding a cloud if a white ball is drawn.

Example 12.8.4. Cauchy Distribution

Suppose that $X_1, X_2, \ldots, X_n$ are independent variates having a Cauchy distribution centred at θ, as in Example 9.8.5. The set of order statistics $X_{(1)}, X_{(2)}, \ldots, X_{(n)}$ is minimally sufficient for θ and no further reduction of the data is possible without loss of information. The use of sufficient statistics produces no reduction of the dimension of the sample space in this case. However, it is possible to find $n-1$ ancillary statistics, and hence reduce the dimension from n to 1 by conditioning on them.

Let us define $U_i \equiv X_i - \theta$ for $i = 1, 2, \ldots, n$. Then, by (6.1.11), U_i has a Cauchy distribution centred at 0. We may write $X_i \equiv \theta + U_i$, where U_i has a distribution not depending upon θ. Now consider the $n-1$ statistics

$$A_i \equiv X_{(i+1)} - X_{(i)} \quad \text{for} \quad i = 1, 2, \ldots, n-1.$$

Since $X_{(i)} \equiv \theta + U_{(i)}$, we have

$$A_i \equiv [\theta + U_{(i+1)}] - [\theta + U_{(i)}] \equiv U_{(i+1)} - U_{(i)},$$

and hence the distribution of $A_1, A_2, \ldots, A_{n-1}$ does not depend upon θ.

Let T be any statistic such that the transformation from $X_{(1)}, X_{(2)}, \ldots, X_{(n)}$ to $T, A_1, A_2, \ldots, A_{n-1}$ is one-to-one. For instance, we could take $T \equiv X_{(i)}$ for any i, or $T \equiv \bar{X}$. Then $(T, A_1, \ldots, A_{n-1})$ is a set of sufficient statistics for θ, and $A_1, \ldots, A_{n-1}$ are ancillary statistics. All of the information about θ is carried by the conditional distribution of T given the observed values $a_1, a_2, \dots, a_{n-1}$. This distribution, which may be found by numerical integration, forms the basis for inferences about θ.

The interpretation of the ancillary statistics $A_1, A_2, \dots, A_{n-1}$ as measures of precision is not as clearcut in this example. They give information about the shape of the likelihood function, but not about its location on the axis. For instance, if $n = 2$, the likelihood function has a unique maximum at $\bar{x}$ when a_1 is small, but is bimodal with a relative minimum at $\bar{x}$ when a_1 is large. The value of A_1 indicates the shape of the likelihood function, but to determine its position (and hence the magnitude of θ), we also require the value of an additional statistic such as $\bar{X}$.

*12.9   Power

We now return to the problem of selecting an appropriate discrepancy measure or test statistic D for a significance test. Our discussion of this has been informal. In justifying a particular test statistic D, we have argued that the ranking of outcomes which it produced was intuitively reasonable.

In this section we briefly introduce a theory of test statistics based on the concept of the power or sensitivity of a test statistic against an alternative hypothesis. Power comparisons may help in determining which of several possible test statistics is more likely to detect departures of a particular type from the hypothesis being tested.

Let H_0 denote the hypothesis to be tested, and let H_1 denote another hypothesis. H_0 and H_1 will be called the null hypothesis and the alternative hypothesis, respectively. The alternative hypothesis is chosen to represent the particular type or types of departure from H_0 that one is anxious to detect in the test.

Initially we assume that H_0 and H_1 are simple hypotheses, so that the probability of a typical outcome x can be computed numerically under both H_0 and H_1. Let D be the discrepancy measure (test statistic) for a test of H_0. The significance level of outcome x in relation to H_0 is

$$SL(x) = P\{D \geq d \mid H_0 \text{ is true}\}$$

where $d = D(x)$ is the observed value of D.

---

* This section may be omitted on first reading.

A real number α is called an <u>achievable significance level</u>
if there exists an outcome x such that $SL(x) = \alpha$. It follows from
the results on coverage frequency in Section 13.1 that

$$P\{SL \leq \alpha | H_0 \text{ is true}\} = \alpha \qquad (12.9.1)$$

for every achievable α. If D is a continuous variate, then every
α between 0 and 1 is achievable. However, if D is a discrete
variate, there will be only a discrete set of achievable significance
levels. Two test statistics are said to be <u>comparable</u> if they have
the same set of achievable significance levels.

The <u>size α critical region</u> of a test is the set C_α of
outcomes x such that $SL(x) \leq \alpha$; that is,

$$C_\alpha = \{x; \ SL(x) \leq \alpha\}.$$

It follows from (12.9.1) that, for any achievable α,

$$P\{X \in C_\alpha | H_0 \text{ is true}\} = \alpha. \qquad (12.9.2)$$

A test statistic defines a nested set of critical regions. If α_1
and α_2 are two achievable significance levels with $\alpha_1 < \alpha_2$, then
$C_{\alpha_1} \subset C_{\alpha_2}$.

The <u>size α power</u> (or sensitivity) of a test statistic D
with respect to the simple alternative hypothesis H_1 is defined by

$$K_\alpha = P\{SL \leq \alpha | H_1 \text{ is true}\} = P\{X \in C_\alpha | H_1 \text{ is true}\}.$$

For instance, $K_{0.05}$ is the probability that a test of H_0 using D
will produce a significance level of 5% or less if H_1 is true. If
$K_{0.05}$ is near 1, the test statistic D is said to be powerful or
sensitive against H_1, because if H_1 were true the test would almost
always give evidence that H_0 was false.

Now let D and D' be two comparable statistics for test-
ing H_0, having power K_α and K'_α against H_1. Then D is said to
be <u>more powerful</u> than D' against the alternative hypothesis H_1 if
$K_\alpha \geq K'_\alpha$ for all achievable significance levels α. A statistic D
is called <u>most powerful</u> for testing H_0 against H_1 if it is more
powerful than any comparable statistic D'.

The following theorem, which is called the Neyman-Pearson
Fundamental Lemma, yields a most powerful test statistic when both H_0
and H_1 are simple hypotheses.

Theorem 12.9.1. Let H_0 and H_1 be simple hypotheses, and let $f_0(x)$ and $f_1(x)$ denote the probability of a typical outcome x under H_0 and under H_1, respectively. Then the statistic

$$D(x) = f_1(x)/f_0(x) \qquad (12.9.3)$$

is most powerful for testing H_0 against H_1.

Proof. Let α be an achievable significance level for D. The size α critical region has the form

$$C_\alpha = \{x; \ D(x) \geq d_\alpha\}$$

where d_α is a positive constant chosen so that $P(X \in C_\alpha | H_0) = \alpha$. Note that

$$f_1(x) \geq d_\alpha f_0(x) \qquad \text{for} \quad x \in C_\alpha; \qquad (12.9.4)$$

$$f_1(x) < d_\alpha f_0(x) \qquad \text{for} \quad x \in \overline{C}_\alpha. \qquad (12.9.5)$$

Now let C'_α be the size α critical region for any comparable statistic D'. Then the sample space is partitioned into four regions $C_\alpha C'_\alpha$, $C_\alpha \overline{C}'_\alpha$, $\overline{C}_\alpha C'_\alpha$, and $\overline{C}_\alpha \overline{C}'_\alpha$. We denote their probabilities under H_0 by p_{ij} and their probabilities under H_1 by q_{ij} as in Table 12.9.1.

Table 12.9.1
Probabilities under the Null Hypothesis and under the Alternative Hypothesis for the Four Regions of the Sample Space Defined by Two Size α Critical Regions

H_0	C'_α	$\overline{C}'_\alpha$	Total	H_1	C'_α	$\overline{C}'_\alpha$	Total
C_α	p_{11}	p_{12}	α	C_α	q_{11}	q_{12}	K_α
$\overline{C}_\alpha$	p_{21}	p_{22}	$1-\alpha$	$\overline{C}_\alpha$	q_{21}	q_{22}	$1-K_\alpha$
Total	α	$1-\alpha$	1	Total	K'_α	$1-K'_\alpha$	1

Since C_α and C'_α are size α critical regions, we have

$$p_{11} + p_{12} = P(X \in C_\alpha | H_0) = \alpha;$$

$$p_{11} + p_{21} = P(X \in C'_\alpha | H_0) = \alpha,$$

and hence $p_{12} = p_{21}$. The size α power is

$$K_\alpha = P(X \in C_\alpha | H_1) = q_{11} + q_{12};$$

$$K'_\alpha = P(X \in C'_\alpha | H_1) = q_{11} + q_{21}$$

and the difference in power is

$$K_\alpha - K'_\alpha = q_{12} - q_{21}.$$

Now since $C_\alpha \overline{C}'_\alpha$ is a subset of C_α, (12.9.4) gives

$$q_{12} = \sum f_1(x) \geq d_\alpha \sum f_0(x) = d_\alpha p_{12}$$

where the sums are taken over $x \in C_\alpha \overline{C}'_\alpha$. Similarly, since $\overline{C}_\alpha C'_\alpha$ is a subset of $\overline{C}_\alpha$, (12.9.5) gives

$$q_{21} = \sum f_1(x) < d_\alpha \sum f_0(x) = d_\alpha p_{21}.$$

Now, since $p_{12} = p_{21}$, we have

$$K_\alpha - K'_\alpha = q_{12} - q_{21} > d_\alpha p_{12} - d_\alpha p_{21} = 0.$$

This result holds for all comparable statistics D' and achievable significance levels α, and hence the theorem follows.

Example 12.9.1. Let $X \sim N(\mu, 1)$ and consider a test of the simple null hypothesis $H_0 : \mu = \mu_0$ versus the simple alternative hypothesis $H : \mu = \mu_1$. The likelihood ratio for $\mu = \mu_1$ versus $\mu = \mu_0$ is

$$D(x) = f_1(x)/f_0(x) = \exp\{-\frac{1}{2}(x - \mu_1)^2 + \frac{1}{2}(x - \mu_0)^2\}$$

$$= \exp\{x(\mu_1 - \mu_0) + \frac{1}{2}(\mu_0^2 - \mu_1^2)\},$$

and this is the most powerful statistic for testing H_0 against H_1.

If $\mu_1 > \mu_0$, large values of D correspond to large values of X. The size α critical region has the form $X \geq b_\alpha$ where b_α is chosen so that $P(X \geq b_\alpha | H_0) = \alpha$. Since $X \sim N(\mu_0, 1)$ under H_0, we have $b_\alpha = \mu_0 + z_\alpha$ where z_α is the value exceeded with probability α in a $N(0,1)$ distribution. The size α power is

$$K_\alpha = P(X \geq b_\alpha | H_1).$$

Since $X \sim N(\mu_1, 1)$ under H_1, we have

$$K_\alpha = P(Z \geq b_\alpha - \mu_1) = 1 - F(\mu_0 + z_\alpha - \mu_1)$$

where F is the $N(0,1)$ c.d.f.

If $\mu_1 < \mu_0$, large values of D correspond to small values of X, and the size α critical region will have the form $X \leq \mu_0 - z_\alpha$. The size α power is then found to be

$$K_\alpha = P(X \leq \mu_0 - z_\alpha | H_1) = F(\mu_0 - z_\alpha - \mu_1).$$

Since $F(-z) = 1 - F(z)$, both cases are covered by the formula

$$K_\alpha = F\{|\mu_0 - \mu_1| - z_\alpha\}.$$

In accepting only large values of X as evidence against $H: \mu = \mu_0$, we achieve maximum power against departures on the high side $(\mu_1 > \mu_0)$, but we lose the ability to detect departures on the low side $(\mu_1 < \mu_0)$. In general, the cost of increased sensitivity to one particular type of departure will be a decrease in sensitivity to other types. The likelihood ratio statistic (12.5.3) ranks outcomes according to the magnitude of $|X - \mu_0|$. This statistic is not most powerful for testing $\mu = \mu_0$ against any alternative value μ_1, but it does have reasonably high power for departures in both directions.

Composite alternative hypothesis

Now consider a slightly more general problem in which H_0 is simple but H_1 is composite. Suppose that a typical outcome x has probability $f(x;\theta)$ where θ is a real-valued parameter. The null hypothesis is taken to be $H_0: \theta = \theta_0$. The alternative hypothesis has the form $H_1: \theta \in \Omega_1$ where Ω_1 is a set of possible parameter values; for instance $H_1: \theta \neq \theta_0$ and $H_1: \theta > \theta_0$ have this form.

Given any particular value $\theta_1 \in \Omega_1$, a most powerful statistic for testing $\theta = \theta_0$ versus $\theta = \theta_1$ is given by

$$D(x) = f(x;\theta_1)/f(x;\theta_0). \qquad (12.9.6)$$

If this statistic defines the same ranking of outcomes (i.e. the same critical regions) for all $\theta_1 \in \Omega_1$, then D is said to be _uniformly most powerful_ for testing H_0 versus H_1.

In Example 12.9.1, one obtains the same ranking of outcomes from smallest (most favourable) to largest (least favourable) whenever $\mu_1 > \mu_0$. Hence there exists a uniformly most powerful statistic for testing $H_0: \mu = \mu_0$ versus $H_1: \mu > \mu_0$. Similarly, there exists a uniformly most powerful statistic for testing $H_0:\mu = \mu_0$ versus $H_1:\mu < \mu_0$. However there is no uniformly most powerful statistic for testing $H_0:\mu = \mu_0$ versus $H_1:\mu \neq \mu_0$ because a different ranking is obtained for $\mu_1 < \mu_0$ than for $\mu_1 > \mu_0$. The following example shows that a similar result holds in general for a wide class of problems.

Example 12.9.2. In Example 9.8.6 we defined the exponential family of distributions. We showed that the likelihood function based on a sample size n is

$$L(\theta) = [A(\theta)]^n e^{tc(\theta)}$$

where $T \equiv \sum d(X_i)$ is a sufficient statistic for θ. The likelihood ratio for $\theta = \theta_1$ versus $\theta = \theta_0$ is

$$D = \left[\frac{A(\theta_1)}{A(\theta_0)}\right]^n e^{t[c(\theta_1)-c(\theta_0)]}.$$

This test statistic is most powerful for testing $\theta = \theta_0$ against the simple alternative hypothesis $\theta = \theta_1$. If $c(\theta_1) > c(\theta_0)$, critical regions have the form $T \geq t_\alpha$, whereas if $c(\theta_1) < c(\theta_0)$ they have the form $T \leq t'_\alpha$. Consider a test of the simple hypothesis $H_0:\theta = \theta_0$ versus the composite alternative hypothesis $H_1:\theta \in \Omega_1$. There exists a uniformly most powerful statistic for testing H_0 against H_1 if $c(\theta_1) - c(\theta_0)$ has the same sign for all $\theta_1 \in \Omega_1$, but not if $c(\theta_1) - c(\theta_0)$ changes sign. $c(\theta)$ is called the natural parameter of the distribution; see Problem 9.8.10.

Discussion.

1. We have seen that there generally will not exist a statistic which is uniformly most powerful for testing $H_0:\theta = \theta_0$ against a two-sided alternative $\theta \neq \theta_0$. In fact, uniformly most powerful tests will rarely exist except in simple textbook examples. To make further progress in defining a theoretically optimum test, additional restrictions must be placed on the types of test to be considered. The restrictions usually suggested seem arbitrary and unconvincing. The situation is even less satisfactory when both the null and alternative hypotheses are composite, and we shall not give details here.

2. Although power considerations will not identify an optimum test statistic except in very special cases, a comparison of power may still be helpful in choosing between two test statistics D and D'. Given a statistic D for testing $H_0:\theta = \theta_0$ against $H_1:\theta = \theta_1$, one can

determine the size α power as a function of θ_1,

$$K_\alpha(\theta_1) = P(SL \le \alpha | \theta = \theta_1).$$

A graphical comparison of the power functions $K_\alpha(\theta)$ and $K'_\alpha(\theta)$ for selected values of α may suggest that one of the statistics is pre-ferable.

3.   A test of significance has two ingredients: a reference set and a test statistic. The choice of the reference set takes precedence over power comparisons. For instance, if there exists an ancillary statistic A, a conditional reference set is appropriate. Power would then be computed from the conditional distribution of outcomes given the observed value of A.

4.   Another use to which power has been put is the determination of sample size. Suppose that we intend to test $H: \theta = \theta_0$ using a test statistic D, and that we want to be 90% sure of obtaining a signi-ficance level of 5% or less if in fact $\theta = \theta_1$. Then the sample size n should be selected so that $K_{0.05}(\theta_1) = 0.9$. For this use of power, the point made in the preceding paragraph concerning the need to condi-tion on ancillary statistics does not apply. For another approach to experimental planning, see the discussion of expected information in Section 13.5.

5.   Power comparisons are not likely to be very useful unless one can be quite specific with respect to the alternative hypothesis.   In many applications of significance tests, one will have only a vague idea concerning the types of departure that may occur. One would like to avoid building elaborate models to explain these until the need for them has been demonstrated in a test of significance.

CHAPTER 13. INTERVALS FROM SIGNIFICANCE TESTS

Suppose that we have a significance test for the hypothesis $\theta = \theta_0$ where θ_0 is any particular value of the unknown parameter θ. The set of values θ_0 such that the significance level is α or more is called a $100(1-\alpha)\%$ confidence interval for θ. Some examples and properties of confidence intervals are discussed in Section 1.

In Section 2 we use the large sample approximation to the distribution of the likelihood ratio statistic to obtain approximate confidence intervals. The relationship between these and likelihood intervals (Section 9.4) is discussed.

A likelihood function is called normal in the parameter θ if the log likelihood function is a quadratic function of θ. In this case likelihood and approximate confidence intervals will have the form $\hat{\theta} \pm z/\sqrt{\hat{I}}$ where $\hat{I}$ is the observed information and z is a quantile of $N(0,1)$. It follows from large-sample properties of $\hat{\theta}$ that all likelihood functions are approximately normal in large samples. However, with finite samples, the normality of the likelihood function depends upon the choice of the parameter θ, and a suitable parameter transformation $\theta = g(\phi)$ may lead to much more accurate normal approximations. These points are illustrated in Section 3.

In Section 4 we suppose that several independent experiments give information about the same parameter θ. If the likelihood functions are all nearly normal in θ, it is very easy to test for homogeneity or consistency across experiments, and if appropriate, to combine or pool the information about θ.

Finally, in Section 13.5, we define Fisher's measure of expected information. Some of its properties are discussed, and its use in planning experiments is illustrated.

13.1 Confidence Intervals

In Chapter 9 we considered the use of the relative likelihood
function in determining which values of an unknown parameter θ are
plausible in the light of the data. If $R(\theta_0)$ is near 1, then θ_0
is a plausible value in the sense that when $\theta = \theta_0$, the data are
nearly as probable as they can be under the model. If $R(\theta_0)$ is near
0, then θ_0 is not a plausible value because there exist other values
of θ which make the data much more probable. A table or graph of
$R(\theta)$ or its logarithm enables us to compare plausibilities of possible
parameter values. One or two likelihood intervals or regions can be
given as a summary of the information about θ when presentation of
the entire relative likelihood function would be inconvenient.

Tests of significance are designed to assess the consistency
of a particular hypothesis with the data rather than to compare possible
values of a parameter. Nevertheless, tests of significance are often
used to determine ranges of "reasonable" values for parameters in the
light of the data. Suppose that we have a test of significance for
the hypothesis $\theta = \theta_0$, where θ_0 is a particular value of θ. We
imagine that the test is performed for each possible parameter value
θ_0. The significance level will depend upon which parameter value is
tested: $SL = SL(\theta_0)$. If $SL(\theta_0)$ is near 1, the hypothesis $\theta = \theta_0$
is consistent with the data, and hence θ_0 is a "reasonable" value of
θ. However, if $SL(\theta_0)$ is near 0, the hypothesis $\theta = \theta_0$ is
strongly contradicted by the data, and θ_0 is not a "reasonable"
value. A table or graph of $SL(\theta)$ can thus be used to compare possible
parameter values.

Regions or intervals of parameter values can be constructed
from $SL(\theta)$ in the same way that likelihood regions or intervals are
obtained from $R(\theta)$. These might be called "significance regions" or
"consonance regions"; see Kempthorne and Folks, *Probability, Statistics,
and Data Analysis,* Iowa State Univ. Press (1971). Although these names
are perhaps more descriptive, the term commonly used is "confidence
region".

The set of parameter values such that $SL(\theta) \geq \alpha$ will be
called a 100(1-α)% confidence region for θ. Usually this set will
be an interval of real values which we call a 100(1-α)% confidence
interval for θ.

A 95% confidence interval or region for θ consists of
all parameter values such that $SL(\theta) \geq 0.05$. Any parameter value θ_0
which lies outside the 95% confidence region is said to be contradicted by the data at the 5% level, since $SL(\theta_0) < 0.05$. Values in-

side the 95% confidence interval are consistent with the data at the 5% level in the particular test of significance used.

Note that the confidence interval depends upon the particular significance test used to define it. If a different discrepancy measure were used, the 95% confidence interval for θ could change substantially. An advantage of likelihood intervals is that they do not depend upon the choice of a discrepancy measure.

As a general rule, it would seem best to use likelihood ratio tests in defining confidence intervals. However other tests (e.g. two-tail tests) which require less computation will sometimes be used.

The above definition of confidence intervals is non-standard. In the usual approach, a confidence interval need not be associated with a test of significance. Reasons for the more restrictive definition used here will be presented at the end of this section.

<u>Example 13.1.1</u>. Confidence interval for a normal mean

Suppose that $X_1, X_2, \ldots, X_n$ are independent $N(\mu, \sigma^2)$ variates where σ^2 is known. In Section 12.5 we showed that the significance level in a likelihood ratio test of $H: \mu = \mu_0$ is given by

$$SL(\mu_0) = P\{|Z| \geq |z_{obs}|\}$$

where $Z \sim N(0,1)$ and $z_{obs} = \sqrt{n}(\bar{x} - \mu_0)/\sigma$. From Table B1, we have

$$P\{|Z| \geq 1.960\} = 0.05.$$

Thus $SL(\mu_0) \geq 0.05$ if and only if $|z_{obs}| \leq 1.960$. Hence the 95% confidence interval for μ consists of those parameter values such that

$$-1.960 \leq \frac{\sqrt{n}(\bar{x} - \mu)}{\sigma} \leq 1.960.$$

Solving for μ now gives

$$\bar{x} - 1.96 \, \frac{\sigma}{\sqrt{n}} \leq \mu \leq \bar{x} + 1.96 \, \frac{\sigma}{\sqrt{n}}.$$

Similarly, since $P\{|Z| \geq 2.576\} = 0.01$, the 99% confidence interval for μ is given by

$$\bar{x} - 2.576 \, \frac{\sigma}{\sqrt{n}} \leq \mu \leq \bar{x} + 2.576 \, \frac{\sigma}{\sqrt{n}}.$$

These intervals are derived from the likelihood ratio test, but a two-tail test would yield the same intervals in this case.

For instance, in Example 12.5.4 we had $n = 50$, $\bar{x} = 37.25$ and $\sigma^2 = 49.35$. The 95% confidence interval for μ is then $35.30 \leq \mu \leq 39.20$, and the 99% confidence interval for μ is $34.69 \leq \mu \leq 39.81$. Any value of μ outside the 99% confidence interval would give rise to a significance level less than 1% in likelihood ratio test, and hence is strongly contradicted by the data.

Example 13.1.2. Confidence interval for an exponential mean

Suppose that $X_1, X_2, \ldots, X_n$ are independent exponential variates with mean θ. A two-tail test of $H : \theta = \theta_0$ was considered in Example 12.7.2. The significance level is

$$SL(\theta_0) = \begin{cases} 2 \cdot P\{\chi^2_{(2n)} \geq u\} & \text{if } u > m \\ 2 \cdot P\{\chi^2_{(2n)} \leq u\} & \text{if } u \leq m \end{cases}$$

where $u = 2t/\theta_0$, $t = \sum x_i$, and m is the median of $\chi^2_{(20)}$. From Table B4 we can determine values a, b such that

$$P\{\chi^2_{(2n)} \leq a\} = .025 = P\{\chi^2_{(2n)} \geq b\}.$$

Then $SL(\theta_0) \geq 0.05$ if and only if $a \leq u \leq b$, and hence the 95% confidence interval for θ is given by

$$a \leq \frac{2t}{\theta} \leq b \longleftrightarrow \frac{2t}{b} \leq \theta \leq \frac{2t}{a}.$$

For instance, Table B4 gives

$$P\{\chi^2_{(20)} \leq 9.591\} = P\{\chi^2_{(20)} \geq 34.17\} = 0.025.$$

Hence the 95% significance interval for θ in Example 12.7.2 is

$$\frac{2 \times 288}{34.17} \leq \theta \leq \frac{2 \times 288}{9.591} \longleftrightarrow 16.86 \leq \theta \leq 60.06.$$

Any value of θ outside this interval would lead to a significance level less than 5% in a two-tail test.

The likelihood ratio test for $H : \theta = \theta_0$ was considered in Section 12.5, and based on this test, the 95% confidence interval for θ is found to be $16.35 \leq \theta \leq 57.82$. Much more arithmetic is required for the likelihood ratio test than for the two-tail test; one also requires either extensive tables of the χ^2 distribution or a computer programme to evaluate the χ^2 integral. Perhaps the easiest computational method is to calculate $SL(\theta_0)$ as in Example 12.5.5 for several values of θ_0, then prepare a graph and determine the interval from it. This is similar to the method used to obtain likelihood intervals in Section 9.4.

Example 13.1.3. Confidence interval for a binomial proportion

Suppose that X has a binomial distribution with index $n = 10$ and proportion p. We wish to find a 95% confidence interval for p on the basis of the observation $X = 6$.

The confidence interval will depend upon the method of ranking outcomes for the test. For simplicity, we consider the ranking by distance from the mean, $D \equiv |X - np|$. The significance level is then

$$SL(p) = P\{|X - 10p| \geq |6 - 10p|\},$$

which we compute by summing the binomial probabilities

$$f(x) = \binom{10}{x} p^x (1 - p)^{10-x}.$$

The range of summation will depend upon the value of p being tested. For instance,

$$SL(.2) = P\{|X - 2| \geq 4\} = f(6) + f(7) + \ldots + f(10) = 0.006;$$

$$SL(.3) = P\{|X - 3| \geq 3\} = f(0) + f(6) + \ldots + f(10) = 0.059;$$

$$SL(.4) = P\{|X - 4| \geq 2\} = f(0) + f(1) + f(6) + \ldots + f(10) = 0.334.$$

In each case, the hypothesized value of p is used in computing the probabilities.

The significance level is plotted as a function of p in Figure 13.1.1. Note the discontinuities in the function at $p = .3, .35$, .4, etc. To see why these occur, consider what happens for p near .3. If p is less than .3, then $|6 - 10p| > 3$ and $|0 - 10p| < 3$. Hence $f(0)$ is not included in computing the significance level when $p < .3$. As p approaches .3 from below, $SL(p)$ approaches 0.030. When $p = 0.3$, we have $|6 - 10p| = |0 - 10p| = 3$, so that $X = 0$ now gives as great a discrepancy as $X = 6$. Hence $f(0)$ must be included in the sum, and $SL(p)$ jumps at $p = 0.3$ by the amount

$$f(0) = \binom{10}{0}(.3)^0(.7)^{10} = 0.028.$$

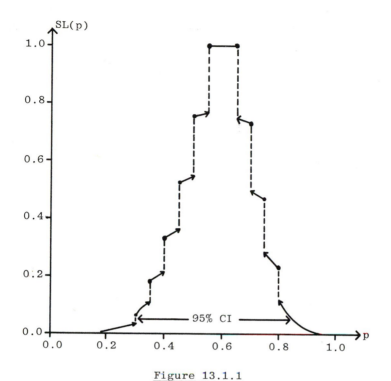

Figure 13.1.1
Significance Level as a Function of p in Example 13.1.3,
Showing a 95% Confidence interval for p.

Similarly, SL(p) approaches 0.108 as p approaches 0.35 from
below, but at p = 0.35, the extra term f(1) comes into the sum, and
SL(p) increases at p = 0.35 by the amount

$$f(1) = \binom{10}{1}(.35)^1(.65)^9 = 0.072.$$

From the graph, we see that SL(p) ≥ 0.05 for
0.300 ≤ p ≤ 0.850, and this is the 95% confidence interval for p.
Other confidence intervals can be obtained from the graph in a similar
way.

Approximate confidence intervals for a binomial proportion
will be considered in Sections 13.2 and 13.3.

Coverage frequency

Suppose that the experiment is repeated a large number of

times, and that θ has the same unknown value in all repetitions. At each repetition we obtain a 95% confidence region for θ from a significance test with discrepancy measure D. We shall show that exactly 95% of these regions will contain or cover the true value of θ if D is continuous, and at least 95% will cover the true value if D is discrete.

The true value, θ_0 say, belongs to the 95% confidence region if and only if a test of $H:\theta = \theta_0$ gives SL ≥ 0.05. If D is a continuous variate, then there exists a variate value d such that $P(D \geq d) = 0.05$ exactly. In this case we have SL ≥ 0.05 if and only if $D \leq d$, so that

$$P(SL \geq 0.05) = P(D \leq d) = 1 - P(D > d) = 0.95$$

because $P(D > d) = P(D \geq d)$ for a continuous variate. In the continuous case, exactly 95% of the regions will cover the true value.

If D is a discrete variate, there usually will not exist a value d such that $P(D \geq d) = 0.05$ exactly. Let d be the variate value for which $P(D \geq d) \geq 0.05$ but $P(D > d) < 0.05$. Then SL ≥ 0.05 if and only if $D \leq d$, and hence

$$P(SL \geq 0.05) = P(D \leq d) = 1 - P(D > d) > 0.95.$$

The exact coverage probability depends upon θ_0 in the discrete case, but it is at least 95%.

Another approach

We now give the standard definition of confidence intervals, which differs from that given at the beginning of the section.

Let V_1, V_2 be variates such that $V_1 \leq V_2$ always, and consider the random interval $[V_1, V_2]$. This interval is called a 100(1-α)% confidence interval for θ if

$$P(V_1 \leq \theta \leq V_2) = 1 - \alpha \qquad (13.1.1)$$

for all values of θ. The confidence coefficient, or coverage probability of the interval is $1 - \alpha$. The interval $[V_1, V_2]$ is called a conservative 100(1-α)% confidence interval for θ if

$$P(V_1 \leq \theta \leq V_2) \geq 1 - \alpha \qquad (13.1.2)$$

for all values of θ.

The parameter θ is considered to be fixed at its true value in (13.1.1), and the probability statement refers to V_1 and V_2. The confidence coefficient is the relative frequency with which the random interval $[V_1,V_2]$ would include the true value of θ in repetitions of the experiment. Suppose that we fix θ at some particular value θ_0, and then repeat the experiment over and over, each time computing the observed values v_1 and v_2. Then, over the long run, a fraction $1 - \alpha$ of the intervals $[v_1,v_2]$ would include θ_0, and a fraction α of the intervals would not include θ_0. In general, it would not be proper to conclude that a particular observed interval $[v_1,v_2]$ had probability $1 - \alpha$ of including θ_0. For instance, it can happen that a 95% confidence interval includes all possible values of θ, and therefore certainly includes the true value.

We have shown that intervals $[V_1,V_2]$ constructed from a significance test satisfy (13.1.1) in the continuous case and (13.1.2) in the discrete case. Thus, intervals obtained from significance tests are confidence intervals according to the usual definition.

When confidence intervals are obtained from a significance test, the significance level provides a ranking of parameter values. Parameter values inside the confidence interval have larger significance levels, and hence are "more reasonable" than values outside the interval. There is a nesting of confidence intervals in this case: all parameter values belonging to the 95% interval (SL $\geq .05$) must also belong to the 99% interval (SL ≥ 0.01).

An interval $[V_1,V_2]$ which merely satisfies (13.1.1) does not have these properties. No ranking of parameter values is implied, and one cannot conclude that values inside the interval are preferable to values outside. Furthermore, there may exist parameter values which belong to the 95% interval but do not belong to the 99% interval.

Because of these difficulties, we have adopted a more restrictive definition which requires that confidence intervals be obtained from significance tests.

Problems for Section 13.1

1. When an automatic shear is set to cut plates to length μ, the lengths actually produced are normally distributed about μ with standard deviation 1.6 inches. The average length of 15 plates cut at one setting was 125.77 inches. Obtain a 90% confidence interval for μ.

2. In Problem 12.5.2, obtain a 95% confidence interval for the expected survival time of patients receiving the new treatment.

†3. (a) Suppose that $X_1, X_2, \ldots, X_n$ are independent $N(\mu, \sigma^2)$ variates, where μ is known. Show that $\sum(X_i - \mu)^2$ is sufficient for σ, and that $\sum(X_i - \mu)^2/\sigma^2$ has a χ^2 distribution with n degrees freedom.

   (b) In a check of the accuracy of their measurement procedures, fifteen engineers are asked to measure a precisely known distance of 3727 feet between two markers. Their results are as follows:

3727.75	3726.43	3728.04	3729.21	3726.30
3728.15	3724.25	3726.29	3724.90	3727.51
3726.85	3728.50	3725.94	3727.69	3726.09

   Assuming that their measurements are independent $N(3727, \sigma^2)$, obtain the 95% confidence interval for σ^2 based on a two-tail test.

13.2 Approximate Confidence Intervals

Large-sample properties of the likelihood function are often used to generate approximate significance tests and confidence intervals in problems where exact solutions are not feasible. In this section we obtain confidence intervals directly from the likelihood ratio statistic using the approximation (12.6.1). In Section 13.3, we consider situations where it is appropriate to give approximate intervals of the form $\hat{\theta} \pm z/\sqrt{\hat{I}}$, where z is a quantile of $N(0,1)$ and $\hat{I}$ is a value computed from the data. In such situations, the relevant information about θ is conveniently summarized by a pair of statistics, $\hat{\theta}$ and $\hat{I}$, from which approximate confidence intervals of any desired size can be obtained. Furthermore, we shall see in Section 13.4 that information about θ from several independent experiments can then be combined very easily.

Let $D \equiv -2 \log R$ be the likelihood ratio statistic for a test of $H: \theta = \theta_0$, where θ is a real-valued parameter. Since only one parameter is fixed by the hypothesis, there is just one degree of freedom for the test. Under appropriate regularity conditions (see Section 12.6), D has approximately a $\chi^2_{(1)}$ distribution in large samples. The significance level in an approximate likelihood ratio test is

$$SL(\theta_0) = P\{\chi^2_{(1)} \geq d(\theta_0)\}$$

where $d(\theta_0)$ is the observed value of D under the hypothesis $\theta = \theta_0$.
Since $P\{\chi_{(1)} \geq 3.841\} = 0.05$ from Table B4, it follows that
$SL(\theta_0) \geq 0.05$ if and only if $d(\theta_0) \leq 3.841$. Thus the set of values
of θ such that $d(\theta) \leq 3.841$ is an approximate 95% confidence interval for θ. Similarly, the set of parameter values for which
$d(\theta) \leq 6.635$ is an approximate 99% confidence interval for θ.

Since $d(\theta) = -2 \cdot r(\theta)$, we have $d(\theta) \leq 3.841$ if and only
if $r(\theta) \geq -1.92$, where $r(\theta)$ is the log relative likelihood function
of θ. Since $e^{-1.92} = 0.147$, the approximate 95% confidence interval is just the 14.7% likelihood interval, and it can be read from a
graph of the relative likelihood function as in Section 9.4.

The following table shows the log relative likelihood r
and the relative likelihood R corresponding to various significance
levels in an approximate likelihood ratio test.

$SL(\theta_0)$	0.24	0.1	0.05	0.032	0.01	0.0024
$r(\theta_0)$	−0.69	−1.35	−1.92	−2.30	−3.32	−4.61
$R(\theta_0)$	0.5	0.258	0.147	0.1	0.036	0.01

This correspondence holds approximately in large samples under appropriate regularity conditions. There need be no simple relationship
between significance levels and relative likelihoods in problems with
small samples and/or non-regular distributions.

Similar results can be obtained for the case of two unknown
parameters by using tables of the χ^2 distribution with two degrees
of freedom.

Example 13.2.1. Exponential mean

Let $X_1, X_2, \ldots, X_n$ be independent exponential variates with
mean θ. From Section 12.5, the likelihood ratio statistic for testing a value of θ is

$$D \equiv 2n\left[\frac{\hat{\theta}}{\theta} - 1 - \log \frac{\hat{\theta}}{\theta}\right]$$

where $\hat{\theta} \equiv \frac{1}{n} \sum X_i$. The 95% confidence interval for θ based on an
approximate likelihood ratio test consists of those parameter values
for which

$$-2 \cdot r(\theta) = 2n\left[\frac{\hat{\theta}}{\theta} - 1 - \log \frac{\hat{\theta}}{\theta}\right] \leq 3.841.$$

In Example 9.5.1 we considered n = 10 lifetimes with aver-
age value $\hat{\theta} = 28.8$. The approximate 95% confidence interval is
given by

$$-2 \cdot r(\theta) = 20 \left[\frac{28.8}{\theta} - 1 - \log \frac{28.8}{\theta} \right] \le 3.841.$$

The log relative likelihood function is plotted in Figure 9.5.1, and from this graph we find that $r(\theta) \ge -1.92$ for $16.42 \le \theta \le 57.47$. For comparison, the exact 95% confidence interval based on the likelihood ratio test is $16.35 \le \theta \le 57.82$; see Example 13.1.2.

Note that this procedure can be used to obtain approximate confidence intervals when some of the lifetimes are censored as in Example 9.6.1. Exact confidence intervals are difficult to derive in this case. They will depend not only on the observed failure and censoring times, but also on hypothetical censoring times for items which failed. (The hypothetical censoring time for an item is the time at which observation would presumably have ceased if the item had not failed at an earlier time.)

Example 13.2.2. Binomial proportion

If X has a binomial (n,p) distribution, the likelihood ratio statistic for testing a value of p is given by

$$D \equiv 2X \log \frac{X}{np} + 2(n - X) \log \frac{n - X}{n(1-p)}$$

from Example 12.5.1. The approximate 95% confidence interval for p contains those values of p such that

$$-2r(p) = 2x \log \frac{x}{np} + 2(n - x) \log \frac{n - x}{n(1-p)} \le 3.841.$$

For $n = 10$ and $x = 6$ as in Example 13.1.3, we have

$$r(p) = 6 \log \frac{0.6}{p} + 4 \log \frac{0.4}{1-p} \ge -1.92.$$

The log relative likelihood function may be evaluated at several p-values and plotted. From the graph, we find that $r(p) \ge -1.92$ for $0.300 \le p \le 0.854$, and this is the approximate 95% confidence interval for p. The exact 95% confidence interval based on distance from the mean was found to be $0.300 \le p \le 0.850$; see Example 13.1.3.

Problems for Section 13.2

†1. Two hundred people were selected at random from a large population and were asked their opinions on capital punishment. Sixty percent indicated that they were in favour. Find approximate 95% and

99% confidence intervals for p, the proportion of people in the population who favour capital punishment.

2. Obtain approximate 95% confidence intervals for the sex ratio in problems 11.3.6 (a),(b), and (c).

3. (a) Let $X_1, X_2, \ldots, X_n$ be independent Poisson variates with mean μ. Show that the likelihood ratio statistic for testing a value of μ is

$$-2r(\mu) = n\hat{\mu}\left[\frac{\hat{\mu}}{\mu} - 1 - \log \frac{\hat{\mu}}{\mu}\right].$$

(b) Suppose that the number of accidents per month at a busy intersection has a Poisson distribution with mean μ. Over a 10-month period, there were a total of 53 accidents. Obtain an approximate 95% confidence interval for μ.

13.3 Intervals from Normal Likelihoods

A likelihood function will be called normal in the parameter θ if the log likelihood function is a quadratic function of θ. If we can find a parameter θ such that the likelihood function is normal in θ, then approximate confidence intervals for θ will have the form $\hat{\theta} \pm z/\sqrt{\hat{I}}$, where $\hat{I}$ is a value computed from the data, and z is obtained from tables of $N(0,1)$. We shall see that, in sufficiently large samples, all likelihoods are approximately normal.

Let $L(\theta;x)$ denote the likelihood function of a continuous parameter θ based on a typical outcome x. Suppose that the MLE $\hat{\theta}$ exists and is a point of relative maximum of the likelihood function, and that $\log L$ has a Taylor's series expansion at $\theta = \hat{\theta}$:

$$\log L(\theta;x) = c_0 + (\theta - \hat{\theta})c_1 + \frac{(\theta - \hat{\theta})^2}{2!} c_2 + \frac{(\theta - \hat{\theta})^3}{3!} c_3 + \ldots$$

where c_i is the value of the ith derivative of $\log L$ at $\theta = \hat{\theta}$:

$$c_i = \frac{\partial^i \log L}{\partial \theta^i}\Bigg|_{\theta=\hat{\theta}}.$$

The first derivative of the log likelihood function is called the score function of θ:

$$S(\theta;x) = \frac{\partial}{\partial\theta} \log L(\theta;x). \tag{13.3.1}$$

The information function of θ is defined to be minus the second derivative:

$$I(\theta;x) = -\frac{\partial^2 \log L(\theta;x)}{\partial\theta^2} = -\frac{\partial S(\theta;x)}{\partial\theta} . \tag{13.3.2}$$

Since $\hat{\theta}$ is assumed to be a relative maximum, we have

$$c_1 = S(\hat{\theta};x) = 0; \qquad c_2 = -I(\hat{\theta};x) > 0.$$

The value of the information function at $\theta = \hat{\theta}$ is called the observed information for θ, and will be denoted by $\hat{I}$ or $\hat{I}_\theta$:

$$\hat{I} = \hat{I}_\theta = I(\hat{\theta};x) = -\frac{\partial^2 \log L}{\partial\theta^2}\Bigg|_{\theta=\hat{\theta}} . \tag{13.3.3}$$

Since $c_0 = \log L(\hat{\theta};x)$, we have

$$\log R(\theta;x) = \log L(\theta;x) - \log L(\hat{\theta};x)$$

$$= -\frac{(\theta-\hat{\theta})^2}{2!}\hat{I} + \frac{(\theta-\hat{\theta})^3}{3!}c_3 + \frac{(\theta-\hat{\theta})^4}{4!}c_4 + \ldots \tag{13.3.4}$$

Normal likelihoods

The likelihood function will be called normal in the parameter θ if the information function $I(\theta;x)$ depends only on x. If this is so, then $c_3 = c_4 = \ldots = 0$, and the log relative likelihood is a quadratic function of θ:

$$\log R(\theta;x) = -\frac{1}{2}\hat{I}(\theta - \hat{\theta})^2 . \tag{13.3.5}$$

The likelihood ratio statistic then has the form

$$D \equiv \hat{I}(\theta - \hat{\theta})^2 .$$

Using the approximation (12.6.1) as in Section 13.2, we obtain

$$\hat{I}(\theta - \hat{\theta})^2 \le 3.841 \iff \theta \in \hat{\theta} \pm 1.96/\sqrt{\hat{I}} \tag{13.3.6}$$

as the approximate 95% confidence interval for θ.

Note that $(\hat{\theta},\hat{I})$ is a pair of sufficient statistics for θ. The MLE $\hat{\theta}$ describes the location of the normal likelihood function on the axis, while $\hat{I}$ measures its spread. The larger the value of $\hat{I}$, the narrower likelihood intervals and approximate confidence intervals for θ will be. The observed information $\hat{I}$ is thus a measure of the informativeness or precision of the experiment with respect to θ, and it plays the role of an ancillary statistic in inferences about θ; see Section 12.8.

<u>Example 13.3.1</u>. Let $X_1, X_2, \ldots, X_n$ be independent $N(\mu, \sigma^2)$ variates where σ is known. From Section 12.5, the likelihood function of μ is

$$L(\mu) = \exp\{- \frac{n}{2\sigma^2} (\hat{\mu} - \mu)^2\}.$$

The information function of μ is

$$I(\mu;x) = - \frac{\partial^2 \log L}{\partial \mu^2} = \frac{n}{\sigma^2}$$

which does not depend upon μ. Hence L is normal in the parameter μ. In this case I does not depend upon the x_i's either, and hence there is a single sufficient statistic $\hat{\mu}$. Also, since the distribution of D is exactly $\chi^2_{(1)}$, (13.3.6) gives an exact 95% confidence interval

$$\mu \in \hat{\mu} \pm 1.96/\sqrt{\hat{I}} = \hat{\mu} \pm 1.96\sqrt{\sigma^2/n} \ .$$

<u>Example 13.3.2</u>. In Example 12.8.2, we assumed that a measurement X was distributed as $N(\mu, 1)$ if $Z = 0$ and as $N(\mu, 100)$ if $Z = 1$, where $P(Z = 0) = P(Z = 1) = \frac{1}{2}$. The likelihood function of μ was shown to be

$$L(\mu) = \exp\{- \frac{1}{2}(x - \mu)^2/10^{2z}\}.$$

The MLE is $\hat{\mu} = x$, and the information function of μ is

$$I(\mu;x) = - \frac{\partial^2 \log L}{\partial \mu^2} = 10^{-2z}.$$

Since I does not depend upon μ, the likelihood function is normal in μ, and $\hat{I} = I = 10^{-2z}$. By (13.3.6), the approximate 95% confidence interval for μ is

$$\mu \in \hat{\mu} \pm 1.96/\sqrt{\hat{I}} = x \pm 1.96 \times 10^z = \begin{cases} x \pm 1.96 & \text{if} \quad z = 0; \\ x \pm 19.6 & \text{if} \quad z = 1. \end{cases}$$

This is, in fact, an exact 95% confidence interval for μ. The conditional distribution of $D \equiv (X - \mu)^2/10^{2z}$ given Z is exactly $\chi^2_{(1)}$ for $Z = 0$ and also for $Z = 1$, so the unconditional distribution of D is also exactly $\chi^2_{(1)}$.

In this example, $\hat{\mu}$ is not a sufficient statistic by it-

self. We also need to know $\hat{I}$ (or equivalently Z) which measures
the informativeness or precision of the experiment with respect to μ.
In Example 12.8.2 we noted that Z was an ancillary statistic, and
argued that inferences about μ should be made conditionally on the
observed value of Z (or equivalently $\hat{I}$). The 95% confidence interval obtained above does reflect the presence of an ancillary statistic, and it can be interpreted as a conditional 95% confidence interval for μ given the value of Z.

Approximate normal likelihoods

Suppose that the information function $I(\theta;x)$ is nearly
constant with respect to θ over the region of plausible parameter
values. Then the cubic and higher terms in (13.3.4) will close to
zero, and

$$\log R(\theta;x) \approx -\frac{1}{2}\hat{I}(\hat{\theta} - \theta)^2 \tag{13.3.7}$$

over the region of plausible θ-values. The score function is then
approximately linear in θ for any given x:

$$S(\theta;x) \approx \hat{I}(\hat{\theta} - \theta).$$

We can still give $\hat{\theta} \pm 1.96/\sqrt{\hat{I}}$ as a 95% confidence interval for θ,
but this now depends upon two approximations: (12.6.1) and (13.3.7).
One can check the accuracy of (13.3.7) by tabulating or plotting both
$\log R(\theta)$ and the normal approximation. If the agreement is poor, confidence intervals of the form $\theta \in \hat{\theta} \pm z/\sqrt{\hat{I}}$ are not appropriate.

Now suppose that the likelihood function is based on a
sample of size n, and that regularity conditions similar to those
described at the beginning of Section 12.6 are satisfied. Then it can
be shown that, as $n \to \infty$, the cubic and later terms in (13.3.4) be-
come negligible in comparison with the quadratic term when $|\theta - \hat{\theta}|$ is
not too large, and hence (13.3.7) holds. Thus, under some mild regularity conditions, all likelihood functions are approximately normal
as $n \to \infty$.

Under these same conditions it can be shown that, as $n \to \infty$,
$\hat{\theta}$ becomes approximately sufficient for θ, and the distribution of
$\hat{\theta}$ approaches normality, with $E(\hat{\theta}) \approx \theta$ and

$$\text{var}(\hat{\theta}) \approx \hat{I}^{-1} \approx I_E(\theta)^{-1} \approx I_E(\hat{\theta})^{-1}$$

where $I_E(\theta)$ is the expected information function; see Section 13.5.

These are equivalent as $n \to \infty$, but they need not agree closely in finite samples. In general, I seems to be the preferable choice for inference problems because it will reflect the presence of ancillary statistics as in Example 13.3.2, but I_E will not.

Example 13.3.3. If X has a binomial (n,p) distribution, then

$$\log L(p) = x \log p + (n - x)\log(1 - p)$$

and the information function of p is

$$I(p;x) = -\frac{\partial^2 \log L}{\partial p^2} = \frac{x}{p^2} + \frac{n - x}{(1-p)^2}.$$

Since $\hat{p} = \frac{x}{n}$, the observed information for p is

$$\hat{I} = I(\hat{p};x) = \frac{n}{\hat{p}} + \frac{n}{1 - \hat{p}} = \frac{n}{\hat{p}(1-\hat{p})}.$$

The approximate 95% confidence interval for p is thus

$$p \in \hat{p} \pm 1.96\sqrt{\frac{\hat{p}(1-\hat{p})}{n}}.$$

From Section 6.8, $X \approx N(np,np(1 - p))$ for n large, and hence $\hat{p} \equiv \frac{X}{n} \approx N(p,\frac{p(1-p)}{n})$. If we estimate the variance by $\hat{p}(1 - \hat{p})/n$, this distribution yields the above 95% confidence interval. If we do not estimate the variance, an approximate 95% confidence interval will be given by

$$-1.96 \le (\hat{p} - p)\sqrt{\frac{n}{p(1-p)}} \le 1.96.$$

The endpoints of the interval may be found as roots of a quadratic equation.

Example 13.3.4. Let $X_1, X_2, \ldots, X_n$ be independent exponential variates with mean θ. The likelihood function of θ is

$$L(\theta) = \theta^{-n}\exp\{-\frac{1}{\theta} \sum x_i\} = \theta^{-n}\exp\{-\frac{n\hat{\theta}}{\theta}\}$$

and the information function of θ is

$$I(\theta) = -\frac{\partial^2 \log L}{\partial \theta^2} = -\frac{n}{\theta^2} + \frac{2n\hat{\theta}}{\theta^3}.$$

The observed information for θ is

$$\hat{I} = I(\hat{\theta}) = n/\hat{\theta}^2$$

and (13.3.5) gives the approximate 95% confidence interval

$$\theta \in \hat{\theta} \pm 1.96\hat{\theta}/\sqrt{n}.$$

For $n = 10$ and $\hat{\theta} = 28.8$ as in Example 9.5.1, we obtain $10.95 \le \theta \le 46.65$, whereas an exact likelihood ratio test yields $16.35 \le \theta \le 57.82$. The agreement is very poor. The log likelihood function is highly skewed (see Figure 9.5.1), whereas the approximation (13.3.7) is symmetrical about $\theta = \hat{\theta}$. The likelihood function does become normal in θ as $n \to \infty$, but n must be very much larger than 10 before the approximation (13.3.7) becomes sufficiently accurate to use. However, we shall see that the situation may be greatly improved by means of a suitable transformation of the parameter.

Parameter transformations

Consider a one-to-one parameter transformation, $\theta = g(\phi)$ say. By the invariance property (Section 9.7), the likelihood function of ϕ is $L(g(\phi))$, and $\hat{\theta} = g(\hat{\phi})$. If L is normal in θ, it will not be normal in ϕ unless g is a linear transformation. Conversely, if L is not normal in θ, we may be able to choose a non-linear transformation $\theta = g(\phi)$ such that L is normal or nearly normal in the new parameter ϕ. One can then obtain confidence intervals for ϕ and transform them back into confidence intervals for θ.

The normal approximation (13.3.7) is obtained by ignoring the cubic and higher terms in (13.3.4). If we can find a transformation $\theta = g(\phi)$ which reduces the size of the cubic term in relation to the quadratic term, the accuracy of the normal approximation should improve.

Example 13.3.4 (continued). Consider the family of power transformations, $\theta = \phi^a$ where $a \neq 0$. The log likelihood function is

$$-n \log \theta - n\hat{\theta}/\theta = -na \log \phi - n\hat{\phi}^a/\phi^a.$$

We differentiate with respect to ϕ and evaluate at $\phi = \hat{\phi}$ to obtain

$$\hat{I}_\phi = -\left.\frac{\partial^2 \log L}{\partial \phi^2}\right|_{\phi=\hat{\phi}} = \frac{na^2}{\hat{\phi}^2}; \quad c_3 = \left.\frac{\partial^3 \log L}{\partial \phi^3}\right|_{\phi=\hat{\phi}} = \frac{na^2(a+3)}{\hat{\phi}^3}.$$

Note that $c_3 = 0$ for $a = -3$; that is, for $\theta = \phi^{-3}$. Hence, if we work with the parameter $\phi = \theta^{-1/3}$, the coefficient of the cubic term in the Tyalor's series expansion of $\log R(\phi)$ about $\phi = \hat{\phi}$ will be zero. The likelihood function should by approximately normal in the new parameter $\phi = \theta^{-1/3}$. The approximate 95% confidence interval for ϕ is then

$$\phi \in \hat{\phi} \pm 1.96/\sqrt{\hat{I}_\phi} = \hat{\phi}(1 \pm 1.96/3\sqrt{n}).$$

Substituting $\phi = \theta^{-1/3}$ and $\hat{\phi} = \hat{\theta}^{-1/3}$ gives

$$\theta^{-1/3} \in \hat{\theta}^{-1/3}(1 \pm 1.96/3\sqrt{n}) \longleftrightarrow \theta \in \hat{\theta}(1 \pm 1.96/3\sqrt{n})^{-3}$$

as the corresponding interval for θ. For $n = 10$ and $\hat{\theta} = 28.8$ this gives $16.39 \le \theta \le 57.67$, which agrees very well with the exact 95% confidence interval $16.35 \le \theta \le 57.82$.

      Table 13.3.1 compares $\log R(\theta)$ with the approximate normal likelihoods for θ and for ϕ over the range of parameter values for which $\log R \ge -5$. Use of the transformation $\theta = \phi^{-3}$ leads to a very substantial improvement in the accuracy of the normal approximation.

Table 13.3.1

Comparison of Approximate Normal Likelihoods

θ	$\phi = \theta^{-1/3}$	$\log R(\theta)$	$-\frac{1}{2}\hat{I}_\theta(\theta - \hat{\theta})^2$	$-\frac{1}{2}\hat{I}_\phi(\phi - \hat{\phi})^2$
12	0.437	−5.25	− 3.40	−5.17
15	0.405	−2.68	− 2.30	−2.65
20	0.368	−0.75	− 0.93	−0.75
25	0.342	−0.10	− 0.17	−0.10
40	0.292	−0.49	− 1.51	−0.48
60	0.255	−2.14	−11.74	−2.12
80	0.232	−3.82	−31.60	−3.75
100	0.215	−5.33	−61.12	−5.19

Comments

      The preceding example shows that the normality of the likelihood function depends not just on the sample size, but also on the choice of parameter. With a poor choice of parameter, a sample of size $n = 1,000,000$ may be too small, whereas with a good choice we may be able to apply large-sample results for $n = 5$ or less. A large sample size is thus no guarantee that large-sample methods can safely be applied. It is necessary to compare the actual likelihood function with the normal approximation.

If one can find a parameter θ such that the approximation (13.3.7) is quite accurate, then in general the distribution of

$$Z \equiv (\hat{\theta} - \theta)\sqrt{\hat{I}}$$

will be very close to $N(0,1)$. Since $-2 \log R \approx Z^2$, the approximation (12.6.1) will also be quite accurate. On the other hand, if there is no parameter θ for which (13.3.7) holds, then the approximation (12.6.1) will not be very good either. For an interesting discussion of large-sample methods, see D.A. Sprott, Application of Maximum Likelihood Methods to Finite Samples, *Sankhya B* 37 (1975), 259-270.

Problems for Section 13.3

†1. Use the two methods described in Example 13.3.3 to obtain approximate 95% and 99% confidence intervals for p in Problem 13.2.1.

2. Suppose that X has a binomial (n,p) distribution. Show that the ratio of the cubic term to the quadratic term in (13.3.4) is

$$\frac{2}{3} (\hat{p} - p)(1 - 2\hat{p})/\hat{p}(1 - \hat{p}).$$

Under what conditions will the likelihood function be nearly normal in p?

3. (a) Let $X_1, X_2, \ldots, X_n$ be independent Poisson variates with mean μ, as in Problem 13.2.3, and consider transformations of the form $\mu = \phi^a$ where $a \neq 0$. Find the cubic term in the expansion of $\log R$ in powers of $\phi - \hat{\phi}$, and show that it is zero for $a = 3$.

   (b) For the numerical example in Problem 13.2.3(b), obtain an approximate 95% confidence for ϕ where $\phi = \mu^{1/3}$, and transform it into an interval for μ. Use a table or graph to show that a much more accurate approximation to $\log R$ is obtained by using the transformed parameter.

4. Find approximate 95% confidence intervals for θ_H and θ_C in Problem 9.6.3. Use a table or graph to check the accuracy of the normal approximation.

5. Obtain approximate 95% confidence intervals for θ in Problems 9.1.4(b) and 9.1.5(b), and investigate the accuracy of the normal approximations to the likelihood function.

†6. Let $\hat{I}_\theta$ and $\hat{I}_\phi$ denote the observed information for parameters θ and ϕ, respectively, where $\theta = g(\phi)$. Show that $\hat{I}_\phi = \hat{I}_\theta \cdot [g'(\hat{\phi})]^2$ where $g'(\phi) = \dfrac{dg(\phi)}{d\phi}$.

13.4   Combining Normal Likelihoods

Suppose that m experiments yield independent likelihood functions $L_1, L_2, \ldots, L_m$ and estimates $\hat{\theta}_1, \hat{\theta}_2, \ldots, \hat{\theta}_m$ for the same parameter θ. In general, the overall MLE $\tilde{\theta}$ based on the data from all m experiments cannot be found just from the individual estimates. It is necessary to construct the overall likelihood function as a product

$$L(\theta) = L_1(\theta)L_2(\theta) \ldots L_m(\theta),$$

and then maximize to determine $\tilde{\theta}$.

There is a considerable simplification when each of the likelihood functions $L_i(\theta)$ is normal or approximately normal in θ. Then all or nearly all of the information about θ from the ith experiment is summarized by the two statistics $(\hat{\theta}_i, \hat{I}_i)$. We shall see that the MLE and approximate confidence intervals for θ based on all m experiments can be obtained directly from these summary statistics.

If $L_1, L_2, \ldots, L_m$ are all normal in θ, then

$$L_i(\theta) = \exp\{-\tfrac{1}{2}\hat{I}_i(\theta - \hat{\theta}_i)^2\} \quad \text{for} \quad i = 1, 2, \ldots, m$$

and the overall likelihood function is

$$L(\theta) = k \prod_{i=1}^{m} L_i(\theta) = k \exp\{-\tfrac{1}{2}\sum_{i=1}^{m}\hat{I}_i(\theta - \hat{\theta}_i)^2\}.$$

Setting the derivative of $\log L$ equal to zero gives

$$\tilde{\theta} = \sum\hat{I}_i\hat{\theta}_i / \sum\hat{I}_i$$

as the combined or pooled estimate of θ. This is a weighted average of $\hat{\theta}_1, \hat{\theta}_2, \ldots, \hat{\theta}_m$, with $\hat{\theta}_i$ being weighted according to the observed information $\hat{I}_i$. An estimate $\hat{\theta}_i$ from a precise experiment ($\hat{I}_i$ large) will have a greater influence in determining $\tilde{\theta}$ than will an estimate from a less informative experiment.

The overall information function of θ is

$$I = -\frac{\partial^2 \log L}{\partial \theta^2} = \sum_{i=1}^{m}\hat{I}_i$$

which is not a function of θ. Hence the overall likelihood function L is again normal in θ:

$$L(\theta) = \exp\{-\tfrac{1}{2}I(\theta - \tilde{\theta})^2\}.$$

Approximate confidence intervals for θ based on all m experiments will thus have the form

$$\theta \in \tilde{\theta} \pm z/\sqrt{I}$$

where $I = \sum \hat{I}_i$, $\tilde{\theta} = \frac{1}{I} \sum \hat{I}_i \hat{\theta}_i$, and z is the appropriate quantile of $N(0,1)$.

Test for homogeneity

Before pooling estimates as above, it is prudent to check that the data are consistent with a single common value of θ in all m experiments. We thus consider a likelihood ratio test of the hypothesis of homogeneity

$$H: \theta_1 = \theta_2 = \ldots = \theta_m,$$

where θ_i denotes the parameter value in the ith experiment. The maximized likelihood ratio for H is

$$R = \frac{L_1(\tilde{\theta}_1) L_2(\tilde{\theta}_2) \ldots L_m(\tilde{\theta}_m)}{L_1(\hat{\theta}_1) L_2(\hat{\theta}_2) \ldots L_m(\hat{\theta}_m)}$$

where $\hat{\theta}_i$ is as before, and $\tilde{\theta}_i$ is the MLE of θ_i under the hypothesis. Since the θ_i's are all equal under H, we have

$$\tilde{\theta}_1 = \tilde{\theta}_2 = \ldots = \tilde{\theta}_m = \tilde{\theta}.$$

Since H decreases the number of unknown parameters from m to 1, there are $m-1$ degrees of freedom for the test, and (12.6.1) gives $-2 \log R \approx \chi^2_{(m-1)}$.

Now suppose that $L_i(\theta_i)$ is normal (or approximately normal) in θ_i, so that

$$L_i(\theta_i) = \exp\{-\frac{1}{2} \hat{I}_i (\theta_i - \hat{\theta}_i)^2\}.$$

Then, since $L_i(\hat{\theta}_i) = 1$, we have

$$R = \prod_{i=1}^{m} L_i(\tilde{\theta}) = \exp\{-\frac{1}{2} \sum_{i=1}^{m} \hat{I}_i (\tilde{\theta} - \hat{\theta}_i)^2\},$$

and the likelihood ratio statistic is

$$D = -2 \log R = \sum_{i=1}^{m} \hat{I}_i (\tilde{\theta}_i - \hat{\theta}_i)^2.$$

The significance level in an approximate likelihood ratio test for homogeneity is then

$$SL = P(\chi^2_{(m-1)} \geq d)$$

where d is the observed value of D. Note that d can be computed directly from the summary statistics $(\hat{\theta}_i, \hat{I}_i)$.

Using Theorem 7.3.2, one can show that the distribution of D is exactly $\chi^2_{(m-1)}$ if the m independent variates $(\hat{\theta}_i - \theta)\sqrt{\hat{I}_i}$ have exactly $N(0,1)$ distributions. The proof uses the following identity:

$$\sum \hat{I}_i(\hat{\theta}_i - \theta)^2 = \sum \hat{I}_i(\hat{\theta}_i - \tilde{\theta} + \tilde{\theta} - \theta)^2$$
$$= \sum \hat{I}_i(\hat{\theta}_i - \tilde{\theta})^2 + (\tilde{\theta} - \theta)^2 \sum \hat{I}_i.$$

The cross-product term is zero by the definition of $\tilde{\theta}$.

Example 13.4.1. Suppose that three independent experiments give likelihood functions that are approximately normal in the same parameter θ, with summary statistics

$$\hat{\theta}_1 = 9.74 \qquad \hat{\theta}_2 = 8.35 \qquad \hat{\theta}_3 = 10.27$$
$$\hat{I}_1 = 0.563 \qquad \hat{I}_2 = 0.345 \qquad \hat{I}_3 = 0.695.$$

These give approximate 95% confidence intervals

$$9.74 \pm 2.61; \qquad 8.35 \pm 3.34; \qquad 10.27 \pm 2.35.$$

To combine information from all three experiments, we compute

$$I = \sum \hat{I}_i = 1.603; \qquad \tilde{\theta} = \frac{1}{I} \sum \hat{I}_i \hat{\theta}_i = 9.67;$$
$$\sum \hat{I}_i(\hat{\theta}_i - \tilde{\theta})^2 = 0.854.$$

An approximate likelihood ratio test of $H: \theta_1 = \theta_2 = \theta_3$ gives

$$SL = P(\chi^2_{(2)} \geq 0.854) > 0.5$$

from Table B4, so the data are consistent with the hypothesis. Based on all three experiments, the approximate 95% confidence interval is

$$\theta \in \tilde{\theta} \pm 1.96/\sqrt{I} = 9.67 \pm 1.55.$$

As expected, this interval is narrower than that derived from one experiment by itself.

Problems for Section 13.4

1. Seven different dilution series experiments were used to estimate
    a parameter h, called the "hit number". The MLE $\hat{h}$ and observed
    information $\hat{I}$ are given below for each of the seven experiments.

$\hat{h}$	2.028	2.108	1.912	1.675	1.730	1.808	1.889
$\hat{I}$	19.63	25.18	32.34	70.54	64.88	67.63	36.58

    In each case, the likelihood function was approximately normal in
    h.
    (a) Are these results consistent with a common value of h in all
         seven experiments?
    (b) Are the combined results consistent with the theoretical value
         h = 2?

*13.5   Expected Information

Let X be a random variable or a vector of random variables
having probability function or probability density function which depends upon a continuous parameter θ. The likelihood function for θ
based on outcome x is

$$L(\theta;x) = k \cdot f(x;\theta)$$

where k is positive and does not depend upon θ. In Section 13.3 we
defined the score and information functions of θ,

$$S(\theta;x) = \frac{\partial \log L}{\partial \theta} = \frac{\partial \log f}{\partial \theta} \; ;$$

$$I(\theta;x) = -\frac{\partial^2 \log L}{\partial \theta^2} = -\frac{\partial \log S}{\partial \theta} = -\frac{\partial^2 \log f}{\partial \theta} \; .$$

The expected information function of θ, also called Fisher's measure
of expected information, is defined to be the expected value of the information function:

$$I_E(\theta) = E\{I(\theta;X)\} = -E\{\frac{\partial^2 \log f}{\partial \theta^2}\}. \tag{13.5.1}$$

We noted in Section 3 that $S(\hat{\theta};x) = 0$ and $\hat{I} = I(\hat{\theta};x) > 0$.
We now show that similar properties hold when we take expected values.
In what follows, we write all expectations as single sums, although they
will usually be multiple sums or integrals.

---
* This section may be omitted on first reading.

For any value of θ, the total probability in the distribution of X is equal to 1:

$$\sum_x f(x;\theta) = 1 \quad \text{for all} \quad \theta.$$

Now we differentiate with respect to θ, and assume that the order of differentiation and summation can be interchanged, to get

$$\sum_x \frac{\partial}{\partial\theta} f(x;\theta) = 0 \quad \text{and} \quad \sum_x \frac{\partial^2}{\partial\theta^2} f(x;\theta) = 0.$$

But since $S = \dfrac{\partial \log f}{\partial\theta} = \dfrac{1}{f}\dfrac{\partial f}{\partial\theta}$, we have $\dfrac{\partial f}{\partial\theta} = Sf$ and hence

$$\sum_x S(x;\theta)f(x;\theta) = 0.$$

This shows that the expected value of the score function is zero. Also, we have

$$I = -\frac{\partial S}{\partial\theta} = -\frac{\partial}{\partial\theta}\left[\frac{1}{f}\frac{\partial f}{\partial\theta}\right] = \frac{1}{f^2}\left(\frac{\partial f}{\partial\theta}\right)^2 - \frac{1}{f}\frac{\partial^2 f}{\partial\theta^2}$$

from which we obtain

$$\frac{\partial^2 f}{\partial\theta^2} = S^2 f - If.$$

It now follows that

$$\sum_x S(x;\theta)^2 f(x;\theta) - \sum_x I(x;\theta)f(x;\theta) = 0.$$

The first sum is $E(S^2)$ and the second sum is the expected information function. We have thus shown that

$$I_E(\theta) = E(S^2) = \text{var}(S) \qquad\qquad (13.5.2)$$

since $E(S) = 0$. In particular, $I_E(\theta) \geq 0$ for all values of θ.

In Section 13.3 we noted that, when the likelihood function is approximately normal in θ, the observed information $I = I(\hat\theta;x)$ gives a measure of the experiment's informativeness or precision with respect to θ. Under similar conditions, the expected information is a measure of the average precision that one would achieve in many repetitions of the experiment. It will not reflect the presence of ancillary statistics, and so will not necessarily indicate the precision achieved in any particular case. For instance, in Example 13.3.2, the observed information is a function of the ancillary statistic Z:

$$I = 10^{-2z} = \begin{cases} 1 & \text{if } z = 0 \text{ (precise experiment)} \\ 0.01 & \text{if } z = 1 \text{ (imprecise experiment)}. \end{cases}$$

However, in computing expected information, we average over Z:

$$I_E = 1 \times P(Z = 0) + 0.01 \times P(Z = 1) = 0.505.$$

It is the observed information I, rather than the expected information I_E, which measures the precision actually attained.

If $\hat{\theta}$ is a sufficient statistic for θ, there exists no ancillary statistic, and one would expect the observed and expected information functions to give similar results. In Example 13.3.4, $\hat{\theta} \equiv \frac{1}{n} \sum X_i$ is a sufficient statistic for the exponential mean θ. The information function of θ is

$$I(\theta) = -\frac{n}{\theta^2} + \frac{2 \sum X_i}{\theta^3}$$

and since $E(\sum X_i) = \sum E(X_i) = n\theta$, the expected information function is

$$I_E(\theta) = n/\theta^2.$$

Note that $I(\hat{\theta}) = I_E(\hat{\theta})$ in this case.

Planning experiments

Expected information can sometimes be used at the planning stage to determine how large an experiment is needed to achieve a desired level of precision in estimating θ, or to decide which of several experiments is likely to be the most informative. We illustrate this in two examples.

Example 13.5.1. Suppose that n items are to be tested for a preset length of time T, and the number Y which survive is to be recorded. The lifetimes are assumed to be independent exponential variates with mean θ. If T is chosen to be too small, the experiment may terminate before any failures occur. If T is too large, all items may fail long before the experiment ends. In either of these cases, one would expect to learn very little about θ. How large should T be chosen in order to maximize the expected information about θ?

Solution. Since the lifetime X of an item is assumed to have an exponential distribution with mean θ, the probability than an item survives time T is

$$p = P(X > T) = \int_T^\infty \frac{1}{\theta} e^{-x/\theta} dx = e^{-T/\theta}.$$

Then Y, the number surviving, has a binomial (n,p) distribution, and the log likelihood function of θ is

$$\ell(\theta) = y \log p + (n - y)\log(1 - p) \quad \text{where} \quad p = e^{-T/\theta}.$$

Differentiating twice with respect to θ gives

$$I(\theta) = \left[\frac{y}{p^2} + \frac{n - y}{(1-p)^2}\right]\left(\frac{dp}{d\theta}\right)^2 - \left[\frac{y}{p} - \frac{n - y}{1 - p}\right]\frac{d^2 p}{d\theta^2}.$$

Since E(Y) = np, the expected information function is

$$I_E(\theta) = \frac{n}{p(1-p)}\left(\frac{dp}{d\theta}\right)^2 = \frac{npT^2}{(1-p)\theta^4} = \frac{n}{\theta^2} \cdot h(p)$$

where

$$h(p) = \frac{p(\log p)^2}{1 - p} \quad \text{and} \quad p = e^{-T/\theta}.$$

With the aid of a calculator, one can easily show that h(p) reaches a maximum value of 0.648 for p = 0.203. Thus we should try to pick T so that about 20% of the items tested will survive the test period. If we guess T nearly right, the expected information is about $0.64n/\theta^2$; that is, about 64% of the expected information n/θ^2 when all n exponential failure times are observed.

This example is rather artificial because we are not taking costs into account. If a shorter experiment costs less, one might well get "more information per dollar" by choosing a smaller value of T so that more than 20% of items would survive testing.

Example 13.5.2. Two experiments are being considered for obtaining information about a linkage parameter θ, where $0 < \theta < \frac{1}{2}$. Each experiment would involve observing multinomial frequencies X_1, X_2, X_3, X_4 where $\sum_i X_i \equiv n$. For the first experiment, the probabilities are

$$p_1 = p_2 = \theta/2; \quad p_3 = p_4 = (1 - \theta)/2.$$

For the second experiment, the probabilities are

$$q_1 = (\theta^2 - 2\theta + 3)/4; \quad q_2 = q_3 = (2\theta - \theta^2)/4; \quad q_4 = (1 - \theta)^2/4.$$

Which experiment can be expected to yield more information about θ?

Solution. The log-likelihood function based on the multinomial distri-bution is $\sum x_i \log p_i$, and differentiating twice with respect to θ gives

$$I(\theta) = \sum \frac{x_i}{p_i^2}\left(\frac{dp_i}{d\theta}\right)^2 - \sum \frac{x_i}{p_i}\left(\frac{d^2 p_i}{d\theta^2}\right).$$

Since $E(X_i) = np_i$, the expected information function is

$$I_E(\theta) = n \sum \frac{1}{p_i}\left(\frac{dp_i}{d\theta}\right)^2 - n \sum \frac{d^2 p_i}{d\theta^2}.$$

The latter sum is zero because $\sum p_i = 1$.

Since $\dfrac{dp_i}{d\theta} = \pm\dfrac{1}{2}$ and $\dfrac{dq_i}{d\theta} = \pm\dfrac{1-\theta}{2}$, the expected information functions for the two experiments are

$$I_1(\theta) = \frac{n}{4} \sum \frac{1}{p_i} \quad \text{and} \quad I_2(\theta) = \frac{n(1-\theta)^2}{4} \sum \frac{1}{q_i}.$$

The ratio of these two functions,

$$\frac{I_2(\theta)}{I_1(\theta)} = (1-\theta)^2 \frac{\sum 1/q_i}{\sum 1/p_i},$$

is called the <u>expected relative efficiency</u> for the second experiment versus the first, and is tabulated below:

θ	0.0	0.1	0.2	0.3	0.4	0.5
I_2/I_1	1	0.88	0.77	0.65	0.55	0.44

For all $\theta > 0$, the first experiment is more efficient (has larger expected information) than the second, and it is considerably more efficient for θ near $\frac{1}{2}$. If costs were equal, the first experiment would be preferable to the second.

CHAPTER 14. INFERENCES FOR NORMAL DISTRIBUTION PARAMETERS

The normal distribution has many important applications in Statistics. Many types of measurements appear as though they could have come from normal distributions. The Central Limit Theorem (Section 6.7) justifies the assumption of a normal distribution, at least as a first approximation, in many situations.

Another reason for the widespread use of normal distribution models is that the corresponding statistical procedures are relatively simple, and multiparameter problems can be dealt with more easily than for most other distributions. Because of this, there is a tendency to use normal theory whenever possible. Even if the original measurements are decidedly non-normal, it may be possible to find a nonlinear transformation such that normal distribution results can be applied to the transformed measurements. An example of this is the analysis of log-lifetimes as normally distributed measurements in many engineering applications. Quantile plots (Section 11.5) may be useful in checking the assumption of normality for the original or transformed measurements.

Unfortunately, it is not uncommon for normal theory to be applied in situations where the assumptions clearly are not met. For instance, it is not difficult to find examples in the scientific literature where $0 - 1$ variables have been analysed as though they were continuous and normally distributed! There might have been an excuse for this twenty years ago because of computational problems with alternate models. Modern computers have changed this, and many alternate methods of analysis are now available for cases in which normality assumptions would be inappropriate.

In Section 1 we describe some general methods to be used in the remainder of this chapter and in Chapter 15. The analysis of a sample of measurements from a single normal distribution with unknown mean and variance is considered in Section 2, and Section 3 discusses the analysis of paired data (e.g. before and after measurements). Sections 4 and 5 deal with the comparison of means and variances of two independent normal samples, and Section 6 considers these problems for k normal samples.

Tests and confidence intervals for normal means are obtained from standardized normal and t distributions. Inferences for normal variances involve χ^2 and F distributions. The reader might wish to go over the material in Section 6.9 again before continuing with this chapter.

14.1   Introduction

In this chapter and the next one we shall consider several situations involving the analysis of measurements which are assumed to be independent and normally distributed.   Usually the measurements are assumed to have the same unknown variance σ^2. The analysis is similar if the variances are known multiples of a single unknown variance σ^2.

The parameter ϕ which is of primary interest is usually related to the means of the assumed normal distributions. For instance, ϕ might equal a single unknown mean, a difference of two unknown means, or more generally, a linear function of several unknown means.

Under the assumptions described above, the MLE $\hat{\phi}$ will be normally distributed, $\hat{\phi} \sim N(\phi, c\sigma^2)$, where c is a known constant, so that

$$Z \equiv \frac{\hat{\phi} - \phi}{\sqrt{c\sigma^2}} \sim N(0,1). \tag{14.1.1}$$

Furthermore, there will exist a statistic V, say, which carries all of the information about σ^2, where V and $\hat{\phi}$ are distributed independently of one another, and

$$U \equiv \frac{V}{\sigma^2} \sim \chi^2_{(\nu)}. \tag{14.1.2}$$

The degrees of freedom ν will depend upon the particular situation. We shall show that the MLE of σ^2 based on the distribution of V is

$$s^2 \equiv V/\nu. \tag{14.1.3}$$

If σ we known, tests and confidence intervals for ϕ would be based on (14.1.1). Either a two-tail test or a likelihood ratio test for $H: \phi = \phi_0$ would give

$$SL(\phi_0) = P\{|Z| \geq |z_{obs}|\} \quad \text{where} \quad z_{obs} = \frac{\hat{\phi} - \phi_0}{\sqrt{c\sigma^2}}.$$

Since $P\{|Z| \geq 1.960\} = 0.05$ from Table B1, it follows that $SL \geq 0.05$ if and only if $|z_{obs}| \leq 1.960$. Hence the 95% confidence interval for ϕ is given by

$$-1.960 \leq \frac{\hat{\phi} - \phi}{\sqrt{c\sigma^2}} \leq 1.960 \longleftrightarrow \phi \in \hat{\phi} \pm 1.960\sqrt{c\sigma^2}. \tag{14.1.4}$$

The special case where $X_1, X_2, \ldots, X_n$ are independent $N(\mu, \sigma^2)$ with σ known was considered in Sections 12.5, 12.7 and 13.1. In that case we have $\phi = \mu$, $\hat{\phi} \equiv \hat{\mu} \equiv \overline{X}$, and $c = \frac{1}{n}$.

If σ is unknown, we can no longer use (14.1.1) as it stands, but must incorporate information about σ from (14.1.2). It turns out that, in this case, we need merely replace σ^2 in (14.1.1) by the estimate $S^2 \equiv V/\nu$. The resulting variate T has a t-distribution (Student's distribution) with ν degrees of freedom:

$$T \equiv \frac{\hat{\phi} - \phi}{\sqrt{cS^2}} \sim t_{(\nu)}. \qquad (14.1.5)$$

This result follows immediately from (6.9.11), because

$$\frac{\hat{\phi} - \phi}{\sqrt{cS^2}} \equiv \frac{\hat{\phi} - \phi}{\sqrt{c\sigma^2}} \div \sqrt{\frac{S^2}{\sigma^2}} \equiv Z \div \sqrt{\frac{U}{\nu}}$$

where $Z \sim N(0,1)$ and $U \sim \chi^2_{(\nu)}$, independently of Z.

When σ is unknown, tests and confidence intervals for ϕ are obtained from (14.1.5) rather than (14.1.1). The t-distribution, like the standardized normal, is symmetrical about the origin. Hence the two-tail significance level of $H: \phi = \phi_0$ is

$$SL = P\{|T| \geq |t_{obs}|\} \quad \text{where} \quad t_{obs} = \frac{\hat{\phi} - \phi_0}{\sqrt{cS^2}}.$$

Table B3 is used to evaluate this probability. To construct a 95% confidence interval for ϕ, we find the value "a" such that $P\{|t_{(\nu)}| \geq a\} = 0.05$ from the column headed $F = .975$ in Table B3. Then $SL \geq 0.05$ if and only if $|t_{obs}| \leq a$, and the 95% confidence interval for μ is given by

$$-a \leq \frac{\hat{\phi} - \phi}{\sqrt{cS^2}} \leq a \quad \longleftrightarrow \quad \phi \in \hat{\phi} \pm a\sqrt{cS^2}. \qquad (14.1.6)$$

We noted in Section 6.9 that $t_{(\nu)} \to N(0,1)$ as $\nu \to \infty$. When ν is large, (14.1.5) will yield approximately the same results as (14.1.1) with $\sigma^2 = s^2$. This justifies the procedure used in Examples 12.5.4 and 13.1.1.

Inferences about σ^2

Inferences about σ^2 will be based on the distribution of V. Since $V \equiv \sigma^2 U$ where $U \sim \chi^2_{(\nu)}$, the p.d.f. of V can be found by a simple change of variables. By (6.9.1), U has p.d.f.

$$f(u) = k_\nu u^{\frac{\nu}{2}-1} e^{-\frac{u}{2}} \quad \text{for} \quad 0 < u < \infty$$

where k_ν is a constant. Now by (6.1.11), the p.d.f. of V is

$$f(u) \cdot \left| \frac{du}{dv} \right| = k_\nu \left[\frac{v}{\sigma^2} \right]^{\frac{\nu}{2}-1} e^{-v/2\sigma^2} \cdot \frac{1}{\sigma^2}$$

$$= k_\nu v^{\frac{\nu}{2}-1} \sigma^{-\nu} e^{-v/2\sigma^2} \quad \text{for} \quad 0 < v < \infty.$$

The likelihood function of σ based on the distribution of V is

$$L(\sigma) = \sigma^{-\nu} e^{-v/2\sigma^2} \quad \text{for} \quad \sigma > 0. \tag{14.1.7}$$

This function is maximized for $\sigma^2 = v/\nu = s^2$; see (14.1.3).

The likelihood ratio (relative likelihood function) of σ based on the distribution of V is

$$R(\sigma) = \frac{L(\sigma)}{L(s)} = \frac{\sigma^{-\nu} e^{-v/2\sigma^2}}{s^{-\nu} e^{-v/2s^2}} .$$

Using (14.1.2) and (14.1.3) it is easy to show that the likelihood ratio statistic is

$$D \equiv -2 \log R \equiv U - \nu [1 + \frac{U}{\nu}]. \tag{14.1.8}$$

This function can be used to rank values of V for a likelihood ratio test of $H: \sigma = \sigma_0$. Alternatively, one could use a two-tail test, which is computationally easier. Based on the two-tail test, the 95% confidence interval for σ^2 is given by

$$a \le \frac{v}{\sigma^2} \le b \longleftrightarrow \frac{v}{b} \le \sigma^2 \le \frac{v}{a} \tag{14.1.9}$$

where $P(\chi^2_{(\nu)} \le a) = P(\chi^2_{(\nu)} \ge b) = 0.025$.

Tests and confidence intervals for σ^2 can be obtained immediately from the results in Section 12.5, 12.7 and 13.1 for the mean θ of an exponential distribution. The latter results were based on

$2T/\theta \sim \chi^2_{(2n)}$, whereas we now have $V/\sigma^2 \sim \chi^2_{(\nu)}$. To convert results for θ to results for σ^2, we need merely substitute $\theta = \sigma^2$, $n = \frac{\nu}{2}$, and $T \equiv \frac{V}{2}$.

14.2   One-sample Problems

Let $X_1, X_2, \ldots, X_n$ be independent $N(\mu, \sigma^2)$ variates, where both μ and σ are unknown. From Example 10.1.1, the likelihood function of μ and σ is

$$L(\mu, \sigma) = \sigma^{-n} \exp\{-\textstyle\sum (x_i - \mu)^2 / 2\sigma^2\}$$

$$= \sigma^{-n} \exp\{-[\textstyle\sum (x_i - \bar{x})^2 + n(\bar{x} - \mu)^2] / 2\sigma^2\}$$

for $\sigma > 0$ and $-\infty < \mu < \infty$. The latter expression shows that $\bar{X}$ and $\sum (X_i - \bar{X})^2$ are sufficient statistics for μ and σ.

Any value of $\bar{X}$, however large or small, could be the result of an equally large or small value of the unknown parameter μ. The information provided by $\bar{X}$ is intimately linked with μ, and $\bar{X}$ gives no information about σ^2 when μ is unknown. As a result, inferences concerning σ^2 will be based only on the statistic $V \equiv \sum (X_i - \bar{X})^2$.

The distributions of $\bar{X}$ and $\sum (X_i - \bar{X})^2$ can be deduced from Theorem 7.3.2; see Application 2 at the end of Section 7.3. One finds that $\bar{X}$ and $\sum (X_i - \bar{X})^2$ are independent variates, with

$$X \sim N(\mu, \sigma^2/n); \quad \sum (X_i - \bar{X})^2 / \sigma^2 \sim \chi^2_{(n-1)}. \tag{14.2.1}$$

We can now apply the results of Section 14.1 to obtain tests and confidence intervals for μ and σ^2. We have

$$\phi = \mu; \quad \hat{\phi} \equiv \hat{\mu} \equiv \bar{X}; \quad c = \frac{1}{n}; \quad V \equiv \sum (X_i - \bar{X})^2; \quad \nu = n - 1.$$

Tests and confidence intervals for σ^2 are based on

$$U \equiv \sum (X_i - \bar{X})^2 / \sigma^2 \sim \chi^2_{(n-1)}, \tag{14.2.2}$$

and by (14.1.3), σ^2 is estimated by

$$s^2 = \frac{1}{n-1} \sum (x_i - \bar{x})^2 = \frac{1}{n-1} [\textstyle\sum x_i^2 - \frac{1}{n} (\sum x_i)^2]. \tag{14.2.3}$$

If σ were known, inferences for μ would be based on

$$Z \equiv \frac{\bar{X} - \mu}{\sqrt{\sigma^2/n}} \sim N(0,1).$$

For σ unknown, we replace σ^2 by S^2 to get

$$T \equiv \frac{\bar{X} - \mu}{\sqrt{S^2/n}} \sim t_{(n-1)}. \qquad (14.2.4)$$

Example 14.2.1. A standard drug produces blood pressure increases that are normally distributed with mean 5.5 and standard deviation 2.3 units. A new drug is also expected to produce normally distributed increases, but with possibly different values of μ and σ. The new drug was given to $n = 10$ individuals, and produced the following blood pressure increases:

4.4   6.7   5.7   3.8   4.6   3.7   4.6   5.1   4.3   2.1

Is there evidence that $\mu \neq 5.5$? that $\sigma \neq 2.3$?

Solution. We first compute the following values from the data:

$$\sum x_i = 45.0, \qquad \bar{x} = 4.5$$

$$\sum x_i^2 = 216.1, \qquad \sum(x_i - \bar{x})^2 = \sum x_i^2 - \frac{1}{n}(\sum x_i)^2 = 13.6$$

$$s^2 = \frac{1}{n-1}\sum(x_i - \bar{x})^2 = 1.51, \qquad s = 1.23.$$

Since σ is unknown, a test of the hypothesis $\mu = 5.5$ will be based on (14.2.4). The observed T-value is

$$t_{obs} = \frac{4.5 - 5.5}{\sqrt{1.51/10}} = -2.57$$

and we compare this value with tables of $t_{(9)}$. Since the t-distribution is symmetrical about the origin, the significance level in a two-tail test is

$$SL = P\{|t_{(9)}| \geq 2.57\} \approx 0.03$$

from Table B3. There is evidence that $\mu \neq 5.5$.

Since μ is unknown, a test of the hypothesis $\sigma = 2.3$ will be based on (14.2.2). The observed U-value is

$$u_{obs} = 13.6/(2.3)^2 = 2.57,$$

and we compare this value with tables of $\chi^2_{(9)}$. It lies on the lower tail of the distribution, and hence the two-tail significance level is

$$SL = 2P\{\chi^2_{(9)} \leq 2.57\} \approx 2(0.02) = 0.04$$

from Table B4. There is evidence that $\sigma \neq 2.3$. The new drug produces a lower mean increase, but it is also less variable in its effect than the old one. This is an advantage, because the effect of the new drug on an individual can be more accurately predicted.

Example 14.2.2. Eight plastic gears were tested at 21°C until they failed. The times to failure (in millions of cycles) were as follows:

2.37   2.01   2.47   2.20   1.87   2.32   2.00   2.86

A common assumption in such situations is that the logarithms of the failure times are independent $N(\mu, \sigma^2)$. The natural logarithms of the $n = 8$ failure times listed above are

0.863   0.698   0.904   0.788   0.626   0.842   0.693   1.051

Obtain 95% confidence intervals for μ and σ.

Solution. Here $n = 8$, and the log failure times give $\bar{x} = 0.8081$, $\sum(x_i - \bar{x})^2 = 0.131$, and $s^2 = 0.01876$. Since σ is unknown, confidence intervals for μ will be obtained from (14.2.4). From Table B3 we find that $P\{|t_{(7)}| \geq 2.365\} = 0.05$. The significance level in a two-tail test of $H:\mu = \mu_0$ will be at least 5% if and only if $|t_{obs}| \leq 2.365$. Hence the 95% confidence interval for μ is given by

$$-2.365 \leq \frac{\bar{x} - \mu}{\sqrt{s^2/n}} \leq 2.365 \longleftrightarrow \mu \in \bar{x} \pm 2.365\sqrt{s^2/n}.$$

Upon substitution for $n, \bar{x}$, and s^2 we find the 95% confidence interval for μ to be $0.693 \leq \mu \leq 0.923$. Any value of μ outside this interval is contradicted by the data at the 5% level in a two-tail test.

Since μ is unknown, inferences about σ are based on (14.2.7). Table B4 gives

$$P\{\chi^2_{(7)} \leq 1.690\} = P\{\chi^2_{(7)} \geq 16.01\} = 0.025.$$

The significance level in a two-tail test for σ will be 5% or more if and only if $1.690 \leq u_{obs} \leq 16.01$, where $u_{obs} = \sum(x_i - \bar{x})^2/\sigma^2$.

Upon substituting $\sum(x_i - \bar{x})^2 = 0.1313$ and rearranging, we get

$$\frac{0.1313}{16.01} \leq \sigma^2 \leq \frac{0.1313}{1.690} .$$

Finally we take square roots to get $0.091 \leq \sigma \leq 0.279$ as the required 95% confidence interval. Any value of σ outside this interval would lead to a significance level less than 5% in a two-tail test.

Problems for Section 14.2

1. The following are the initial velocities in metres per second of seven projectiles fired from the same gun:

    451   447   454   450   454   449   452

    Assuming that velocity is normally distributed, obtain a 90% confidence interval for the mean velocity μ.

†2. Under a special diet, twelve rats made the following weight gains (in grams) from birth to age three months:

    55.3   54.8   65.9   60.7   59.4   62.0   62.1   58.7   64.5   62.3   67.6   61.2

    Assuming that the weight gains are independent normal, obtain 95% confidence intervals for the mean and for the variance.

†3. A manufacturer wishes to determine the mean breaking strength μ of string "to within a pound", which we interpret as requiring that the 95% confidence interval for μ should have length at most 2 pounds. If ten preliminary measurements gave $\sum(x_i - \bar{x})^2 = 80$, how many additional measurements would you advise the manufacturer to make?

4. Sixteen packages are randomly selected from the production of a detergent packaging machine. Their weights (in grams) were as follows:

    287   293   295   295   297   298   299   300
    300   302   302   303   306   307   308   311

    It may be assumed that weights are independent $N(\mu, \sigma^2)$.

    (a) Determine 95% confidence intervals for μ and σ.

    (b) Assuming that μ and σ are equal to their estimates, find an interval which contains the weight of a new randomly chosen package with probability 0.95.

14.3 Analysis of Differences

Suppose that an experiment yields n pairs of related measurements (A_i, B_i) for $i = 1, 2, \ldots, n$. For instance, B_i might represent a measurement on an individual before treatment, while A_i represents a measurement on the same individual after treatment. There is a natural grouping or blocking of the data into pairs, and we are likely to find greater similarity between two measurements A_i, B_i which belong to the same pair than between measurements A_i, B_j which belong to different pairs. It would be improper to ignore this pairing in the analysis. It would be equally improper to introduce such pairing artificially when it does not exist in the experiment - for instance, by randomly pairing observations with one another. Methods appropriate for situations in which there is no natural pairing will be considered in Section 4.

In the analysis of paired data, we will be interested in determining whether there is proof of a treatment effect. If there is, we would wish to estimate its size. The analysis is usually performed on the n differences,

$$X_i \equiv B_i - A_i, \quad i = 1, 2, \ldots, n,$$

and these are assumed to be independent.

If the treatment has no effect, each non-zero difference has a 50% probability of being positive. Given that there are m non-zero differences, the number of positive differences should have a binomial distribution with parameters m and $p = \frac{1}{2}$. A test of the hypothesis $p = \frac{1}{2}$ can be performed as in Example 12.1.1, and evidence against this hypothesis would be evidence that the treatment does have an effect.

In the preceding paragraph, we assumed only the independence of the X_i's, and did not make use of their numerical values. In some situations, we may be willing to make stronger assumptions - for instance, that the X_i's have a common normal distribution $N(\mu, \sigma^2)$. The normality assumption can be checked by means of a quantile plot as in Example 11.5.3. If the treatment has no effect, then $\mu = 0$. We can use (14.2.4) to test the hypothesis $\mu = 0$, or to obtain confidence intervals for the mean treatment effect μ.

Example 14.3.1. The following table gives the average number of man-hours per month lost due to accidents in each of eight factories over a period of one year before and after the introduction of an industrial safety programme.

Factory	i	1	2	3	4	5	6	7	8
Before	B_i	48.5	79.2	25.3	19.7	130.9	57.6	88.8	62.1
After	A_i	28.7	62.2	28.9	0.0	93.5	49.6	86.3	40.2
Difference	X_i	19.8	17.0	-3.6	19.7	37.4	8.0	2.5	21.9

Do these observations give proof that the safety programme has had an effect?

Solution.   (a)   First we consider a test of significance based only on the signs of the X_i's. There are $m = 8$ nonzero differences. If the programme had no effect, the probability of observing y positive differences out of 8 would be $\binom{8}{y}(\frac{1}{2})^8$. The expected number of positive differences is 4, and we take the discrepancy measure to be $|Y - 4|$. We have observed $y = 7$, and hence

$$SL = P(Y = 0, 1, 7, \text{ or } 8) = 18(\tfrac{1}{2})^8 = 0.070.$$

The sign test does not give conclusive evidence that the programme has had an effect.

Note that most of the positive differences are fairly large, whereas the only negative difference is rather small. The sign test does not take this into account, because it makes no use of the numerical values of the X_i's. The same result would have been obtained if all of the positive differences were +100 and the negative difference was -0.1.

(b)   Now let us assume that the differences are independent values from $N(\mu, \sigma^2)$. A quantile plot shows this to be not an unreasonable assumption. We have

$$n = 8, \quad \bar{x} = 15.34, \quad s^2 = 164.12,$$

so that under the hypothesis $\mu = 0$, the observed value of the T-statistic (14.2.4) is

$$t_{obs} = \frac{15.34 - 0}{\sqrt{164.12/8}} = 3.39.$$

From Table B3, the two-tail significance level is

$$SL = P\{|t_{(7)}| \geq 3.39\} \approx 0.01.$$

This test gives strong evidence that the safety programme has reduced the average number of man-hours lost. Using (14.2.4), a 95% confidence interval for the mean reduction is found to be $4.63 \leq \mu \leq 26.05$.

Discussion

The main advantage of the analysis of differences is that it permits differences in the means for the n experimental units (e.g. factories) to be eliminated from the analysis. Suppose that, before treatment, the expected value for the ith unit is $E(B_i) = \mu_i$, and that the effect of the treatment is to add a constant amount δ to all the means. Then the expected value for the ith unit after treatment is $E(A_i) = \mu_i + \delta$. The expected value of the difference is

$$E(X_i) = E(A_i) - E(B_i) = \delta,$$

and μ_i cancels out. If the μ_i's differ substantially from one another and they are not eliminated by taking differences, the effect of the treatment may not show up.

For instance, consider what happens if we ignore the pairing in the preceding example, and compare the eight "before" measurements with the eight "after" measurements. The sample mean and variance of the "before" measurements are

$$\bar{B} = 64.01, \quad s_B^2 = \frac{1}{n-1} \sum (B_i - \bar{B})^2 = 1295,$$

and for the "after" measurements we obtain

$$\bar{A} = 48.67, \quad s_A^2 = \frac{1}{n-1} \sum (A_i - \bar{A})^2 = 977.$$

To determine whether the programme was effective, we might consider the difference in sample means, $\bar{B} - \bar{A}$. If before and after measurements are independent, then

$$\text{var}(\bar{B} - \bar{A}) = \text{var}(\bar{B}) + \text{var}(\bar{A})$$

$$= \frac{1}{n} \sigma_B^2 + \frac{1}{n} \sigma_A^2 \approx \frac{1295 + 977}{8} = 284,$$

and hence the standard deviation of $\bar{B} - \bar{A}$ is about $\sqrt{284} = 16.85$. The observed difference is 15.34, and this lies within one standard deviation of zero. Therefore, we could not claim to have proof that the programme was effective. The variance estimate is inflated by the large differences among factories, and this masks the real effect of the programme.

In general, we should try to set up the experiment in such a way that extraneous effects like the μ_i's can be eliminated in comparing treatments. The Design of Experiments is a major branch of statistics which deals with such problems.

Problems for Section 14.3

†1. The following table gives the additional hours of sleep gained by
each patient in the use of two drugs:

Patient	1	2	3	4	5	6	7	8	9	10
Drug A	0.7	-1.6	-0.2	-1.2	-0.1	3.4	3.7	0.8	0.0	2.0
Drug B	1.9	0.8	1.1	0.1	-0.1	4.4	5.5	1.6	4.6	3.4

Perform two different tests of the hypothesis that the two drugs
are equally effective, and state the assumptions upon which the
analysis depends in each case.

2. The following table gives the results of a series of measurements
of the corrosion of coated and uncoated underground pipes:

Soil type:	1	2	3	4	5	6	7	8
Coated :	15.6	21.0	22.6	56.8	13.2	20.9	8.6	31.2
Uncoated :	10.9	46.7	25.7	69.7	36.7	20.4	29.4	10.2

Soil type:	9	10	11	12
Coated :	25.4	8.5	11.2	35.8
Uncoated :	71.6	42.8	23.9	49.2

Obtain a 99% confidence interval for the mean difference in the
amounts of corrosion for the two types of pipe.

3. Two analysts carried out simultaneous measurements of the percentage
of ammonia in a plant gas on nine successive days to find the extent of the bias, if any, between their results. Their measurements
were:

Day	1	2	3	4	5	6	7	8	9
Analyst A	4	37	35	43	34	36	48	33	33
Analyst B	18	37	38	36	47	48	57	28	42

Obtain a 95% confidence interval for the mean difference in their
measurements. On what assumptions does your analysis depend?

†4. Six automobiles of different models were used to compare two brands
of tires. Each car was fitted with tires of brand A and driven
over a difficult course until one of its tires could no longer be
used. Tires of brand B were then fitted to the same cars, and
the procedure was repeated. The following are the observed mileages
to tire failure in thousands of miles:

Car	1	2	3	4	5	6
Brand A	18	23	16	27	19	17
Brand B	15	22	16	21	15	16

(a) Test whether these data are consistent with the hypothesis
that the mean lifetimes for the two brands are equal.

(b) What factor, other than difference in tire quality, might
     account for the lower mileage achieved with brand B? Suggest
     an improvement in the design of the experiment which would
     have helped to eliminate this source of bias.

14.4 Comparison of Two Means

        In many statistical applications, a set of measurements
$X_1, X_2, \ldots, X_n$ is to be compared with another set of similar measurements $Y_1, Y_2, \ldots, Y_m$ taken under somewhat different conditions. For
instance, we may wish to compare the lifetimes of plastic gears at two
different operating temperatures, or the weight gains of animals on
two different diets. We wish to determine whether the two sets of
measurements show "real" differences - that is, differences not attributable to chance. If real differences do exist, we want to describe
them simply and estimate their magnitudes.

        If the conditions under which the two samples are taken are
radically different, the probability distributions of the X_i's and
Y_i's will likely have different means and variances, and possibly different shapes as well. For instance, at very high operating temperatures, plastic gears will fail due to melting, while at very low temperatures they become quite brittle and tend to fracture. There is no
reason to expect that the distributions of lifetimes will be similar
at these two extremes.

        In most practical situations, relatively modest changes in
conditions are involved. Then one would not expect the distribution
of the X_i's to be much different from the distribution of the Y_i's.
In fact, it is often the case that the two distributions have nearly
the same shapes and variances, and that the effect of a small change in
conditions shows up primarily as a difference in the means. Models in
which sets of measurements are assumed to have distributions of the
same shape (usually normal) with equal variances are very common in
statistical work.

        In this section, we consider tests for the equality of two
means, and estimation of the difference in means. The analysis is
based on the assumption that all $n + m$ measurements are independent,
and that they come from two possibly different normal distributions -

        Sample 1: $X_1, X_2, \ldots, X_n \sim N(\mu_1, \sigma_1^2)$
        Sample 2: $Y_1, Y_2, \ldots, Y_m \sim N(\mu_2, \sigma_2^2)$.

The analysis will depend upon what is assumed about the variances. First we suppose that both n and m are large, so that the variances can be taken as known. Then we consider the most common case in practice, in which the variances are assumed to be equal. This analysis is easily extended to the case in which the variance ratio $\lambda = \sigma_1^2/\sigma_2^2$ is known. Finally, we briefly discuss the problem of comparing means when the variance ratio is unknown.

The assumption of normality can be checked by means of quantile plots. If there were substantial departures, we would try to devise a better model, for which different methods of analysis would be required. Failing this, we could make use of "nonparametric" methods, which do not involve the assumption of a particular distributional form for the X_i's and Y_i's. For instance, we might use a nonparametric test based on the theory of runs, as described at the end of Section 2.6. Although such methods can be used to demonstrate that real differences exist, they are not usually very helpful in describing the nature of the differences and estimating their magnitudes.

Throughout this section and the next one, we are assuming that there does not exist a natural pairing of the X_i's with the Y_i's. If such a pairing does exist in the experiment, it should be taken into account in the analysis (see Section 3).

Preliminary Results

The sufficient statistics from the first sample are $\overline{X}$ and $V_1 \equiv \sum(X_i - \overline{X})^2$, and those from the second sample are $\overline{Y}$ and $V_2 \equiv \sum(Y_i - \overline{Y})^2$. These four statistics are distributed independently of one another as follows:

$$\overline{X} \sim N(\mu_1, \sigma_1^2/n); \qquad \overline{Y} \sim N(\mu_2, \sigma_2^2/m)$$
$$V_1/\sigma_1^2 \sim \chi^2_{(n-1)}; \qquad V_2/\sigma_2^2 \sim \chi^2_{(m-1)}.$$

The two variance estimates are $s_1^2 = \frac{1}{n-1} V_1$ and $s_2^2 = \frac{1}{m-1} V_2$.

Let $\phi = \mu_1 - \mu_2$ be the difference between the two means. The hypothesis of equal means, $\mu_1 = \mu_2$, is then equivalent to $H: \phi = 0$. The estimate of ϕ is $\hat{\phi} \equiv \overline{X} - \overline{Y}$. Since $\overline{X}$ and $\overline{Y}$ are independent normal variates, (6.6.7) gives

$$\hat{\phi} \sim N(\phi, \frac{1}{n}\sigma_1^2 + \frac{1}{m}\sigma_2^2). \tag{14.4.1}$$

Also, by (6.9.3), a sum of independent χ^2 variates has a χ^2 distribution, and hence

$$\frac{V_1}{\sigma_1^2} + \frac{V_2}{\sigma_2^2} \sim \chi_{(n+m-2)}^2 . \qquad (14.4.2)$$

Comparing means when the variances are known

From (14.4.1) we have

$$Z \equiv \frac{\hat{\phi} - \phi}{\sqrt{\frac{1}{n}\sigma_1^2 + \frac{1}{m}\sigma_2^2}} \sim N(0,1) \qquad (14.4.3)$$

and inferences about ϕ can be based on this result when σ_1^2 and σ_2^2 are known. If the variances are unknown but n and m are both large, we can replace σ_1^2 and σ_2^2 by their estimates s_1^2 and s_2^2 and use (14.4.3). This procedure gives results that are sufficiently accurate for most practical purposes provided that n and m are both at least 30.

Example 14.4.1. Two fabrics were tested to determine their wearing qualities. A piece of fabric was rubbed until a hole appeared, and the number of rubs was recorded. This was done for $n = 100$ pieces of Fabric A, and for $m = 50$ pieces for Fabric B. The assumption of normality appeared reasonable from inspection of the data, and the following values were computed:

Fabric A: $n = 100$ $\bar{x} = 1867.1$ $s_1^2 = 164.1$
Fabric B: $m = 50$ $\bar{y} = 2034.2$ $s_2^2 = 429.2$

Obtain a 95% confidence interval for the difference in the mean number of rubs necessary to produce a hole.

Solution. Since both n and m are large, we use (14.4.3) with $\sigma_1^2 = 164.1$ and $\sigma_2^2 = 429.2$. Since $P\{|Z| \geq 1.960\} = 0.05$ from Table B1, the 95% confidence interval for ϕ is given by

$$\phi \in \hat{\phi} \pm 1.960\sqrt{\frac{1}{n}\sigma_1^2 + \frac{1}{m}\sigma_2^2} = -167.1 \pm 6.3;$$

that is, $-173.4 \leq \mu_1 - \mu_2 \leq -160.8$.

Comparing means when the variances are equal

Let $\sigma_1^2 = \sigma_2^2 = \sigma^2$, say. Then (14.4.1) gives

$$\hat{\phi} \sim N(\phi, (\frac{1}{n} + \frac{1}{m})\sigma^2)$$

and hence

$$Z \equiv \frac{\hat{\phi} - \phi}{\sqrt{c\sigma^2}} \sim N(0,1) \tag{14.4.4}$$

where $c = \frac{1}{n} + \frac{1}{m}$. If σ is unknown, we need a statistic which carries all of the information about σ. To find this, we examine the likelihood function of μ_1, μ_2 and σ. From Section 14.2, the likelihood function based on the X_i's is

$$L(\mu_1, \sigma) = \sigma^{-n} \exp\{-[v_1 + n(\overline{x} - \mu_1)^2]/2\sigma^2\}$$

and that based on the Y_i's is similar. Their product is

$$L(\mu_1, \mu_2, \sigma) = \sigma^{-n-m} \exp\{-[v + n(\overline{x} - \mu_1)^2 + m(\overline{y} - \mu_2)^2]/2\sigma^2\}$$

where $V \equiv V_1 + V_2$. We now see that $\overline{X}, \overline{Y}$ and V are sufficient statistics for μ_1, μ_2 and σ. One can argue, as in Section 14.2, that $\overline{X}$ and $\overline{Y}$ give no information about σ when μ_1 and μ_2 are unknown, and hence that inferences about σ should be based only on V.

Now if we take $\sigma_1^2 = \sigma_2^2 = \sigma^2$ in (14.4.2), we find that

$$U \equiv \frac{V}{\sigma^2} \equiv \frac{V_1}{\sigma^2} + \frac{V_2}{\sigma^2} \sim \chi^2_{(n+m-2)}.$$

By (14.1.3), the variance estimate based on both samples is

$$s^2 = \frac{v}{\nu} = \frac{v_1 + v_2}{n + m - 2} = \frac{(n-1)s_1^2 + (m-1)s_2^2}{(n-1) + (m-1)}.$$

The combined (pooled) variance estimate is an average of the two sample variances s_1^2 and s_2^2, weighted according to their degrees of freedom.

Now, proceeding as in Section 14.1, we replace σ^2 in (14.4.4) by the variance estimate s^2 to get

$$T \equiv \frac{\hat{\phi} - \phi}{\sqrt{cs^2}} \sim t_{(n+m-2)} \tag{14.4.5}$$

where $c = \frac{1}{n} + \frac{1}{m}$. Tests and confidence intervals for $\phi = \mu_1 - \mu_2$ are based on this result when $\sigma_1^2 = \sigma_2^2$.

Example 14.4.2. In Example 11.5.3 we considered 24 measurements of
the lengths of cuckoos' eggs, of which $n = 9$ were found in reed-warblers' nests and $m = 15$ were found in wrens' nests. The quantile
plots given in Figure 11.5.3 indicate that each set of measurements
is in reasonable agreement with a normal distribution. Because the
two lines are nearly parallel, the two standard deviations are approximately equal. However, it appears that the means are different. We
shall assume normality and equal variances, and use the results derived above to test the hypothesis $\mu_1 = \mu_2$.

The following values are computed from the measurements recorded in Example 11.5.3.

Sample 1 (reed-warblers' nests)
$$n = 9 \qquad \overline{x} = 22.20 \qquad s_1^2 = 0.4225$$

Sample 2 (wrens' nests)
$$m = 15 \qquad \overline{y} = 21.12 \qquad s_2^2 = 0.5689$$

Hence $\hat{\phi} = \overline{x} - \overline{y} = 1.08$, and the pooled variance estimate is

$$s^2 = \frac{8s_1^2 + 14s_2^2}{8 + 14} = 0.5157$$

with $8 + 14 = 22$ degrees of freedom.

Since $\phi = \mu_1 - \mu_2$, we wish to test the hypothesis that
$\phi = 0$. Under this hypothesis, the observed value of the T-statistic
(14.4.5) is

$$t_{obs} = \frac{1.08 - 0}{\sqrt{(\frac{1}{9} + \frac{1}{15})(.5157)}} = 3.57.$$

The two-tail significance level is then

$$SL = P\{|t_{(22)}| \geq 3.57\} \approx .002$$

from Table B3. The data provide very strong evidence that $\mu_1 \neq \mu_2$.

Example 14.4.3. The log-lifetimes of $n = 8$ plastic gears at 21°C
were analysed in Example 14.2.2. The following are the log-lifetimes
of four additional gears run at 30°C:

$$0.364 \qquad 0.695 \qquad 0.558 \qquad 0.359$$

We assume that the log-lifetimes are normally distributed with the
same variance σ^2, but with possibly different means μ_1 (at 21°C)

and μ_2 (at 30°C). We wish to determine a 95% confidence interval for $\phi = \mu_1 - \mu_2$.

First we compute the following values from the data:

Sample 1 (21°C) – values from Example 14.2.2
$$n = 8 \qquad \overline{x} = 0.8081 \qquad s_1^2 = 0.01876$$

Sample 2 (30°C)
$$m = 4 \qquad \overline{y} = 0.4940 \qquad s_2^2 = 0.02654.$$

Hence $\hat{\phi} = \overline{x} - \overline{y} = 0.3140$, and the combined variance estimate is

$$s^2 = \frac{7s_1^2 + 3s_2^2}{7 + 3} = 0.02109$$

with 10 degrees of freedom. From Table B3, we find that $P\{|t_{(10)}| \geq 2.228\} = 0.05$, and thus the 95% confidence interval is

$$\phi \in \hat{\phi} \pm 2.228\sqrt{(\tfrac{1}{n} + \tfrac{1}{m})s^2} = 0.3140 \pm 0.1981;$$

that is, $0.1159 \leq \mu_1 - \mu_2 \leq 0.5121$.

Note. Since the normal distribution is symmetrical, the mean log-lifetime μ_i is also the median log-lifetime. The median lifetime will be simply $m_i = e^{\mu_i}$, but from Problem 6.6.12, the mean lifetime is $\exp\{\mu_i + \tfrac{1}{2}\sigma_i^2\}$. Thus the ratio of median lifetimes is

$$m_1/m_2 = e^{\mu_1 - \mu_2}$$

and a 95% confidence interval for m_1/m_2 is given by

$$e^{0.1159} \leq e^{\mu_1 - \mu_2} \leq e^{0.5121} \longleftrightarrow 1.123 \leq m_1/m_2 \leq 1.669.$$

This is also a 95% confidence interval for the ratio of the means when $\sigma_1^2 = \sigma_2^2$, as we are assuming here, but not otherwise.

Comparing means when the variance ratio is known

A similar analysis is possible when the variances are unequal provided that the variance ratio $\lambda = \sigma_1^2/\sigma_2^2$ is known. Let σ^2 denote $\mathrm{var}(X_i)$, so that $\sigma_2^2 = \mathrm{var}(Y_i) = \sigma^2/\lambda$. Then

$$\hat{\phi} \equiv \overline{X} - \overline{Y} \sim N(\phi, \tfrac{1}{n}\sigma^2 + \tfrac{1}{\lambda m}\sigma^2)$$

and hence

$$Z \equiv \frac{\hat{\phi} - \phi}{\sqrt{c\sigma^2}} \sim N(0,1)$$

where now $c = \frac{1}{n} + \frac{1}{\lambda m}$. The likelihood function is

$$L(\mu_1, \mu_2, \sigma) = \sigma^{-n-m} \exp\{-[v + n(\bar{x} - \mu_1)^2 + m(\bar{y} - \mu_2)^2]/2\sigma^2\}$$

where $V \equiv V_1 + \lambda V_2$. Now (14.4.2) gives

$$U \equiv \frac{V}{\sigma^2} \equiv \frac{V_1}{\sigma^2} + \frac{V_2}{\sigma^2/\lambda} \sim \chi^2_{(n+m-2)}.$$

The pooled variance estimate becomes

$$s^2 = \frac{v}{\nu} = \frac{v_1 + \lambda v_2}{n + m - 2} = \frac{(n-1)s_1^2 + (m-1)\lambda s_2^2}{(n-1) + (m-1)}.$$

This is again a weighted average of two estimates of σ^2: s_1^2 from the first sample and λs_2^2 from the second sample. Replacing σ^2 by S^2 now gives

$$T \equiv \frac{\hat{\phi} - \phi}{\sqrt{cS^2}} \sim t_{(n+m-2)}. \tag{14.4.6}$$

When λ is known but σ is unknown, inferences about ϕ are based on this T-statistic. Our previous results for equal variances are obtained by taking $\lambda = 1$.

One use of the results for general λ is in checking the extent to which conclusions based on the assumption of equal variances ($\lambda = 1$) are dependent upon this assumption. One can repeat the calculation of the significance level or confidence interval for several values of λ in order to see whether the conclusions would be seriously affected if the variances were not equal. A suitable range of values for λ can be found from the results of the next section.

Example 14.4.2 (continued). Table 14.4.1 shows the calculation of observed T-values under the hypothesis $\mu_1 = \mu_2$ for several values of the variance ratio $\lambda = \sigma_1^2/\sigma_2^2$. The values of λ were selected to cover the 90% confidence interval for λ; see Example 14.5.1. Our previous analysis under the assumption of equal variances corresponds to $\lambda = 1$. The significance level in a two-tail test of $H: \mu_1 = \mu_2$ is

240

Table 14.4.1
Observed T-values under $H: \mu_1 = \mu_2$ for
Several Values of the Variance Ratio λ

λ	0.25	0.5	0.75	1.0	1.5	2.0	2.5
c	0.378	0.244	0.200	0.178	0.156	0.144	0.138
s^2	0.244	0.335	0.425	0.516	0.697	0.878	1.059
t_{obs}	3.56	3.77	3.70	3.57	3.28	3.03	2.83

$P\{|t_{(22)}| \geq |t_{obs}|\}$. Since $P\{|t_{(22)}| \geq 2.819\} = 0.01$ from Table B3, the significance level is less than 1% over the entire range of λ-values considered. Our previous conclusion, that $\mu_1 \neq \mu_2$, does not depend critically on the assumption of equal variances.

Comparing means when the variance ratio is unknown

The problem of comparing the means of two normal distributions when the variance ratio $\lambda = \sigma_1^2/\sigma_2^2$ is unknown is called the Behrens-Fisher problem. This problem is controversial. A variety of solutions and approximate solutions have been proposed, none of which has gained universal acceptance.

For any given value of λ, we can compute the significance level in a test of $H: \mu_1 = \mu_2$ as in the preceding example. The problem is to combine the results for different λ-values to obtain a single overall significance level, taking into account the plausibilities of the various λ-values. This can be done by averaging the significance level over a Bayesian or fiducial distribution for λ (see Chapter 16), but many statisticians do not accept this solution.

Problems for Section 14.4

1. The following are measurements of ultimate tensile strength (UTS) for twelve specimens of insulating foam of two different densities.

High density 98.5 105.5 111.6 114.5 126.5 127.1
Low density   79.7   84.5   85.2   98.0 105.2 113.6

Assuming normality and equality of variances, obtain 95% confidence intervals for the common variance and for the difference in mean UTS.

†2. Twenty-seven measurements of yield were made on two industrial pro-

cesses, with the following results:

Process 1: $n = 11$ $\bar{x} = 6.23$ $s_1^2 = 3.79$
Process 2: $m = 16$ $\bar{y} = 12.74$ $s_2^2 = 4.17$

Assuming that the yields are normally distributed with the same variance, find 95% confidence intervals for the means μ_1 and μ_2, and for the difference in means $\mu_1 - \mu_2$.

3. An experiment to determine the effect of a drug on the blood glucose concentration of diabetic rats gave the following results:

Control rats   2.05   1.82   2.00   1.94   2.12
Treated rats   1.71   1.37   2.04   1.50   1.69   1.83

The hypothesis of interest is that the drug has no effect on the mean concentration. Test this hypothesis under the assumption of equal variances. Repeat the test for $\lambda = 0.1, 0.25, 0.5$ and 0.75, where $\lambda = \sigma_1^2/\sigma_2^2$. What can be concluded about the effect of the drug?

†4. For what value of λ is the denominator of (14.4.6) minimized?

14.5 Comparison of Two Variances

As in the preceding section, we consider $n + m$ independent measurements from two possibly different normal distributions –

Sample 1: $X_1, X_2, \ldots, X_n$ from $N(\mu_1, \sigma_1^2)$;
Sample 2: $Y_1, Y_2, \ldots, Y_m$ from $N(\mu_2, \sigma_2^2)$.

We suppose that the means μ_1 and μ_2 are unknown, and consider the comparison of the variances σ_1^2 and σ_2^2.

Because variance is related to the measurement scale, the comparison of variances is most naturally made in terms of the ratio $\lambda = \sigma_1^2/\sigma_2^2$. A change in the scale of measurement (e.g. from pounds to ounces, or from inches to feet) will not change λ, but it will change the difference in variances $\sigma_1^2 - \sigma_2^2$, or the difference in standard deviations $\sigma_1 - \sigma_2$. There are situations, such as in the analysis of variance with random effects, where one is interested in the difference of two variances. However, the analysis is more difficult in this case.

When μ_1 and μ_2 are unknown, inferences about σ_1^2 and σ_2^2 are based on the statistics $V_1 \equiv \sum(X_i - \bar{X})^2$ and $V_2 \equiv \sum(Y_i - \bar{Y})^2$.

These are independent variates, with

$$U_1 \equiv V_1/\sigma_1^2 \sim \chi^2_{(n-1)}; \quad U_2 \equiv V_2/\sigma_2^2 \sim \chi^2_{(m-1)}.$$

The variance estimates are $S_1^2 \equiv \frac{1}{n-1} V_1$ and $S_2^2 \equiv \frac{1}{m-1} V_2$.

The ratio U_1/U_2, or any function of this ratio, will have a distribution that depends only on the ratio $\lambda = \sigma_1^2/\sigma_2^2$. The distribution which has been tabulated is that of the ratio of mean squares $\frac{1}{n-1} U_1$ and $\frac{1}{m-1} U_2$. A mean square was defined in Section 6.9 to be a χ^2 variate divided by its degrees of freedom. By (6.9.8), a ratio of independent mean squares has an F-distribution.

In the present case, the ratio of mean squares is

$$\frac{U_1 \div (n-1)}{U_2 \div (m-1)} \equiv \frac{\sigma_2^2}{\sigma_1^2} \cdot \frac{V_1 \div (n-1)}{V_2 \div (m-1)} \equiv \frac{1}{\lambda} \frac{S_1^2}{S_2^2}$$

which has an F-distribution with $n-1$ numerator and $m-1$ denominator degrees of freedom:

$$\frac{1}{\lambda} (S_1^2/S_2^2) \sim F_{n-1,m-1}. \qquad (14.5.1)$$

Tests and confidence intervals for λ may be based on this result.

If the variances are equal $(\lambda = 1)$, the observed value of the F-statistic is

$$F_{obs} = s_1^2/s_2^2$$

and we compare this value with tables of $F_{n-1,m-1}$. Both large values (>1) and small values (<1) of F would give evidence that $\sigma_1^2 \neq \sigma_2^2$, and hence a two-tail test is appropriate. Another possibility is to use a likelihood ratio test based on the distribution of S_1^2/S_2^2, but this presents computational problems.

As we noted in Section 6.9, tables of the F-distribution list only variate values greater than 1. We can save a little arithmetic by arranging to put the larger variance estimate in the numerator when we compute F_{obs}. The two-tail significance level is then $2 \cdot P(F_{n-1,m-1} \geq F_{obs})$, where $n-1$ is the degrees of freedom for the larger estimate and $m-1$ is the degrees of freedom for the smaller estimate.

To construct a 95% confidence interval for λ, we consult Table B5 to obtain values a, b such

$$P(F_{m-1,n-1} \geq a) = P(F_{n-1,m-1} \geq b) = 0.025.$$

Then, by (6.9.7),

$$P(F_{n-1,m-1} \leq \frac{1}{a}) = 0.025.$$

The 95% confidence interval based on a two-tail test is given by

$$\frac{1}{a} \leq \frac{1}{\lambda} (s_1^2/s_2^2) \leq b \longleftrightarrow \frac{1}{b} (s_1^2/s_2^2) \leq \lambda \leq a(s_1^2/s_2^2).$$

Example 14.5.1. Test the hypothesis $\sigma_1^2 = \sigma_2^2$ in Example 14.4.2, and obtain a 90% confidence interval for the variance ratio $\lambda = \sigma_1^2/\sigma_2^2$.

Solution. The observed sample variances are

$$s_1^2 = 0.4225 \quad \text{with} \quad 8 \quad \text{degrees of freedom};$$
$$s_2^2 = 0.5689 \quad \text{with} \quad 14 \quad \text{degrees of freedom}.$$

We divide the larger estimate by the smaller one to obtain

$$F_{obs} = s_2^2/s_1^2 = \frac{0.5689}{0.4225} = 1.35.$$

There are 14 numerator and 8 denominator degrees of freedom, and the two-tail significance level for the hypothesis $\sigma_1^2 = \sigma_2^2$ is

$$SL = 2P\{F_{14,8} \geq 1.35\} > 0.2$$

from Table B5. There is no evidence that the variances are different. To construct a 90% confidence interval, we consult the table of 95th percentiles of the F-distribution. Using linear interpolation in the table, we find that

$$P(F_{8,14} \geq 2.71) = 0.05 = P(F_{14,8} \geq 3.25).$$

The 90% confidence interval for λ is then given by

$$\frac{1}{2.71} (s_1^2/s_2^2) \leq \lambda \leq 3.25(s_1^2/s_2^2) \longleftrightarrow 0.27 \leq \lambda \leq 2.41.$$

Effects of Non-Normality

All of the methods obtained in this Chapter are based on the assumption that the measurements are normally distributed. In applications, the distribution of the measurements will not be exactly normal, but only approximately so. Therefore, it is desirable that the statistical procedures used should not be too seriously affected by small departures from normality. Procedures which are not seriously affected by small changes in the assumptions are called <u>robust</u>. There has been much research into the robustness of estimation procedures and tests of significance in recent years.

The methods which we have discussed for making inferences about means are quite robust, and may safely be used when there are moderate departures from normality. However, the methods for making inferences about variances, and in particular the F-test described in this section, are quite sensitive to departures from normality. The reason for this is that one extreme observation can have a very substantial effect on the sum of squares $\sum(x_i - \bar{x})^2$. The probability of obtaining extreme values, and consequently the distribution of $\sum(X_i - \bar{X})^2$, depends very much on what the tails of the actual distribution are like. A large F-value could be indicative of departures from normality rather than unequal variances. For this reason, it is essential to check the normality assumption by means of quantile plots as in Example 11.5.3.

For an alternate test of variance equality which is less sensitive to departures from normality, see "Non-normality and Tests on Variances" by G.E.P. Box in *Biometrika 40* (1953), pages 318-335.

Problems for Section 14.5

1. Check the assumptions of normality and equal variances in problem 14.4.1.

†2. Find a 90% confidence interval for the ratio of the variances in problem 14.4.3.

3. A common final examination was written by 182 honours mathematics students, of whom 61 were in the co-operative programme. The results were as follows:

Co-op students	n = 61	$\bar{x}$ = 68.30	s_1 = 10.83
Others	m = 121	$\bar{y}$ = 65.93	s_2 = 15.36

Assuming that the examination marks are normally distributed, determine whether the means and variances are significantly different for the two groups.

†4. A standard kilogram weight is weighed four times on each of two sets of scales, with the following results:

First set : 1.003   0.994   1.001   1.002
Second set: 0.990   0.987   1.009   1.004

Assuming that all measurements are independent normal <u>with known</u> <u>mean</u> 1, obtain a 90% confidence interval for the ratio of the two variances.

5. (a) Let $\overline{X}$ be the average of n independent variates having an exponential distribution with mean θ_1, and let $\overline{Y}$ be the average of m independent variates having an exponential distribution with mean θ_2. Show that

$$\overline{X}\theta_2/\overline{Y}\theta_1 \sim F_{(2n,2m)}.$$

(b) The time from the treatment of a particular disease until the recurrence of the disease is known to have an exponential distribution. Two new treatments were administered to a total of 12 patients, and the number of months to recurrence was observed.

Treatment 1: 9   186   25   6   44   115
Treatment 2: 1   18   6   25   14   45

Obtain a 90% confidence interval for θ_1/θ_2, the ratio of the two mean recurrence times. Is there conclusive evidence that one treatment is superior to the other?

14.6 <u>k-sample Problems</u>

Now suppose that we have n independent normally distributed measurements in k samples. Suppose that there are n_1 measurements in the first sample, n_2 in the second sample,..., and n_k in the kth sample, where $n_1 + n_2 + \ldots + n_k = n$. The n_i measurements in the ith sample, which we denote by $Y_{i1}, Y_{i2}, \ldots, Y_{in_i}$, are assumed to have mean μ_i and variance σ_i^2; that is,

$$Y_{i1}, Y_{i2}, \ldots, Y_{in_i} \sim N(\mu_i, \sigma_i^2)$$

independently of one another and the other $k-1$ samples. The ith sample yields sufficient statistics

$$\overline{Y}_i \equiv \frac{1}{n_i} \sum_{j=1}^{n_i} Y_{ij}; \qquad V_i \equiv \sum_{j=1}^{n_i} (Y_{ij} - \overline{Y}_i)^2$$

where $\overline{Y}_i$ and V_i are independent and

$$\overline{Y}_i \sim N(\mu_i, \sigma_i^2/n_i); \qquad V_i/\sigma_i^2 \sim \chi_{(n_i-1)}^2.$$

The variance estimate from the ith sample is $s_i^2 = \frac{1}{n_i - 1} v_i$ with $n_i - 1$ degrees of freedom.

Analysis when the variances are assumed equal

We now assume that the k variances are equal:

$$\sigma_1^2 = \sigma_2^2 = \ldots = \sigma_k^2 = \sigma^2.$$

We shall see that the methods described in Section 14.1 can then be applied in a straightforward way. As we noted at the beginning of Section 14.4, equal-variance models are often used in statistical applications. A similar analysis can be used when the variances are unequal but the variance ratios are known; that is, when $\sigma_i^2 = a_i\sigma^2$ where σ^2 may be unknown but the a_i's are all known.

Let ϕ be any linear function of the μ_i's:

$$\phi = b_1\mu_1 + b_2\mu_2 + \ldots + b_k\mu_k$$

where the b_i's are known constants. Then ϕ is estimated by

$$\hat{\phi} \equiv b_1\bar{Y}_1 + b_2\bar{Y}_2 + \ldots + b_k\bar{Y}_k.$$

Since the $\bar{Y}_i$'s are independent normal variates, (6.6.7) gives

$$\hat{\phi} \sim N(\textstyle\sum b_i\mu_i, \sum b_i^2\sigma^2/n_i);$$

that is, $\hat{\phi} \sim N(\phi, c\sigma^2)$ where $c = \sum b_i^2/n_i$ is a known constant.

To apply the methods of Section 14.1, we also require a statistic V which carries all of the information about σ. This can be found, as in Section 14.4, by writing down the likelihood function as a product of k factors, one for each sample:

$$L(\mu_1, \ldots, \mu_k, \sigma) = \prod_{i=1}^{k} \sigma^{-n_i} \exp\{-[v_i + n_i(\bar{y}_i - \mu_i)^2]/2\sigma^2\}$$

$$= \sigma^{-n} \exp\{-[v + \sum_{i=1}^{k} n_i(\bar{y}_i - \mu_i)^2]/2\sigma^2\}$$

where $V \equiv \sum V_i$. Since the V_i's are independent and $V_i/\sigma^2 \sim \chi^2_{(n_i-1)}$, (6.9.3) implies that V/σ^2 is χ^2 with $\sum(n_i - 1) = n - k$ degrees of freedom:

$$U \equiv \frac{V}{\sigma^2} \sim \chi^2_{(n-k)}.$$

The V_i's are independent of the $\bar{Y}_i$'s, and hence V is independent of $\hat{\phi}$. The pooled variance estimate is

$$s^2 = \frac{v}{n-k} = [\sum_{i=1}^{k} (n_i - 1)s_i^2] \div [\sum_{i=1}^{k} (n_i - 1)].$$

As in Section 14.1, we have

$$T \equiv \frac{\hat{\phi} - \phi}{\sqrt{cs^2}} \sim t_{(n-k)}$$

and this result is used for inferences about ϕ when σ is unknown.

Test for equality of the means

We continue to assume equality of variances, and construct a test for the hypothesis that the k means are equal:

$$H: \mu_1 = \mu_2 = \ldots = \mu_k = \mu.$$

A maximized likelihood ratio (Section 12.5) will be used to suggest an appropriate test statistic.

The likelihood function of $\mu_1, \mu_2, \ldots, \mu_k$ and σ is given above. It is maximized for $\hat{\mu}_i = \bar{y}_i$ and $\hat{\sigma}^2 = v/n$. (Note that $\hat{\sigma}^2$, the MLE obtained from the joint likelihood function, is different from s^2, the MLE based on the distribution of V.) The unrestricted maximum of the likelihood is

$$L(\hat{\mu}_1, \hat{\mu}_2, \ldots, \hat{\mu}_k, \hat{\sigma}) = \hat{\sigma}^{-n} \exp\{- \frac{n}{2}\}.$$

Under $H: \mu_1 = \mu_2 = \ldots = \mu_k = \mu$, the likelihood function becomes

$$\sigma^{-n} \exp\{- [v + \sum n_i (\bar{y}_i - \mu)^2]/2\sigma^2\}$$

and the MLE's are found to be

$$\tilde{\mu} = \frac{\sum n_i \bar{y}_i}{\sum n_i} = \frac{1}{n} \sum\sum y_{ij} = \bar{y};$$

$$\tilde{\sigma}^2 = \frac{1}{n}[v + \sum n_i (\bar{y}_i - \bar{y})^2].$$

The maximum of the likelihood under H is thus

$$L(\tilde{\mu}, \tilde{\mu}, \ldots, \tilde{\mu}, \tilde{\sigma}) = \tilde{\sigma}^{-n} \exp\{- \frac{n}{2}\}.$$

By (12.5.2), the maximized likelihood ratio is

$$R = \frac{L(\tilde{\mu}, \ldots, \tilde{\mu}, \tilde{\sigma})}{L(\hat{\mu}_1, \ldots, \hat{\mu}_k, \hat{\sigma})} = \left[\frac{\tilde{\sigma}}{\hat{\sigma}}\right]^{-n} = \left[1 + \frac{\sum n_i(\bar{y}_i - \bar{y})^2}{V}\right]^{-\frac{n}{2}}.$$

Any strictly decreasing function of R may be selected as the test statistic for a likelihood ratio test. It will be convenient here to choose

$$D \equiv \frac{n-k}{k-1} \cdot \frac{\sum n_i(\bar{Y}_i - \bar{Y})^2}{V} \equiv \frac{Q \div (k-1)}{V \div (n-k)} \qquad (14.6.1)$$

where

$$Q \equiv \sum_{i=1}^{k} n_i(\bar{Y}_i - \bar{Y})^2. \qquad (14.6.2)$$

Since Q is a function of the $\bar{Y}_i$'s and V is a function of the V_i's, the numerator and denominator of D are independent. We shall show that, if H is true, then $Q/\sigma^2 \sim \chi^2_{(k-1)}$. It then follows by (6.9.8) that D has an F-distribution with $k-1$ numerator and $n-k$ denominator degrees of freedom. Since only large values of D (i.e. small values of R) give evidence against the hypothesis, the significance level is

$$SL = P(D \geq d) = P(F_{k-1, n-k} \geq d)$$

where d is the observed value of D. In this case, a one-tail F-test is appropriate rather than a two-tail test as in Section 14.5.

To obtain the distribution of Q, we note that under the hypothesis, the k variates

$$U_i \equiv \sqrt{n_i}(\bar{Y}_i - \mu)/\sigma$$

are independent $N(0,1)$. Define $Z_1 \equiv \sum a_i U_i$ where $a_i = \sqrt{n_i/n}$. Since $\sum a_i^2 = \sum n_i/n = 1$, Theorem 7.3.2 implies that $\sum U_i^2 - Z_1^2$ is distributed as $\chi^2_{(k-1)}$. But

$$\sum U_i^2 \equiv \frac{\sum n_i(\bar{Y}_i - \mu)^2}{\sigma^2} \equiv \frac{\sum n_i(\bar{Y}_i - \bar{Y} + \bar{Y} - \mu)^2}{\sigma^2}$$

$$\equiv \frac{\sum n_i(\bar{Y}_i - \bar{Y})^2}{\sigma^2} + \frac{(\bar{Y} - \mu)^2 \sum n_i}{\sigma^2} + \frac{2(\bar{Y} - \mu)\sum n_i(\bar{Y}_i - \bar{Y})}{\sigma^2}.$$

The last term is zero by the definition of $\bar{Y}$, the second last term is easily seen to be Z_1^2, and the first term is Q/σ^2. Hence it follows that

$$Q/\sigma^2 \equiv \sum U_i^2 - Z_1^2 \sim \chi^2_{(k-1)}.$$

Test for Equality of Variances

A test for the equality of variances can also be constructed from a maximized likelihood ratio. Suppose that $V_1, V_2, \ldots, V_k$ are independent variates with $V_i/\sigma_i^2 \sim \chi^2_{(\nu_i)}$, so that σ_i^2 is estimated by $s_i^2 = v_i/\nu_i$. The likelihood function of $\sigma_1, \sigma_2, \ldots, \sigma_k$ can be written down as a product of k factors like (14.1.7), and its maximum value is

$$L(s_1, s_2, \ldots, s_k) = \prod_{i=1}^{k} s_i^{-\nu_i} e^{-\nu_i/2}.$$

Under $H: \sigma_1 = \sigma_2 = \ldots = \sigma_k$, the MLE of the variance is found to be $s^2 = v/\nu$ where $V \equiv \sum V_i$ and $\nu = \sum \nu_i$, and the maximum value of the likelihood function is

$$L(s, s, \ldots, s) = s^{-\nu} e^{-\nu/2}.$$

The likelihood ratio statistic (12.5.3) is now found to be

$$D \equiv -2 \log R \equiv \nu \log s^2 - \sum \nu_i \log s_i^2. \tag{14.6.3}$$

The exact distribution of D is quite complicated, so we consider the approximation (12.6.1). There are k unknown parameters initially, and only one under the hypothesis, so there will be $k - 1$ degrees of freedom for the test. Hence

$$SL \approx P(\chi^2_{(k-1)} \geq d)$$

where d is the observed value of D.

When the degrees of freedom ν_i are small, this approximation underestimates the significance level. A correction was suggested by M.S. Bartlett (*Journal of the Royal Statistical Society Supplement, vol. 4* (1937), page 137), and the corrected test is known as Bartlett's test for homogeneity of variance. Unfortunately, like the F-test in the preceding section, it is quite sensitive to departures from normality, and a small significance level may be due to one or two extreme observations rather than to a difference in the variances.

Example 14.6.1. The following observations arose as part of a study on

a pulse-jet pavement breaker using nozzles of three different diame-
ters. The measurement is the penetration (in millimeters) of a con-
crete slab produced by a single discharge.

Nozzle	Penetration									$\bar{y}_i$	s_i^2
Small	67	47.5	46	62.5	49	53.5	42	55.5	39	51.33	85.00
Medium	88	60	72	73.5	62	72.5	73.5	44	54.5	66.67	167.38
Large	83	53	87	71	78	51.5	68	58	61	67.83	167.63

There are $n_i = 9$ observations in each of the $k = 3$ samples, so
$n = 27$ and each variance estimate s_i^2 has $n_i - 1 = 8$ degrees of
freedom.   The overall mean is

$$\bar{y} = \sum n_i \bar{y}_i / \sum n_i = 61.94$$

and the pooled variance estimate is

$$s^2 = \frac{v}{n-k} = \frac{\sum (n_i - 1) s_i^2}{\sum (n_i - 1)} = 140.00.$$

First consider a test of $H: \mu_1 = \mu_2 = \mu_3$ with the variances assumed
equal.   We find that

$$Q = \sum n_i (\bar{y}_i - \bar{y})^2 = 1356.60$$

and hence the observed value of (14.6.1) is

$$d = \frac{1356.60 \div 2}{140.00} = 4.84.$$

From Table B5, the significance level is

$$SL = P\{F_{2,24} \geq 4.84\} \approx 0.04.$$

There is some evidence of a difference in the means.   The small nozzle
gives a lower mean penetration than the other two.
    To check the equal-variance assumption, we compute (14.6.3)
where in this case $v = 24$ and $v_1 = v_2 = v_3 = 8$. We find that $d = 1.12$,
and

$$SL \approx P(\chi^2_{(2)} \geq 1.12) > 0.5.$$

There is no evidence of heterogeneity in the variances. (A small significance level here could be due either to unequal variances or to
departures from normality.)

Problems for Section 14.6

†1. Several chemical analyses of samples of a product were performed
on each of four successive days, and the following table gives the
percentage impurity found in each analysis.

Day 1: 2.6 2.6 2.9 2.0 2.1 2.1 2.0
Day 2: 3.1 2.9 3.1 2.5
Day 3: 2.6 2.2 2.5 2.2 1.2 1.2 1.8
Day 4: 2.5 2.4 3.0 1.5 1.7

(a) Assuming equal variances, test whether there is a difference
in the mean percentage impurity over the four days.

(b) Check the equal-variance assumption.

2. Three laboratories each carried out five independent determinations
of the nicotine content of a brand of cigarettes. Their findings,
in milligrams per cigarette, were as follows:

Laboratory A: 16.3   15.6   15.5   16.7   16.2
Laboratory B: 13.5   17.4   16.9   18.2   15.6
Laboratory C: 14.1   13.2   14.3   12.9   12.8

Are there real differences among the results produced by the three
laboratories?

3. Three series of measurements on g, the acceleration due to
gravity, were reported by Preston-Thomas et al., *Canadian Journal of
Physics 38* (1960), page 845. Given a measurement y, define
$x = (y - 980.61) \times 10^4$. The 85 observed x-values were as follows:

August 1957				August 1958				December 1959			
-11	22	23	89	32	31	54	42	22	30	30	24
41	31	29	62	18	39	33	59	30	24	32	24
60	31	07	38	48	46	44	51	22	16	23	06
40	56	14	57	43	47	40	41	24	28	18	23
44	40	23		45	36	36	42	27	27	32	35
				39	39	48	32	30	23	22	18
				35	31	36	25	17	23	22	21
				30	24	39	32	21	17	30	29

Examine these data with respect to normality, equality of variances,
and equality of means. If it is appropriate to do so, give a 95%
confidence interval for g on the basis of all 83 measurements.
If it is not, give three separate confidence intervals.

Review Problems for Chapter 14

1. Two experiments were carried out to determine μ, the mean increase in blood pressure due to a certain drug. Six different
subjects were used, three in each experiment, and the following

increases were observed:

> Experiment 1:   4.5   5.6   4.9
> Experiment 2: -1.2   9.8   21.4

Indicate, with reasons, which experiment produces stronger evidence that the drug does have an effect on blood pressures. Which experiment points to the greater effect?

†2. Fourteen men were used in an experiment to determine which of two drugs produces a greater increase in blood pressure. Drug 1 was given to seven of the men chosen at random, and drug 2 was given to the remaining seven. The observed increases in blood pressure are:

> Drug 1: 0.7 -0.2 3.4 3.7 0.8 0.0 2.0
> Drug 2: 1.9   1.1 4.4 5.5 1.6 4.6 3.4

(a) Are these data consistent with the hypothesis of equal variances in blood pressure for the two drugs?

(b) Assuming the variances to be equal, obtain a 95% confidence interval for the difference in mean blood pressure increase $\mu_2 - \mu_1$, and for the common variance σ^2.

(c) What changes would you make in the conduct of the experiment and the analysis if the fourteen men consisted of seven sets of identical twins?

3. The following are yields (in pounds) of 16 tomato plants grown on 8 separate uniform plots of land. One plant in each plot was treated with fertilizer A and the other with fertilizer B.

Plot	1	2	3	4	5	6	7	8
Fertilizer A	4.0	5.7	4.0	6.9	5.5	4.6	6.5	8.4
Fertilizer B	4.8	5.5	4.4	4.8	5.9	4.2	4.4	6.3

Are these data consistent with the hypothesis that the two fertilizers are equally effective in increasing yield?

4. The spectrochemical analysis of 11 samples of nickel showed the following percentage impurities:

1.94 1.99 1.98 2.03 2.03 1.96 1.95 1.96 1.96 1.92 2.00

The following results were obtained by routine chemical analysis of 10 additional samples:

1.99 1.98 1.94 2.06 2.02 1.97 2.02 2.01 2.00 1.98

Show that the means and variances of the two sets of measurements are not significantly different, and use both sets of data to obtain a 95% confidence interval for the mean percentage impurity.

†5. Ten steel ingots chosen at random from a large shipment gave the

following hardness measures:

71.7 71.1 68.0 69.6 69.1 69.4 68.8 70.4 69.3 68.2

(a) Are these observations consistent with the manufacturer's claim that the mean hardness is 70?

(b) Obtain a 95% confidence interval for the variance of hardness.

(c) Approximately how many ingots would it be necessary to measure in order that the 95% confidence interval for the mean would have length 0.4 units?

6. An experiment was performed to compare two different methods of measuring the phosphate content of material. Ten samples were chosen so that the material within a sample was relatively homogeneous. Each sample was then divided in half, one half being analysed by method A and the other half by method B.

Sample	1	2	3	4	5	6	7	8	9	10
Method A	55.6	62.4	48.9	45.5	75.4	89.6	38.4	96.8	92.5	98.7
Method B	58.4	66.3	51.2	46.1	74.3	92.5	40.2	97.3	94.8	99.0

Find a 95% confidence interval for the mean difference in phosphate content as measured by the two methods, and state the assumptions upon which your analysis depends.

†7. In a progeny trial, the clean fleece weights of 9 ewe lambs from each of four sires were as follows:

Sire 1: 2.74 3.50 3.22 2.98 2.97 3.47 3.47 3.68 4.22
Sire 2: 3.88 3.36 4.29 4.08 3.90 4.71 4.25 3.41 3.84
Sire 3: 3.28 3.92 3.66 3.47 2.94 3.26 3.57 2.62 3.76
Sire 4: 3.52 3.54 4.13 3.29 3.26 3.04 3.77 2.88 2.90

Test the hypothesis that the mean fleece weight does not depend upon the sire.

CHAPTER 15. FITTING A STRAIGHT LINE

In this chapter we consider the problem of fitting a straight line to n observed points (x_i, y_i), $i = 1, 2, \ldots, n$. This problem can arise in many different contexts, and the appropriate analysis will depend upon the situation. We suppose that the aim is to use the value of one variable x (the independent variable) to explain or predict the value of the second variable y (the dependent variable). The values of x are treated as known constants in the analysis. The y_i's are assumed to be independent values from a normal distribution whose mean is a linear function of x. A more detailed description of the model and various important generalizations of it are given in Section 1.

In Section 2, we consider the estimation of parameters in the straight line model by the method of maximum likelihood. Tests of significance and confidence intervals are derived in Section 3. Section 4 discusses methods for checking the adequacy of the straight line model.

15.1 Linear Models

Suppose that n measurements $Y_1, Y_2, \ldots, Y_n$ are to be made on the same variable under various different conditions. For instance, one might record the gasoline mileage achieved by a car for several driving speeds, weather conditions, brands of gasoline, etc. We wish to formulate a model which describes or explains the way in which the variable y depends upon these conditions.

The special case in which measurements are taken under just two different conditions was considered in Sections 3,4 and 5 of the preceding chapter. The comparison of k samples was considered in Section 14.6. Another special case is considered in this chapter. We now suppose that each measurement Y_i has associated with it a value x_i of another variable, and that the x-values (e.g. driving speeds) can help to explain or predict the corresponding Y-values (gasoline mileages). x is called the independent variable or explanatory variable, and Y is called the dependent variable or response variable. A straight line model relating Y to x is developed in the example below, and its analysis is discussed in the remaining sections of this chapter.

Both the straight line model considered here, and the models considered in the last chapter, are examples of normal linear models.

These are defined following the example. Linear models, both normal and non-normal, are very widely used in modeling the dependence of a measurement Y on various conditions. They are quite flexible, yet relatively easy to deal with mathematically and computationally.

Example 15.1.1. In Examples 14.2.2 and 14.4.3 we considered lifetime data for plastic gears tested at 21°C and at 30°C. In the same experiment, gears were tested at seven other temperatures as well. The complete data set is given in Table 15.1.1, and the 40 log-lifetimes are plotted against temperature in Figure 15.1.1.

Note that there is considerable variability in the log-lifetimes Y of gears tested at the same temperature x. We assume that the log-lifetimes of gears tested at temperature x are normally distributed with mean μ and variance σ^2, where μ and σ may depend upon x. The analysis given in Example 14.2.2 and 14.4.3 depended on this assumption.

Table 15.1.1

Log-lifetimes of plastic gears at nine operating temperatures

Temperature x (C)	Number tested	y = natural logarithm of lifetime (in millions of cycles)			
-16	4	1.690	1.779	1.692	1.857
0	4	1.643	1.584	1.585	1.462
10	4	1.153	0.991	1.204	1.029
21	8	0.863	0.698	0.904	0.788
		0.626	0.842	0.693	1.051
30	4	0.364	0.695	0.558	0.359
37	4	0.412	0.425	0.574	0.649
47	4	0.116	0.501	0.296	0.099
57	4	-0.355	-0.269	-0.354	-0.459
67	4	-0.736	-0.343	-0.965	-0.705

It appears from Figure 15.1.1 that the amount of variability or spread in the Y-values is about the same at all temperatures. There is no indication that the variance of Y is changing with the temperature in any systematic way. We shall assume that σ is constant (not a function of x) over the range of temperatures being considered. We assumed that the variance of Y was the same at 21°C as at 30°C in the analysis of Example 14.4.3. See the discussion at the beginning of Section 14.4.

The diagram clearly shows that the mean lifetime μ depends upon the temperature. In fact, the dependence of μ on x appears to be roughly linear over the range of temperatures considered. At least as a first approximation, it is reasonable to assume that $\mu = \alpha + \beta x$,

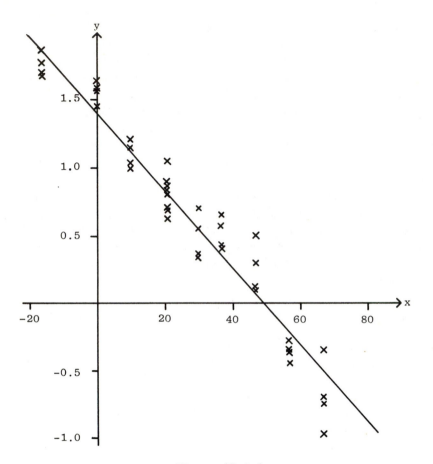

Figure 15.1.1

Scatterplot of Log-lifetimes (y) versus Temperature (x)

where α and β are constants not depending upon x.

The probability model which we propose for the experiment is therefore as follows. The log-lifetime Y of a gear tested at temperature x is normally distributed with mean $\alpha + \beta x$ and variance σ^2,

$$Y \sim N(\alpha + \beta x, \sigma^2).$$
(15.1.1)

The three unknown parameters α, β, and σ are assumed to be constants not depending on x. The x-values are treated as known constants, and the Y-values as independent observations from (15.1.1).

It is sometimes useful to write (15.1.1) in a slightly different form. The <u>error variable</u> ϵ is defined by

$$\epsilon \equiv Y - \alpha - \beta x,$$

and it follows from (6.6.6) that $\epsilon \sim N(0, \sigma^2)$. We may thus rewrite (15.1.1) in the form

$$Y \equiv \alpha + \beta x + \epsilon \quad \text{where} \quad \epsilon \sim N(0, \sigma^2). \qquad (15.1.2)$$

The errors corresponding to different Y-values are assumed to be independent.

The 9 samples of log-lifetimes could also be analysed as a k-sample problem with $k = 9$; see Section 14.6. This analysis would make no use of the x-values (temperatures) associated with the nine samples. By comparing this analysis with one based on the straight line model (15.1.1), one can obtain a test of significance for the adequacy of the straight line model. See Section 15.4 for details.

The normal linear model

The normal linear model is defined by the following two assumptions:

(i) $Y_1, Y_2, \ldots, Y_n$ are independent normal variates with the same variance σ^2;

(ii) the expected values of the Y_i's are known linear functions of a set of unknown parameters $\beta_1, \beta_2, \ldots, \beta_r$; that is,

$$\mu_i = E(Y_i) = a_{i1}\beta_1 + a_{i2}\beta_2 + \ldots + a_{ir}\beta_r \quad \text{for} \quad i = 1, 2, \ldots, n$$

where the a_{ij}'s are known constants.

The a_{ij}'s are selected to describe the conditions under which the measurements are taken. They may be quantitative measurements (e.g. driving speeds or temperatures) or they may be indicator variables (e.g. $a_{i1} = 1$ if Brand A was used for the ith measurement, and $a_{i1} = 0$ if some other brand was used).

A straight line or polynomial model relating E(Y) to some other variable x is a special case of the linear model. For instance, for the straight line model (15.1.1) we have

$$\mu_i = E(Y_i) = \alpha + \beta x_i = a_{i1}\alpha + a_{i2}\beta$$

where $a_{i1} = 1$ and $a_{i2} = x_i$. Similarly, for a second degree polynomial model we have

$$\mu_i = a_{i1}\alpha + a_{i2}\beta + a_{i3}\gamma \quad \text{for} \quad i = 1, 2, \ldots, n$$

where $a_{i1} = 1$, $a_{i2} = x_i$, and $a_{i3} = x_i^2$. Note that it is not necessary for μ_i to be a linear function of x_i. What we require is that μ_i be a linear function of the unknown parameters α, β, and γ.

The two-sample problem with equal variances (Section 14.4) can also be regarded as a special case of the normal linear model. Here measurements are taken under only two different conditions, under treatment A and under treatment B, say. Define $a_{i1} = 1$ if the ith measurement is made with treatment A, and $a_{i1} = 0$ if it is made with treatment B. Define $a_{2i} = 1 - a_{1i}$, and consider the linear model

$$\mu_i = E(Y_i) = a_{i1}\alpha + a_{i2}\beta \quad \text{for} \quad i = 1, 2, \ldots, n.$$

This states merely that measurements with treatment A have mean α, and measurements with treatment B have mean β.

For historical reasons, linear models in which the a_{ij}'s are quantitative measurements are also called (linear) regression models. The straight line model (15.1.1) is called a simple linear regression model, and models involving several explanatory variables are called multiple regression models. The origin of the term "regression" is explained in Section 7.5.

A satisfactory general treatment of normal linear models depends upon results from linear algebra and properties of the multivariate normal distribution. For this reason, the discussion in the following sections is restricted to the straight line model. However, we shall make a few general comments. The MLE's $\hat{\beta}_1, \hat{\beta}_2, \ldots, \hat{\beta}_r$ in the normal linear model are obtained by solving a set of r linear equations. Each estimate $\hat{\beta}_j$ is a linear function of the Y_i's and hence is distributed as $N(\beta_j, c_j\sigma^2)$ where c_j is a known constant depending upon the a_{ij}'s. A similar result holds for any linear function of the $\hat{\beta}_j$'s. The ith residual is defined to be the difference between the measurement y_i and its estimated expected value under the model:

$$\hat{\epsilon}_i = y_i - a_{i1}\hat{\beta}_1 - a_{i2}\hat{\beta}_2 - \ldots - a_{ir}\hat{\beta}_r.$$

It can be shown that the residual sum of squares $\sum \hat{\epsilon}_i^2$ carries all of the information about σ. Furthermore, $\sum \hat{\epsilon}_i^2 / \sigma^2 \sim \chi^2_{(n-r)}$, independently of $\hat{\beta}_1, \hat{\beta}_2, \ldots, \hat{\beta}_p$. Hence the **general** setup is the same as that described in Section 14.1, and the statistical procedures outlined there will apply to the normal linear model. These same procedures will apply if we relax the assumption of equal variances, provided that all of the variance ratios are known.

Non-normal linear models

Linear models can also be formulated for distributions other than the normal. For instance, for binomial data, we can express the success probability p_i, or some suitable function of p_i, as a known linear function of unknown parameters $\beta_1, \beta_2, \ldots, \beta_r$:

$$g(p_i) = a_{i1}\beta_1 + a_{i2}\beta_2 + \ldots + a_{ir}\beta_r \quad \text{for} \quad i = 1, 2, \ldots, n$$

where the a_{ij}'s are known constants. The model developed in Section 10.3 to relate the probability of death to the dose of a drug is of this type, with $g(p)$ being the logistic transform of p:

$$g(p) = \log \frac{p}{1-p} .$$

For Poisson data, the logarithm of the mean is usually taken to be linear in the unknown parameters:

$$\log \mu_i = a_{i1}\beta_1 + a_{i2}\beta_2 + \ldots + a_{ir}\beta_r \quad \text{for} \quad i = 1, 2, \ldots, n.$$

Iterative procedures, such as the Newton-Raphson method, will usually be needed to obtain the MLE's in non-normal linear models, but the calculations do not usually present any real difficulties. However, exact significance tests may be unavailable or else too cumbersome to use. Tests and confidence intervals are usually based on large-sample results, and approximate likelihood ratio tests (Section 12.6) are particularly useful.

15.2 Parameter Estimates

Suppose that we have observed n pairs of measurements (x_i, y_i) for $i = 1, 2, \ldots, n$. The x_i's are treated as known constants, and the y_i's are assumed to be independent observations from the straight line model (15.1.1):

$$Y_i \sim N(\alpha + \beta x_i, \sigma^2) \quad \text{for} \quad i = 1, 2, \ldots, n.$$

The probability density function of Y_i is

$$f_i(y_i) = \frac{1}{\sqrt{2\pi}\, \sigma} \exp\{-(y_i - \alpha - \beta x_i)^2 / 2\sigma^2\}.$$

Since the Y_i's are assumed to be independent, their joint p.d.f. is

$$f(y_1, \ldots, y_n) = \prod_{i=1}^{n} f_i(y_i) = (\frac{1}{\sqrt{2\pi}\, \sigma})^n \exp\{-\sum (y_i - \alpha - \beta x_i)^2 / 2\sigma^2\},$$

and the log likelihood function of α, β and σ is

$$\ell(\alpha,\beta,\sigma) = -n \log \sigma - \frac{1}{2} \sigma^{-2} \sum (y_i - \alpha - \beta x_i)^2 = -n \log \sigma - \frac{1}{2} \sigma^{-2} \sum \epsilon_i^2$$

where $\epsilon_i = y_i - \alpha - \beta x_i$ is the ith error.

The MLE's are chosen to maximize $\ell(\alpha,\beta,\sigma)$. For any fixed σ, maximizing $\ell(\alpha,\beta,\sigma)$ with respect to α and β is equivalent to minimizing the sum of squared errors $\sum \epsilon_i^2$. The maximum likelihood estimates of α and β are thus the same as the least squares estimates. The same is true of the general normal linear model as described in the last section: the MLE's of $\beta_1, \beta_2, \ldots, \beta_r$ are obtained by minimizing $\sum \epsilon_i^2$, and hence are the same as the least squares estimates. This agreement is a result of the assumptions of constant variance and normality. If we permit unequal variances or assume other than a normal distribution, the maximum likelihood estimates will generally differ from the least squares estimates.

To find $\hat{\alpha}$ and $\hat{\beta}$, we find the derivatives of $\sum \epsilon_i^2$ with respect to α and β:

$$\frac{\partial}{\partial \alpha} \sum \epsilon_i^2 = 2 \sum \epsilon_i \frac{\partial \epsilon_i}{\partial \alpha} = 2 \sum \epsilon_i (-1) = -2 \sum \epsilon_i ;$$

$$\frac{\partial}{\partial \beta} \sum \epsilon_i^2 = 2 \sum \epsilon_i \frac{\partial \epsilon_i}{\partial \beta} = 2 \sum \epsilon_i (-x_i) = -2 \sum x_i \epsilon_i .$$

Hence the sum of squared errors is minimized (and the log-likelihood is maximized) for

$$\sum \hat{\epsilon}_i = 0 \quad \text{and} \quad \sum x_i \hat{\epsilon}_i = 0 \qquad (15.2.1)$$

where the ith residual $\hat{\epsilon}_i$ is defined by

$$\hat{\epsilon}_i = y_i - \hat{\alpha} - \hat{\beta} x_i ; \quad i = 1, 2, \ldots, n. \qquad (15.2.2)$$

The ith residual is the difference between the observation y_i and its estimated mean $\hat{\alpha} + \hat{\beta} x_i$. The residuals contain information about the fit of the model to the data. Examination of the residuals may reveal defects in the model, and may indicate necessary modifications; see Section 4.

The residual sum of squares is defined to be the sum of squares of the residuals, $\sum \hat{\epsilon}_i^2$. By (15.2.1), the n residuals $\hat{\epsilon}_1, \hat{\epsilon}_2, \ldots, \hat{\epsilon}_n$ satisfy two linear restrictions, and hence, they have $n - 2$ degrees of freedom. In fact, we shall show that $\sum \hat{\epsilon}_i^2 / \sigma^2$ has a χ^2 distribution with $n - 2$ degrees of freedom. By (14.1.3), the variance estimate based on this distribution is

$$s^2 = \frac{1}{n-2} \sum \hat{\epsilon}_i^2. \qquad (15.2.3)$$

From (15.2.1) and (15.2.2) we have

$$0 = \sum \hat{\epsilon}_i = \sum y_i - n\hat{\alpha} - \hat{\beta} \sum x_i.$$

Dividing by n and rearranging gives

$$\hat{\alpha} = \bar{y} - \hat{\beta}\bar{x}. \qquad (15.2.4)$$

Substitution for $\hat{\alpha}$ in (15.2.2) now gives

$$\hat{\epsilon}_i = y_i - \bar{y} - \hat{\beta}(x_i - \bar{x}) \qquad (15.2.5)$$

and the second equation of (15.2.1) becomes

$$\sum x_i [y_i - \bar{y} - \hat{\beta}(x_i - \bar{x})] = 0.$$

Rearranging and solving gives

$$\hat{\beta} = \frac{\sum x_i(y_i - \bar{y})}{\sum x_i(x_i - \bar{x})} = \frac{S_{xy}}{S_{xx}}. \qquad (15.2.6)$$

The expression in the numerator is called the corrected sum of products. Since $\sum (y_i - \bar{y}) = \sum (x_i - \bar{x}) = 0$, we have

$$S_{xy} = \sum x_i(y_i - \bar{y}) = \sum (x_i - \bar{x})(y_i - \bar{y}) = \sum (x_i - \bar{x})y_i.$$

Similarly, the expression in the numerator is called the corrected sum of squares:

$$S_{xx} = \sum x_i(x_i - \bar{x}) = \sum (x_i - \bar{x})^2.$$

The following formulae are convenient for hand calculation:

$$S_{xx} = \sum x_i^2 - \frac{1}{n}(\sum x_i)^2; \quad S_{yy} = \sum y_i^2 - \frac{1}{n}(\sum y_i)^2;$$

$$S_{xy} = \sum x_i y_i - \frac{1}{n}(\sum x_i)(\sum y_i);$$

$$\sum \hat{\epsilon}_i^2 = S_{yy} - \hat{\beta} S_{xy}. \qquad (15.2.7)$$

The latter result may be proved using (15.2.5). These formulae are susceptible to round off errors, and should not be used for machine calculation.

Example 15.2.1. The following table gives the age (x) and systolic blood pressure (y) for each of n = 12 women:

x	56	42	72	36	63	47	55	49	38	42	68	60
y	147	125	160	118	149	128	150	145	115	140	152	155

We assume that blood pressure Y is normally distributed, with mean $\alpha + \beta x$ and constant variance σ^2. To obtain estimates of α, β, and σ^2, we first compute the following sums:

$$\sum x_i = 628 \qquad \sum x_i^2 = 34416$$
$$\sum y_i = 1684 \qquad \sum y_i^2 = 238822 \qquad \sum x_i y_i = 89894$$

From these we obtain the sample means and corrected sums:

$$\overline{x} = 52.33 \qquad \overline{y} = 140.33$$

$$S_{xx} = 1550.67 \qquad S_{yy} = 2500.67 \qquad S_{xy} = 1764.67$$

Now (15.2.6), (15.2.4), and (15.2.7) and (15.2.3) give

$$\hat{\beta} = S_{xy}/S_{xx} = 1.138; \qquad \hat{\alpha} = \overline{y} - \hat{\beta}\overline{x} = 80.78$$
$$\sum \hat{\epsilon}_i^2 = S_{yy} - \hat{\beta} S_{xy} = 492.47; \qquad s^2 = \frac{1}{n-2} \sum \hat{\epsilon}_i^2 = 49.247.$$

The data are plotted in Figure 15.2.1, and the fitted line

$$y = 80.78 + 1.138x$$

is drawn in for comparison. The straight line seems to give a reasonably good fit to the data.

Example 15.2.2. Consider the lifetime data for plastic gears in Table 15.1.1. We have n = 40 pairs of measurements (x_i, y_i), where x_i is the temperature and y_i is the log-lifetime. Note that we have repeated x-values:

$$x_1 = x_2 = x_3 = x_4 = -16,$$
$$x_5 = x_6 = x_7 = x_8 = \quad 0,$$

and so on. To fit the normal linear model (15.1.1) to these data, we first compute the following sums:

$$\sum x_i = 4(-16) + 4(0) + 4(10) + 8(21) + \ldots = 1096$$
$$\sum x_i^2 = 4(-16)^2 + 4(0)^2 + 4(10)^2 + 8(21)^2 + \ldots = 53816$$

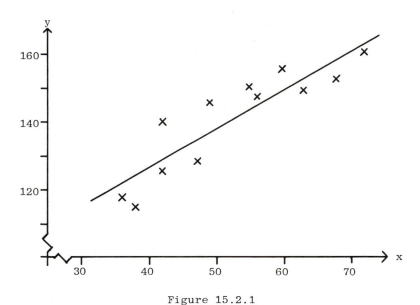

Figure 15.2.1

Scatterplot of Systolic Blood Pressure (y) versus
Age (x), showing the Fitted Regression Line

$$\sum y_i = 24.996 \qquad \sum y_i^2 = 37.506862 \qquad \sum x_i y_i = -15.781$$

We can now compute means, corrected sums of squares, and estimates as
in the preceding example. The fitted line is

$$y = 1.432 - 0.02946x$$

and the residual sum of squares is 1.24663 with 38 degrees of freedom, giving the variance estimate $s^2 = 0.03281$.

15.3 Tests of Significance and Confidence Intervals

As in the preceding section, we assume that $Y_1, Y_2, \ldots, Y_n$
are independent, and that

$$Y_i \sim N(\alpha + \beta x_i, \sigma^2) \quad \text{for} \quad i = 1, 2, \ldots, n$$

where all the x_i's are known constants. Significance tests and confidence intervals will be based upon the following theorem which is proved
at the end of this section.

Theorem 15.3.1. Under the above assumptions, the statistics $\overline{Y}, \hat{\beta}$, and

$\sum \hat{\epsilon}_i^2$ are jointly sufficient for the parameters α, β, and σ. These three statistics are distributed independently of one another as follows:

$$\bar{Y} \sim N(\alpha + \beta\bar{x}, \sigma^2/n), \quad \hat{\beta} \sim N(\beta, \sigma^2/S_{xx}), \quad \sum \hat{\epsilon}_i^2/\sigma^2 \sim \chi^2_{(n-2)}.$$

Inferences about σ

It can be argued that $\sum \hat{\epsilon}_i^2$ carries all of the information about σ when α and β are unknown. Inferences about σ will thus be based on the fact that

$$U \equiv \sum \hat{\epsilon}_i^2/\sigma^2 \sim \chi^2_{(n-2)}. \qquad (15.3.1)$$

By (14.1.3), the MLE of σ^2 based on the distribution of $\sum \hat{\epsilon}_i^2$ is

$$s^2 = \frac{1}{n-2} \sum \hat{\epsilon}_i^2.$$

Inferences about β

Since $\hat{\beta} \sim N(\beta, \sigma^2/S_{xx})$, it follows that

$$Z \equiv \frac{\hat{\beta} - \beta}{\sqrt{\sigma^2/S_{xx}}} \sim N(0,1).$$

If σ^2 is unknown, we replace it by the estimate derived from (15.3.1) to obtain

$$T \equiv \frac{\hat{\beta} - \beta}{\sqrt{s^2/S_{xx}}} \sim t_{(n-2)}. \qquad (15.3.2)$$

Significance tests and confidence intervals for β when σ is unknown are based on this result.

For instance, consider a test of the hypothesis $\beta = 0$ in Example 15.2.1. If $\beta = 0$, the observed T-value is

$$t_{obs} = \frac{\hat{\beta}}{\sqrt{s^2/S_{xx}}} = \frac{1.138}{\sqrt{49.247/1550.67}} = 6.39.$$

Since $n = 12$, there are 10 degrees of freedom, and the two-tail significance level is

$$SL \approx P\{|t_{(10)}| \geq 6.39\} < 0.001$$

from Table B3. There is strong evidence that $\beta \neq 0$. Since $P\{|t_{(10)}| \geq 2.228\} = 0.05$, the 95% confidence interval for the slope is

$$\beta \in \hat{\beta} \pm 2.228\sqrt{s^2/S_{xx}} = 1.138 \pm 0.397.$$

Note that we will be able to determine β quite precisely (i.e. the confidence interval will be quite narrow) if S_{xx} is large. Thus, if we are planning an experiment to obtain information about β, we should select $x_1, x_2, \ldots, x_n$ so that $S_{xx} = \sum(x_i - \bar{x})^2$ is large. To maximize the information, we would need to take half of the x_i's as large as possible, and the other half as small as possible. However if we did this, we would be unable to check the assumption that the dependence of $E(Y)$ on x is linear. As a result, one would usually compromise by taking observations over the whole range of x-values, but with more observations at the extremes than in the middle of the range. It would then be possible to check the fit of the model, and also to make fairly precise statements about β.

Inferences about E(Y)

Given a particular value for x, the expected value of Y is $\mu = E(Y) = \alpha + \beta x$. This is estimated by

$$\hat{\mu} \equiv \hat{\alpha} + \hat{\beta}x \equiv \bar{Y} + (x - \bar{x})\hat{\beta}$$

by (15.2.4). Since $\bar{Y}$ and $\hat{\beta}$ are independent normal variates (Theorem 15.3.1), it follows by (6.6.7) that $\hat{\mu}$ has a normal distribution with mean and variance

$$E(\hat{\mu}) = E(\bar{Y}) + (x - \bar{x})E(\hat{\beta});$$
$$\text{var}(\hat{\mu}) = \text{var}(\bar{Y}) + (x - \bar{x})^2 \text{var}(\hat{\beta}).$$

Upon substituting from Theorem 15.3.1, we obtain

$$E(\hat{\mu}) = (\alpha + \beta\bar{x}) + (x - \bar{x})\beta = \alpha + \beta x = \mu;$$
$$\text{var}(\hat{\mu}) = \sigma^2[\frac{1}{n} + (x - \bar{x})^2/S_{xx}].$$

It follows that

$$Z \equiv \frac{\hat{\mu} - \mu}{\sqrt{\sigma^2[\frac{1}{n} + (x - \bar{x})^2/S_{xx}]}} \sim N(0,1).$$

If σ^2 is unknown, we replace it by the estimate derived from (15.3.1)

266

to obtain

$$T \equiv \frac{\hat{\mu} - \mu}{\sqrt{s^2 [\frac{1}{n} + (x - \bar{x})^2/S_{xx}]}} \sim t_{(n-2)}. \qquad (15.3.3)$$

The 95% confidence interval for μ is then

$$\mu \in \hat{\mu} \pm a\sqrt{s^2 [\frac{1}{n} + (x - \bar{x})^2/S_{xx}]}$$

where $P\{|t_{(n-2)}| \geq a\} = 0.05$.

Note that the width of the 95% confidence interval for μ increases as x gets further away from $\bar{x}$. We can estimate μ less and less precisely as x moves away from the centre of the sample used to fit the model. For instance, let us take $x = 50$ in Example 15.2.1, so that $\mu = \alpha + 50\beta$ is the mean blood pressure of women aged 50. Then $\hat{\mu} = \hat{\alpha} + 50\hat{\beta} = 137.68$, and $a = 2.228$ from tables of $t_{(10)}$. The 95% confidence interval for μ is given by

$$\mu \in 137.68 \pm 2.228\sqrt{49.247[\frac{1}{12} + (50 - 52.33)^2/1550.67]}$$

$$= 137.68 \pm 4.61.$$

For $x = 70$, we obtain $\hat{\mu} = 160.44$, and the 95% confidence interval is

$$\mu \in 160.44 \pm 2.228\sqrt{49.247[\frac{1}{12} + (70 - 52.33)^2/1550.67]}$$

$$= 160.44 \pm 8.34.$$

The 95% confidence interval is much narrower for $x = 50$ than for $x = 70$, because the former value is much closer to the average age in the sample - see Figure 15.3.1. The above confidence intervals for μ are computed under the assumption that μ is a linear function of x. They may be quite misleading if the actual dependence is non-linear. For instance, the confidence interval for mean blood pressure at age 30 in Figure 15.3.1 would be seriously in error if μ were a quadratic function of x. It is always dangerous to extrapolate beyond the range of x-values observed in the sample. Even if the model fits the data very well, there is no guarantee that it will apply over a wider range of x-values.

Inferences about α

The intercept α is the expected value of Y when $x = 0$.

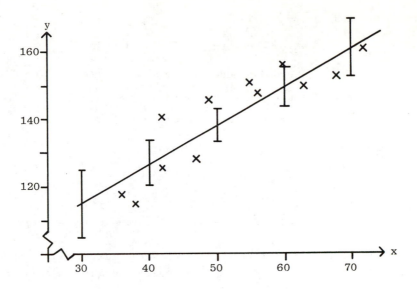

Figure 15.3.1

Scatterplot of Systolic Blood Pressure (y) versus
Age (x), showing 95% Confidence Intervals
for Mean Blood Pressure at Various Ages

Hence we need merely put $x = 0$ in the preceding analysis to get

$$T \equiv \frac{\hat{\alpha} - \alpha}{\sqrt{s^2 [\frac{1}{n} + \bar{x}^2/S_{xx}]}} \sim t_{(n-2)}. \tag{15.3.4}$$

Significance tests and confidence intervals for α can now be obtained. Note, however, that α would not be a parameter of interest in Example 15.2.1, since there is no reason to suppose that the straight line model holds in the neighbourhood of $x = 0$.

Proof of Theorem 15.3.1.

We now give a proof of the results stated in Theorem 15.3.1. We begin by proving an identity.

Lemma 15.3.1. $\sum \epsilon_i^2 = \sum \hat{\epsilon}_i^2 + n(\bar{y} - \alpha - \beta\bar{x})^2 + (\hat{\beta} - \beta)^2 S_{xx}.$

Proof. Upon solving (15.2.5) for y_i and substituting, we obtain

$$\epsilon_i = y_i - \alpha - \beta x_i = \hat{\epsilon}_i + \bar{y} + \hat{\beta}(x_i - \bar{x}) - \alpha - \beta x_i$$
$$= \hat{\epsilon}_i + (\bar{y} - \alpha - \beta\bar{x}) + (\hat{\beta} - \beta)(x_i - \bar{x}).$$

We next square and sum to obtain

$$\sum \epsilon_i^2 = \sum \hat{\epsilon}_i^2 + \sum(\overline{y} - \alpha - \beta\overline{x})^2 + \sum(\hat{\beta} - \beta)^2(x_i - \overline{x})^2$$

$$+ \text{ cross-product terms}$$

$$= \sum \hat{\epsilon}_i^2 + n(\overline{y} - \alpha - \beta\overline{x})^2 + (\hat{\beta} - \beta)^2 S_{xx}$$

$$+ \text{ cross-product terms.}$$

The three cross-product terms are as follows:

$$2\sum\hat{\epsilon}_i(\overline{y} - \alpha - \beta\overline{x}) = 2(\overline{y} - \alpha - \beta\overline{x})\sum\hat{\epsilon}_i$$

$$2\sum\hat{\epsilon}_i(\hat{\beta} - \beta)(x_i - \overline{x}) = 2(\hat{\beta} - \beta)(\sum x_i \hat{\epsilon}_i - \overline{x}\sum\hat{\epsilon}_i)$$

$$2\sum(\overline{y} - \alpha - \beta\overline{x})(\hat{\beta} - \beta)(x_i - \overline{x}) = 2(\overline{y} - \alpha - \beta\overline{x})(\hat{\beta} - \beta)(\sum x_i - n\overline{x}).$$

The first two cross-product terms are zero by the maximum likelihood equations (15.2.1). The third term is also zero because $\overline{x} = \sum x_i/n$, and the identity follows. $\square$

From Section 15.2, the log-likelihood function of α, β, σ is

$$\ell(\alpha, \beta, \sigma) = -n \log \sigma - \frac{1}{2} \sigma^{-2}\sum \epsilon_i^2.$$

By the above lemma, $\sum \epsilon_i^2$ depends on the y_i's only through the values of $\overline{Y}, \hat{\beta}$, and the residual sum of squares $\sum\hat{\epsilon}_i^2$. Hence the log likelihood function depends only on these three functions of the y_i's, and they form a set of sufficient statistics. (The x_i's, like n, are regarded as known constants.)

The distributions of $\overline{Y}, \hat{\beta}$, and $\sum\hat{\epsilon}_i^2$ can now be deduced from Theorem 7.3.3. By the Lemma, we have

$$V \equiv \frac{\sum\hat{\epsilon}_i^2}{\sigma^2} \equiv \frac{\sum\epsilon_i^2}{\sigma^2} - \frac{n(\overline{Y} - \alpha - \beta\overline{x})^2}{\sigma^2} - \frac{(\hat{\beta} - \beta)^2 S_{xx}}{\sigma^2}$$

$$\equiv \sum U_i^2 - Z_1^2 - Z_2^2$$

where

$$U_i \equiv \frac{1}{\sigma}\epsilon_i \equiv \frac{1}{\sigma}(Y_i - \alpha - \beta x_i) \sim N(0,1);$$

$$Z_1 \equiv \frac{\sqrt{n}(\overline{Y} - \alpha - \beta\overline{x})}{\sigma}; \qquad Z_2 \equiv \frac{(\hat{\beta} - \beta)\sqrt{S_{xx}}}{\sigma}.$$

Since $Y_i \equiv \alpha + \beta x_i + \epsilon_i$, we have $\overline{Y} \equiv \alpha + \beta\overline{x} + \overline{\epsilon}$, and hence

$$Z_1 \equiv \frac{\sqrt{n}\,\overline{\epsilon}}{\sigma} \equiv \sqrt{n}\,\overline{U} \equiv \frac{1}{\sqrt{n}}\sum U_i \equiv \sum a_i U_i$$

where $a_i = 1/\sqrt{n}$. Similarly, since

$$S_{xY} \equiv \sum(x_i - \bar{x})Y_i \equiv \sum(x_i - \bar{x})(\alpha + \beta x_i + \epsilon_i)$$

$$\equiv \alpha\sum(x_i - \bar{x}) + \beta\sum(x_i - \bar{x})x_i + \sum(x_i - \bar{x})\epsilon_i$$

it follows that

$$\hat{\beta} \equiv S_{xy}/S_{xx} \equiv \beta + \sum(x_i - \bar{x})\epsilon_i/S_{xx}$$

and hence that

$$Z_2 \equiv \frac{1}{\sigma}\sum(x_i - \bar{x})\epsilon_i/\sqrt{S_{xx}} \equiv \sum b_i U_i$$

where $b_i = (x_i - \bar{x})/\sqrt{S_{xx}}$. Now it is easy to verify that

$$\sum a_i^2 = \sum b_i^2 = 1; \qquad \sum a_i b_i = 0.$$

Hence, by Theorem 7.3.3, Z_1, Z_2 and V are independent, and

$$Z_1 \sim N(0,1), \qquad Z_2 \sim N(0,1), \qquad V \sim \chi^2_{(n-2)}.$$

It follows that $\bar{Y}, \hat{\beta}$, and $\sum\hat{\epsilon}_i^2$ are independent, with

$$\sum\hat{\epsilon}_i^2/\sigma^2 \sim \chi^2_{(n-2)};$$

$$\bar{Y} \equiv \alpha + \beta\bar{x} + \sigma Z_1/\sqrt{n} \sim N(\alpha + \beta\bar{x}, \sigma^2/n);$$

$$\beta \equiv \beta + \sigma Z_2/\sqrt{S_{xx}} \sim N(\beta, \sigma^2/S_{xx})$$

as stated in Theorem 15.3.1.

Problems for Section 15.3

†1. A new procedure is being investigated for measuring calcium content.
    The following table gives the actual calcium content x for each
    of ten samples, and the measurement y given by the new procedure.

x	4.0	8.0	12.5	16.0	20.0	25.0	31.0	36.0	40.0	40.0
y	3.7	7.8	12.1	15.6	19.8	24.5	31.1	35.5	39.4	39.5

(a) Prepare a scatterplot of the data and fit a straight line
     (y on x).
(b) Test the hypothesis that the slope of the line equals 1.
(c) Test the hypothesis that the line passes through the origin.

(d) Obtain a 95% confidence interval for the expected y-value
     when x = 23.25.

2. Observations were made to determine the growth rate of algae cells
    in a culture. Because many natural growth processes are exponen-
    tial, it seems appropriate to relate log volume to time rather than
    volume to time. The log volume of cells in the culture was recor-
    ded on each of 8 consecutive days, with the following results:

x (day)	1	2	3	4	5	6	7	8
y (log volume)	3.538	3.828	4.349	4.833	4.911	5.297	5.566	6.036

(a) Fit a straight line to these data, and show it on a scatter
     diagram. Does the fit of the model appear to be satisfactory?
(b) Obtain a 95% confidence interval for the growth rate (i.e.
     slope).
(c) How would the analysis be affected if the days were numbered
     $0,1,\ldots,7$ rather than $1,2,\ldots,8$?

3. Theory suggests that a linear relationship exists between the
    shearing strength of steel bolts and their diameters. The follow-
    ing table gives the diameter x and strength y for 9 bolts of
    a particular type.

x	1/8	1/4	3/8	1/2	5/8	3/4	7/8	1	3/2
y	47	72	97	126	165	186	233	257	311

(a) Fit a straight line to the data, and test the hypothesis
     that the intercept is zero.
(b) Find a 95% confidence interval for the mean strength of
     bolts with diameter 1.25 units.

†4. Fitting a line through the origin. Suppose that $Y_1, Y_2, \ldots, Y_n$
     are independent, with $Y_i \sim N(\beta x_i, \sigma^2)$ where $x_1, x_2, \ldots, x_n$ are
     given constants.
     (a) Show that the maximum likelihood estimate of β is

$$\hat{\beta} \equiv \sum x_i Y_i / \sum x_i^2.$$

     (b) Prove the identity

$$\sum (Y_i - \beta x_i)^2 \equiv \sum (Y_i - \hat{\beta} x_i)^2 + (\hat{\beta} - \beta)^2 \sum x_i^2.$$

     (c) Use Theorem 7.3.2 to show that $\hat{\beta}$ and $\sum (Y_i - \hat{\beta} x_i)^2$ are in-
          dependently distributed, with

$$\hat{\beta} \sim N(\beta, \sigma^2 / \sum x_i^2); \quad \sum (Y_i - \hat{\beta} x_i)^2 / \sigma^2 \sim \chi^2_{(n-1)}.$$

     (d) Fit a straight line through the origin in Problem 15.3.1, and
          test the hypothesis $\beta = 1$. Explain why the result obtained
          here is different from that obtained in Problem 15.3.1(b).

15.4   Residual Plots and Tests of Fit

The analysis described in the preceding two sections is based on the assumption of a linear relationship between y and x, with independent normal errors. In most applications, we will not know whether this assumption is appropriate, and in fact the aim of the analysis will be to find out how y and x are related. There is no better way to do this than by plotting the data and having a look! Graphs permit us to check the assumptions made in the analysis, and they may suggest modifications to the model and analysis. A graph may reveal something interesting and important that does not show up in the arithmetic. This point is illustrated in the following example, which was given by F.J. Anscombe, Graphs in Statistical Analysis, *The American Statistician* 27 (1973), pages 17–21.

Example 15.4.1. Table 15.4.1 gives four data sets, each consisting of $n = 11$ pairs (x,y). All four sets give approximately the same numerical values:

$$\overline{x} = 9.0 \qquad \overline{y} = 7.5$$
$$S_{xx} = 110 \quad S_{xy} = 55 \qquad S_{yy} = 41.25$$

Fitted line: $y = 3 + 0.5x$

$$\sum \hat{\epsilon}_i^2 = 13.75; \quad s^2 = 1.528.$$

Table 15.4.1

Four Data Sets for Example 15.4.1

Set 1		Set 2		Set 3		Set 4	
x	y	x	y	x	y	x	y
4	4.26	4	3.10	4	5.39	8	6.58
5	5.68	5	4.74	5	5.73	8	5.76
6	7.24	6	6.13	6	6.08	8	7.71
7	4.82	7	7.26	7	6.42	8	8.84
8	6.95	8	8.14	8	6.77	8	8.47
9	8.81	9	8.77	9	7.11	8	7.04
10	8.04	10	9.14	10	7.46	8	5.25
11	8.33	11	9.26	11	7.81	8	5.56
12	10.84	12	9.13	12	8.15	8	7.91
13	7.58	13	8.74	13	12.74	8	6.89
14	9.96	14	8.10	14	8.84	19	12.50

Estimates, tests of significance, and confidence intervals will thus be the same for all four data sets.

Scatterplots of the four data sets are given in Figure 15.4.1, and the fitted line $y = 3 + 0.5x$ is also shown. The points of the

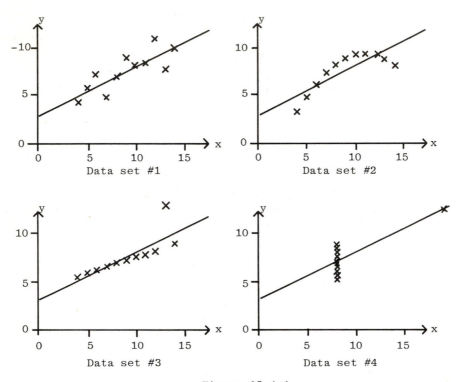

Figure 15.4.1
Scatterplots of the four data sets in Table 15.4.1

first data set appear to be scattered randomly about the fitted line, and the straight line model is quite satisfactory. For the second data set, the dependence of y on x is clearly not linear. The straight line model is inappropriate, and a quadratic polynomial model

$$E(Y) = \alpha + \beta x + \gamma x^2$$

might be tried; see Section 15.1.

In data set number three there is an outlying point at x = 13. It shifts the fitted line upward so that it does not properly fit the remaining ten points either. If we remove this outlier and recalculate, the fitted line is y = 4 + 0.346x, which gives a close fit to the remaining ten points. Both the fitted line and the outlying point should be reported.

The fourth data set shows good agreement with the straight line model, but the estimate of the slope depends entirely upon a single observation. If this observation were found to be in error and deleted, the slope could not be estimated. Furthermore, without measurements at additional values of x, there is no way of determining whether the actual dependence of y on x is even close to being linear. The fact that the analysis depends so heavily on a single observation should be reported along with the numerical results.

In summary, although all four data sets yield the same numerical results for the straight line model (15.1.1), different conclusions are appropriate in the four cases. Examination of a graph is an indispensable part of the statistical analysis.

Residual Plots

Departures from the model may show up more clearly if, instead of plotting y-values, we plot vertical deviations of y-values from the fitted line. The difference between the measurement y_i and the fitted value $\hat{\mu}_i = \hat{\alpha} + \hat{\beta}x_i$ is the ith residual, $\hat{\epsilon}_i = y_i - \hat{\alpha} - \hat{\beta}x_i$. A scatterplot of the residuals $\hat{\epsilon}_i$ versus x-values will convey the same information as a plot of y versus x, but in a more dramatic form. Another slight but convenient modification is to use fitted values $\hat{\mu}_i = \hat{\alpha} + \hat{\beta}x_i$ rather than x-values as abscissae in the plot. The only effect of this is to change the measurement scale on the horizontal axis so that it becomes the same as that on the vertical axis. One can then judge more easily whether a deviation $\hat{\epsilon}_i$ is large enough to be of practical importance. One might, for instance, find definite indications of curvature in a set of data, but decide to use the straight line model anyway because the deviations from it are too small to bother about over the range of x-values being considered.

It is thus recommended that a scatterplot of the residuals $\hat{\epsilon}_i$ versus the fitted values $\hat{\mu}_i$ be prepared. Four such plots are shown in Figure 15.4.2. In the first of these, the n points lie in a horizontal band of roughly constant width about zero, and the model appears to be satisfactory. In (ii), there is more scatter in the residuals as $\hat{\mu}$ increases, indicating a non-constant error variance σ^2. A non-linear transformation of the y_i's might remedy this, or else one could try to formulate a more complicated model in which σ was a function of x.

In the third diagram there is an outlying point. If this is removed and the model is refitted, the new residual plot should resemble (i). Finally, (iv) shows evidence of curvature, and suggests that a second-degree term in x should be added to the model.

If the scatterplot is satisfactory, a quantile plot of the residuals can be made to check the assumption of normality. If the

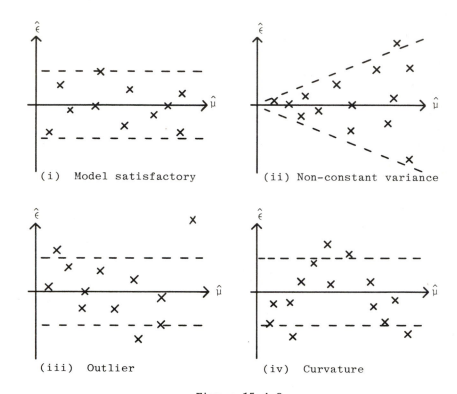

(i) Model satisfactory

(ii) Non-constant variance

(iii) Outlier

(iv) Curvature

Figure 15.4.2
Patterns in Residual Plots

model is correct, $\hat{\epsilon}_i$ is an estimate of the $N(0,\sigma^2)$ error
$\epsilon_i = y_i - \alpha - \beta x_i$. Hence the residuals should resemble a sample of size
n from a normal distribution $N(0,\sigma^2)$. In fact, because of (15.2.1),
the residuals are slightly correlated and their variances are not quite
equal, but these complications can usually be ignored.

Various other residual plots may be useful, depending upon
the situation. For instance, we might wish to plot the residuals in
Example 15.2.1 against the weights of the women if these were available.
In the plastic gear example, we might plot the residuals in the order
that the corresponding measurements were made, with the purpose of
checking whether there was a systematic change in laboratory conditions.

Although residual plots are very useful in statistical analysis, a word of caution is necessary. Even if the model is correct,
random variation will produce patterns in the residuals rather more often than most people would expect. Many beginners spot an "unusual"
pattern in almost every residual plot that they examine! A worthwhile
exercise is to generate random observations from $N(0,1)$ on the com-

puter, and to plot them against arbitrary x-values in various ways. After you have done this several times, you will get an idea of the frequencies with which various patterns can be expected to occur by chance, and be in a better position to judge whether an observed pattern should be taken seriously.

Figure 15.4.3 shows a plot of the residuals against age in Example 15.2.1. With a little imagination, one can see an indication of curvature in this graph, and it would be interesting to examine some additional data. It would also be interesting to know which of the observations in the 40 to 50 age range were for pre-menopausal and which were for post-menopausal women. However, in the absence of such additional information, the fit of the model would be considered adequate. There is a great deal of variability in measurements taken at or near the same age, and this could easily account for the slight curvature observed.

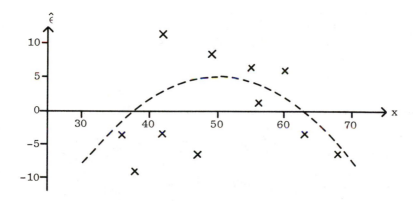

Figure 15.4.3.
Plot of Residuals versus Age in Example 15.2.1

Regression with repeated measurements

When there are repeated measurements of Y at some of the x-values, as in Example 15.1.1, the data can also be analysed as a k-sample problem (Section 14.6), and the adequancy of the straight line model can be tested.

Suppose that there are k distinct values $x_1, x_2, \ldots, x_k$ of the independent variable x, with n_i measurements $Y_{i1}, Y_{i2}, \ldots, Y_{in_i}$

at $x = x_i$, where $n_1 + n_2 + \ldots + n_k = n$. We assume, as in Section 14.6, that the n measurements are independent normal with the same variance σ^2, but with possibly different means $\mu_1, \mu_2, \ldots, \mu_k$ for the k groups. Define

$$\bar{Y}_i \equiv \frac{1}{n_i} \sum_{j=1}^{n_i} Y_{ij}; \qquad S_i^2 \equiv \frac{1}{n_i - 1} \sum_{j=1}^{n_i} (Y_{ij} - n_i)^2.$$

From Section 14.6, the combined variance estimate is $S^2 \equiv \frac{1}{n-k} V$, where

$$V \equiv \sum (n_i - 1) S_i^2 \equiv \sum\sum (Y_{ij} - \bar{Y}_i)^2$$

and $V/\sigma^2 \sim \chi^2_{(n-k)}$.

The above assumptions together with the hypothesis

$$H: \mu_i = \alpha + \beta x_i \quad \text{for} \quad i = 1, 2, \ldots, k$$

define the straight model (15.1.1). When we fit this model, the residual corresponding to observation y_{ij} is

$$\hat{\epsilon}_{ij} = y_{ij} - \hat{\alpha} - \hat{\beta} x_i = (y_{ij} - \bar{y}_i) + (\bar{y}_i - \hat{\alpha} - \hat{\beta} x_i),$$

and hence the residual sum of squares is

$$\sum\sum \hat{\epsilon}_{ij}^2 = \sum\sum (y_{ij} - \bar{y}_i)^2 + \sum\sum (\bar{y}_i - \hat{\alpha} - \hat{\beta} x_i)^2 + 2\sum\sum (y_{ij} - \bar{y}_i)(\bar{y}_i - \hat{\alpha} - \hat{\beta} x_i).$$

The third term on the right is equal to

$$\sum_i (\bar{y}_i - \hat{\alpha} - \hat{\beta} x_i) \sum_j (y_{ij} - \bar{y}_i),$$

and this is zero by the definition of $\bar{y}_i$. The residual sum of squares has thus been split into two parts:

$$\sum\sum \hat{\epsilon}_{ij}^2 = \sum\sum (y_{ij} - \bar{y}_i)^2 + \sum n_i (\bar{y}_i - \hat{\alpha} - \hat{\beta} x_i)^2. \qquad (15.4.1)$$

The first part measures the variability among y_{ij}'s taken at the same x-value, and is called the <u>pure error sum of squares</u>. The second part measures the deviation of the $\bar{y}_i$'s from the fitted line, and is called the <u>lack of fit sum of squares</u>.

We noted earlier that

$$\sum\sum (Y_{ij} - \bar{Y}_i)^2 / \sigma^2 \sim \chi^2_{(n-k)}.$$

It can be shown that, if H is true, then

$$\sum n_i(\overline{Y}_i - \hat{\alpha} - \hat{\beta}x_i)^2/\sigma^2 \sim \chi^2_{(k-2)} \qquad (15.4.2)$$

independently of the pure error sum of squares. By (6.9.8), the ratio of independent mean squares has an F distribution. Hence, if H is true, then

$$\frac{\text{(lack of fit SS)} \div (k-2)}{\text{(pure error SS)} \div (n-k)} \sim F_{k-2,\,n-k}. \qquad (15.4.3)$$

We use this result to test the fit of the model. If $E(Y) \neq \alpha + \beta x$, we expect the lack of fit SS to be large. Only large values of the ratio give evidence against the model, and hence a one-tail test is appropriate.

The independence of the numerator and denominator in (15.4.3) follows from the fact that the numerator depends on only the means $\overline{Y}_1, \overline{Y}_2, \ldots, \overline{Y}_k$, whereas the denominator depends only on the sums of squares $V_i \equiv \sum(Y_{ij} - \overline{Y}_i)^2$. One can deduce (15.4.2) from Theorem 7.3.3 by essentially the same argument used at the end of Section 15.3, using standardized normal variates

$$U_i \equiv \sqrt{n_i}(\overline{Y}_i - \alpha - \beta x_i)/\sigma \quad \text{for} \quad i = 1, 2, \ldots, k.$$

The above test of significance can be derived from a maximized likelihood ratio. One finds, as in Section 14.6, that the maximized likelihood ratio under the hypothesis is

$$R = \left[\frac{\tilde{\sigma}}{\hat{\sigma}}\right]^{-n} = \left[\frac{\sum\sum\hat{\epsilon}_{ij}^2}{v}\right]^{-n/2} = \left[1 + \frac{\sum n_i(\overline{y}_i - \hat{\alpha} - \hat{\beta}x_i)^2}{v}\right]^{-n/2}$$

The F-statistic used for the test is a strictly decreasing function of R. Only large values of the F-statistic (small values of R) give evidence against the hypothesis.

Example 15.4.2. Table 15.4.2 shows the variance estimates computed at each of the $k = 9$ temperatures for the plastic gear data of Example 15.1.1. From the table, the pure error SS is 0.64037 with $n - k = 31$ degrees of freedom. From Example 15.2.2, the residual SS is 1.24663 with $n - 2 = 38$ degrees of freedom. By (15.4.1), the lack of fit SS is

$$1.24663 - 0.64037 = 0.60626,$$

with $k - 2 = 7$ degrees of freedom. The ratio of mean squares is

Table 15.4.2

Sample means and variances for the plastic gear data

x_i	n_i	$\bar{y}_i$	$\sum(y_{ij} - \bar{y}_i)^2$	s_i^2
-16	4	1.755	0.01917	0.0064
0	4	1.569	0.01741	0.0058
10	4	1.094	0.03041	0.0101
21	8	0.808	0.13129	0.0188
30	4	0.494	0.07962	0.0265
37	4	0.515	0.04015	0.0134
47	4	0.253	0.10584	0.0353
57	4	-0.359	0.01814	0.0060
67	4	-0.687	0.19834	0.0661
Total	40	–	0.64037	–

$$F_{obs} = \frac{0.60626 \div 7}{0.64037 \div 31} = 4.19,$$

and the (one-tail) significance level is

$$SL = P\{F_{7,31} \geq 4.19\} < 0.01$$

from Table B5. There is quite strong evidence of lack of fit.

This finding could have been anticipated from an inspection of Figure 15.1.1, since at several temperatures all of the points lie on one side of the fitted line. It is also fairly clear from the diagram that the addition of a quadratic term to the model would not substantially improve the fit. There does not appear to be any simple pattern in the departures from linearity.

A possible explanation of this erratic behaviour can be found in the way that the experiment was performed. In order to save time in setting up equipment, the experimenter usually tested all of the gears at a particular temperature one after the other. Measurements made close together in time are likely to be more similar than those made on widely separated occasions, because it is impossible to maintain absolutely identical conditions over several weeks of testing. As a result, it is likely that the pure error mean square underestimates the variance, and this could account for the large F-value observed. Because of the way in which the experiment was done, there is no valid estimate of the experimental error. (The experiment should have been run in four replications. In the first replication, one gear would be tested at each temperature, with the order of testing being decided at random. This procedure would then be repeated three more times, with a different random order each time.)

An examination of the variance estimates in Table 15.4.2 reveals some tendency for the variance of the log-lifetime to increase with x. The model could be modified to permit σ^2 to depend upon x. For instance, we could take $\sigma^2 = e^{\gamma+\delta x}$. The analysis is then some-what more complicated, and we shall not go into the details here.

Problems for Section 15.4

†1. A new technique for determing the fraction x of a given gas in a mixture of gases was investigated. Eleven gas mixtures with known x were prepared, and each of them was divided into three portions. For each portion, the quantity y of the gas which dissolved in a liquid was recorded.

x=content	y=amount dissolving			x=content	y=amount dissolving		
0.080	2.67	2.68	2.75	0.131	4.46	4.40	4.43
0.082	2.73	2.69	2.62	0.139	4.78	4.80	4.86
0.091	2.88	3.02	3.04	0.164	5.77	5.85	5.82
0.095	3.17	3.28	3.18	0.189	6.56	6.65	6.49
0.096	3.27	3.28	3.08	0.231	7.88	7.97	7.76
0.106	3.51	3.68	3.58				

(a) Compute an estimate of the variance at each value of x, and plot these estimates against x. Is there evidence that the variance of y depends on x?

(b) Assuming that the variance is constant, fit a straight line to the data. Obtain 95% confidence intervals for the slope, and for the expected amount dissolving when the content is x = 0.100.

(c) Use a test of significance and a residual plot to check the fit of the model.

2. Measurements of the ultimate tensile strength (UTS) were made for specimens of insulating foam of five different densities.

Density (x)	Ultimate tensile strength (y)							
4.155	82.8	95.5	97.5	102.8	105.6	107.8	115.7	118.9
3.555	79.7	84.5	85.2	98.0	105.2	113.6		
3.55	71.0	98.2	104.9	106.9	109.6	117.8		
3.23	67.1	77.0	80.3	81.8	83.0	84.1		
4.25	98.5	105.5	111.6	114.5	126.5	127.1		

(a) Fit a straight line to the data, and test the adequacy of the model.

(b) Obtain 95% confidence intervals for the mean UTS of foam at densities 3.2, 3.7, and 4.2.

CHAPTER 16. TOPICS IN STATISTICAL INFERENCE*

        In the preceding chapters, we have based inferences about an
unknown parameter θ on the likelihood function itself (Chapters 9 and
10) or on tests of significance (Chapters 13, 14 and 15).   These tests
were themselves closely related to the likelihood function through the
use of sufficient statistics and the likelihood ratio criterion.   In
Sections 1 and 2 below, we consider two additional methods for making
inferences about an unknown parameter.   With both the fiducial argument
and Bayesian methods, information concerning θ is summarized in a
probability distribution defined on the parameter space.   For Bayesian
methods one requires prior information about θ which is also in the
form of a probability distribution.   For the fiducial argument, θ
must be completely unknown before the experiment.

        In Section 3, we consider the problem of predicting a value
of a random variable Y whose probability distribution depends upon
an unknown parameter θ.   When a Bayesian or fiducial distribution for
θ is available, one can obtain a predictive distribution for Y which
does not depend upon θ.   Section 4 considers the use of predictive
distributions in statistical inference, with particular reference to
the Behrens-Fisher problem.   Finally, in Section 5 we illustrate how
a test of a true hypothesis can be used to obtain intervals of reasonable values for a future observation or an unknown parameter.

16.1   The Fiducial Argument

        Suppose that we have obtained data from an experiment whose
probability model involves a real-valued parameter θ which is completely unknown.   We shall see that, under certain conditions, it is
possible to deduce the probability that θ ≤ k for any specified
parameter value k.   The procedure for obtaining this probability is
called the <u>fiducial argument</u>, and the probability is called a <u>fiducial
probability</u> to indicate the method by which it was obtained.

Probability distributions of constants

        In the fiducial argument, the probability distribution of a
variate U is regarded as a summary of all the available information

---

* This chapter may be omitted on first reading.

about U. This distribution continues to hold until such time as additional information about U becomes available. If U has a certain distribution before an experiment is performed, and if the experiment provides no information about U, then U has the same distribution after the experiment as before.

For example, consider a lottery in which there are N tickets numbered 1,2,...,N, one of which is selected at random. Let U denote the number on the winning ticket. Then

$$P(U = u) = \frac{1}{N} \quad \text{for} \quad u = 1, 2, \ldots, N. \tag{16.1.1}$$

Now suppose that the winning ticket has been chosen, but that the number U has not been announced. A value of U has now been determined, but we have no more information concerning what that value is than we had before the draw. A ticket-holder would presumably feel that he had the same chance of winning as he had before the draw was made. The fiducial argument is based on the assertion that (16.1.1) summarizes the uncertainty about U even after the draw has been made, provided that no information concerning the outcome of the draw is available. After the draw, U is no longer subject to random variation, but is fixed at some unknown value. Now (16.1.1) summarizes all the available information concerning the unknown constant U, and may be called its fiducial distribution.

The fiducial argument does not involve any new "definition" of probability. Instead, it enlarges the domain of application of the usual (long-run relative frequency) notion of probability. Of course, one could take the position (as some people have) that (16.1.1) applies only before the draw, and that, after the draw, no probability statements whatsoever can be made. This position seems unnecessarily restrictive, and if adopted, would rule out many important applications of probability.

Before proceeding with the general discussion, we illustrate the fiducial argument in two examples.

Example 16.1.1. A deck of N cards numbered 1,2,...,N is shuffled and one card is drawn. Let U denote the number on this card. Then U has probability distribution (16.1.1). To this number is added a real number θ which is completely unknown to us. We are not told the value of U or the value of θ, but only the value of their total $T \equiv \theta + U$. What can be said about θ in the light of an observed total t?

The observed total t could have arisen in N different ways:

$$(U = 1, \theta = t - 1), (U = 2, \theta = t - 2), \ldots, (U = N, \theta = t - N).$$

Given t,   there is a one-to-one correspondence between values of U
and possible values of θ. If we knew the value of θ, we could determine which value of U had been obtained. If we knew that θ was an
even integer, then we could deduce whether an odd or even value of U
had been obtained. However, if we know nothing about θ, then the experiment will tell us nothing about U;   the state of uncertainty concerning the value of U will be the same after the experiment as before.   Hence we assume that (16.1.1) also holds when t is known.
But, given t, θ has N possible values t-1, t-2, ..., t-N in one-to-
one correspondence with the possible values of U, and we may write

$$P(\theta = t - u) = P(U = u) = \frac{1}{n}, \quad u = 1, 2, \ldots, N.$$

This probability distribution over the possible values of θ is called
the fiducial distribution of θ.

For instance, suppose that N = 13, and that the observed
total is t = 20.   Then θ has 13 possible values 19, 18, 17, ..., 7,
each with probability $\frac{1}{13}$.   The probability of any subset of θ values
is now obtained by addition.   For example,

$$P(\theta \le 11) = P(\theta = 11) + P(\theta = 10) + \ldots + P(\theta = 7) = \frac{5}{13}.$$

Alternately, we may note that if $\theta \le 11$,   then the observed total 20
must have resulted from a value of U greater than or equal to 9,
and hence

$$P(\theta \le 11) = P(U \ge 9) = \frac{5}{13}.$$

Example 16.1.2. Suppose that $T \sim N(\theta, 1)$ where θ is completely unknown, and that the experiment yields an observed value t. We define
$U \equiv T - \theta$, so that U has a standardized normal distribution. The observed value t arose from some pair of values $(U = u, \theta = t - u)$. Given
t, there is a one-to-one correspondence between possible values of U
and possible values of θ. Since θ is unknown, the experiment will
tell us nothing about which value of U was actually obtained. Consequently, we assume that $U \sim N(0, 1)$ even after t has been observed.

We can now compute probabilities of statements about θ by
transforming them into statements about U. For instance, $\theta \le k$ if
and only if $U \ge t - k$, and hence

$$P(\theta \le k) = P(U \ge t - k) = 1 - F(t - k) = F(k - t) \qquad (16.1.2)$$

where F is the N(0,1) cumulative distribution function. For any
k, the fiducial probability of $\theta \le k$ can be obtained from N(0,1)
tables. For example, if we observe t = 10, the fiducial probability
of $\theta \le 11$ is

$$P(\theta \le 11) = F(11 - 10) = 0.841.$$

Note that probability statements obtained from (16.1.2) are
the same as would be obtained if θ were a random variable having a
normal distribution with mean t and variance 1. We say that given
T = t, the <u>fiducial distribution</u> of θ is N(t,1). This does not
mean that θ is a random variable, but rather that we know precisely
as much about θ as we would about an observation to be taken at random from N(t,1).

From (16.1.2), the cumulative distribution function of the
fiducial distribution of θ is $F(\theta - t)$, where F is the c.d.f. of
N(0,1). Differentiation with respect to θ gives

$$\frac{\partial}{\partial \theta} F(\theta - t) = f(\theta - t) \frac{\partial(\theta - t)}{\partial \theta} = f(\theta - t)$$

where f is the p.d.f. of N(0,1). Hence the fiducial p.d.f. of θ
is

$$f(\theta;t) = \frac{1}{\sqrt{2\pi}} \exp\{-\frac{1}{2}(\theta - t)^2\} \quad \text{for} \quad -\infty < \theta < \infty.$$

This is the p.d.f. of a normal distribution with mean t and variance
1. As a result of the fiducial argument, θ and T have switched
roles, with the observed t now appearing as a "parameter" in the
fiducial distribution of θ.

Sufficient conditions for the fiducial argument

In the preceding two examples, we made use of a quantity
U which was a function of both the data and the parameter, and whose
probability distribution did not depend upon θ. Such a function is
called a <u>pivotal quantity</u>.

The following conditions are sufficient to permit application
of the fiducial argument in the one-parameter case:

C1. There is a single real-valued parameter θ which is completely
     unknown.
C2. There exists a statistic T which is minimally sufficient for θ.
C3. There exists a pivotal quantity $U \equiv U(T,\theta)$ such that
     (a) for each value of θ, $U(t,\theta)$ is a one-to-one function of t;
     (b) for each value of t, $U(t,\theta)$ is a one-to-one function of θ.

If the variate T is continuous, we also require that U be continuous (and hence monotonic) in both t and θ.

The purpose of conditions C2 and C3(a) is to ensure that inferences about θ are based on all of the relevant information contained in the data. C2 can be replaced by the weaker condition that
there exists a set of minimally sufficient statistics (T,A) where T
is real-valued and A is a vector of ancillary statistics (see Section
12.8). We then use the conditional distributions of T and U given
the observed value of A.

Given T = t, there is a one-to-one correspondence between
possible values of θ and possible values of U by C3(b). Since θ
is completely unknown, observing t will give us no information about
which value of U was actually obtained. Hence we assume that the
distribution of U is the same after t has been observed as it was
before observing t. Given t, we can convert statements about θ
into statements about U and hence obtain their (fiducial) probabilities.

The above conditions are quite restrictive. In particular,
C3(a) and (b) imply a one-to-one correspondence between values of T
given θ, and values of θ given T, which will very rarely exist
if T is discrete. Example 16.1.1 is exceptional in that, when t
is known, there are only finitely many possible values for θ.

If the sufficient statistic T is continuous, one can usual-
ly take $U \equiv F(T;\theta)$, where F is the cumulative distribution function
of T. From Section 6.3, U has a uniform distribution between 0
and 1 for each value of θ, and hence is a pivotal quantity. Since
$F(t;\theta) = P(T \leq t)$ is an increasing function of t, C3(a) will also
be satisfied, and only C3(b) needs to be checked. If C3(b) holds, then
$P(\theta \leq k)$ will be equal to either $F(t;k)$ or $1 - F(t;k)$, depending
upon whether $F(t;\theta)$ is an increasing or decreasing function of θ,
and the fiducial p.d.f. of θ is given by

$$f(\theta;t) = \left| \frac{\partial}{\partial \theta} F(t;\theta) \right|.$$

Example 16.1.3. Suppose that the MLE $\hat{\phi}$ is a sufficient statistic
for the unknown parameter φ, and that $\hat{\phi} \sim N(\phi,c)$ where c is a
known constant. Then the standardized variable

$$Z \equiv (\hat{\phi} - \phi)/\sqrt{c}$$

is pivotal and is distributed as N(0,1). It satisfies conditions
3(a) and 3(b). To obtain the fiducial distribution of φ, we assume

that Z is still distributed as N(0,1) when the variate $\hat{\phi}$ is replaced by its observed value. Then we have

$$\phi \equiv \hat{\phi} - Z\sqrt{c}$$

where $\hat{\phi}$ and c are known constants, and (6.6.6) gives

$$\phi \sim N(\hat{\phi},c).$$

Given $\hat{\phi}$, the fiducial distribution of ϕ is normal with mean $\hat{\phi}$ and variance c.

<u>Example 16.1.4.</u> Let $X_1,X_2,\ldots,X_n$ be independent variates having an exponential distribution with unknown mean θ. Then $T \equiv \sum X_i$ is sufficient for θ, and

$$U \equiv 2T/\theta \sim \chi^2_{(2n)}$$

is a pivotal quantity satisfying conditions 3(a) and 3(b). To obtain the fiducial distribution of θ, we replace T by its observed value t and assume that U is still distributed as $\chi^2_{(2n)}$. Statements about θ can now be converted into statements about U, and their probabilities can be obtained from tables of the χ^2 distribution.

The fiducial p.d.f. of θ can be obtained from the p.d.f. of U by standard change of variables methods. By (6.9.1), the p.d.f. of U is

$$f(u) = ku^{n-1}e^{-u/2} \quad \text{for} \quad u > 0$$

where $k = 1/2^n\Gamma(n)$. The fiducial p.d.f. of θ is thus

$$g(\theta;t) = f(u)\cdot\left|\frac{du}{d\theta}\right| = k(\frac{2t}{\theta})^{n-1}e^{-t/\theta}\cdot\frac{2t}{\theta^2}$$

$$= \frac{1}{\theta\Gamma(n)}(\frac{t}{\theta})^n e^{-t/\theta} \quad \text{for} \quad \theta > 0.$$

<u>Example 16.1.5.</u> Consider the situation described in Example 16.1.1, but now suppose that n cards are drawn at random with replacement from the deck. The same unknown θ is added to the number on each card, and we are told the n totals $x_1,x_2,\ldots,x_n$. We wish to make inferences about θ on the basis of the data.

Each X_i can take N equally probable values $\theta+1,\theta+2,\ldots,\theta+N$, so that the probability function of X_i is

$$f(x) = P(X_i = x) = N^{-1} \quad \text{for} \quad x = \theta+1,\theta+2,\ldots,\theta+N.$$

Under random sampling with replacement, the X_i's are independent, and hence their joint probability function is

$$f(x_1)f(x_2)\ldots f(x_n) = N^{-n} \quad \text{for} \quad \theta + 1 \le x_1, x_2, \ldots, x_n \le \theta + N.$$

The likelihood function of θ is thus constant over the range of possible parameter values. We must have $\theta + 1 \le x_{(1)}$ and $\theta + N \ge x_{(n)}$, where $x_{(1)}$ and $x_{(n)}$ are the smallest and largest sample values, so that

$$L(\theta) = 1 \quad \text{for} \quad x_{(n)} - N \le \theta \le x_{(1)} - 1.$$

It follows that $X_{(1)}$ and $X_{(n)}$ are jointly minimally sufficient for θ.

The number of possible parameter values is

$$x_{(1)} - 1 - [x_{(n)} - N - 1] = N - a$$

where $A \equiv X_{(n)} - X_{(1)}$ is the sample range. The larger the value of A obtained, the more precisely we may determine the value of θ. If we observe $A = 0$, there are N equally likely values for θ, but if $A = N - 1$, the value of θ can be determined exactly without error. Thus A is a measure of the experiment's informativeness, and is in fact an ancillary statistic. To see this, we write $X_i \equiv \theta + U_i$, where U_i is the number on the ith card drawn $(i = 1, 2, \ldots, n)$. Then $X_{(1)} \equiv \theta + U_{(1)}$ and $X_{(n)} \equiv \theta + U_{(n)}$, so that

$$A \equiv X_{(n)} - X_{(1)} \equiv U_{(n)} - U_{(1)}.$$

The distribution of A thus depends only on the range of numbers which appear on the n cards drawn, and does not depend upon θ.

We now define a statistic T such that the transformation from $X_{(1)}, X_{(n)}$ to T, A is one-to-one; for instance, we could take $T \equiv X_{(1)}$. Then T, A are jointly sufficient for θ and A is ancillary. Inferences about θ will be based on the conditional distribution of T given the observed value of A. To obtain this distribution, we could first derive the joint probability function of $X_{(1)}$ and $X_{(n)}$ as in Problem 7.2.11, change variables, sum out T to get the probability function of A, and divide to get the required conditional probability function,

$$f(t|a;\theta) = \frac{1}{N-a} \quad \text{for} \quad t = \theta+1, \theta+2, \ldots, \theta + N - a.$$

Given that A = a, the n totals must fall in a range of length a
which lies entirely between $\theta + 1$ and $\theta + N$. There are $N - a$ such
ranges, with lower limits $\theta + 1, \theta + 2, \ldots, \theta + N - a$, and these will be
equally probable.

Now define $U \equiv T - \theta$. The conditional distribution of U
given that A = a is uniform,

$$P(U = u \mid a) = \frac{1}{N - a} \quad \text{for} \quad u = 1, 2, \ldots, N-a, \qquad (16.1.3)$$

and does not depend upon θ. Given A and θ, there is a one-to-one
correspondence between possible values of U and T. Given A and
T, there is a one-to-one correspondence between possible values of U
and θ. Thus, when A is given, the sufficient conditions for the
fiducial argument are satisfied. The fiducial distribution of θ is
obtained by assuming that (16.1.3) continues to hold when T is replaced by its observed value t, and this gives

$$P(\theta = k) = \frac{1}{N - a} \quad \text{for} \quad k = t-1, t-2, \ldots, t - N + a. \qquad (16.1.4)$$

For example, suppose that N = 13, and that we observe the
n = 4 totals 17, 11, 14, 23. Then $t = x_{(1)} = 11$, $x_{(n)} = 23$, and
a = 23 - 11 = 12. Now (16.1.4) implies that $\theta = 10$ with probability
1. In this case the experiment is very informative and completely
determines the value of θ. If we were less fortunate, we might observe totals such as 13, 17, 19, 13. Then t = 13 and a = 6, so that
now (16.1.4) gives

$$P(\theta = k) = \frac{1}{7} \quad \text{for} \quad k = 12, 11, 10, \ldots, 6.$$

There are now seven equally probable values of θ. In the worst possible case, we observe equal totals 18, 18, 18, 18. Then t = 18, a = 0,
and (16.1.4) gives

$$P(\theta = k) = \frac{1}{13} \quad \text{for} \quad k = 17, 16, 15, \ldots, 5$$

so that there are 13 equally probable values of θ.

Two-parameter fiducial distributions

Sometimes a double application of the one-parameter fiducial
argument can be used to obtain a two-parameter fiducial distribution.
However, there are examples where this can be done in two or more different ways, leading to different two-parameter distributions. There
are serious difficulties in extending the fiducial argument beyond the

one-parameter case, and the precise conditions under which this can be done are not known.

16.2 Bayesian Methods

In all of the procedures discussed so far, only the information provided by the experimental data is formally taken into account. However, in some situations we may wish to incorporate information about θ from other sources as well. If this additional information is in the form of a probability distribution for θ, it can be combined with the data using Bayes's Theorem (3.5.1).

Suppose that the probability model for the experiment depends on a parameter θ, and that an event E with probability $P(E;\theta)$ is observed to occur. In addition, suppose that θ is itself a random variable with a known probability distribution, called the prior distribution of θ, with probability or probability density function g, say. The conditional distribution of θ given that E has occurred is called the posterior distribution of θ. The posterior distribution has probability or probability density function given by

$$f(\theta|E) = P(E;\theta)g(\theta)/P(E) \qquad (16.2.1)$$

where P(E) is a normalizing constant:

$$P(E) = \begin{cases} \displaystyle\sum_{\theta\,\in\,\Omega} P(E;\theta)g(\theta) & \text{if } \theta \text{ is discrete;} \\[2ex] \displaystyle\int_{-\infty}^{\infty} P(E;\theta)g(\theta)d\theta & \text{if } \theta \text{ is continuous.} \end{cases} \qquad (16.2.2)$$

The posterior distribution combines the information about θ provided by the experimental data with the information contained in the prior distribution.

The likelihood function of θ based on the observed event E is given by

$$L(\theta;E) = kP(E;\theta)$$

where k does not depend upon θ. Hence we may write

$$f(\theta|E) = cL(\theta;E)g(\theta) \qquad (16.2.3)$$

where c is a constant with respect to θ, and is chosen so that the total probability in the posterior distribution is 1:

$$\frac{1}{c} = \begin{cases} \sum_{\theta \in \Omega} L(\theta;E)g(\theta) & \text{if } \theta \text{ is discrete;} \\ \\ \int_{-\infty}^{\infty} L(\theta;E)g(\theta)d\theta & \text{if } \theta \text{ is continuous.} \end{cases} \qquad (16.2.4)$$

The posterior p.f. or p.d.f. is thus proportional to the product of the likelihood function and the prior p.f. or p.d.f. of θ.

Example 16.2.1. Consider the inheritance of hemophilia as discussed previously in Example 3.5.4. Suppose that a woman has n sons, of whom x are hemophilic and n - x are normal. The probability of this event is

$$P(x;\theta) = \binom{n}{x}\theta^x(1 - \theta)^{n-x} \qquad (16.2.5)$$

where θ is the probability that a particular son will be hemophilic. The problem is to make inferences about θ.

      Given no additional information about θ, inferences would be based on (16.2.5). One could graph the relative likelihood function of θ, or compute confidence intervals. However, it may be possible to extract some information about θ by examining the woman's family tree. For instance, suppose that the woman had normal parents, but she had a brother who was hemophilic. Then her mother must have been a carrier, and she therefore had a 50% chance of inheriting the gene for hemophilia. If she did inherit the gene, then there is a 50% chance that a particular son will inherit the disease ($\theta = \frac{1}{2}$), and if she did not, all of her sons will be normal ($\theta = 0$). (The possibility of a mutation is ignored in order to simplify the example.) The prior probability distribution of θ is thus given by

$$g(0) = P(\theta = 0) = \frac{1}{2}; \qquad g(\tfrac{1}{2}) = P(\theta = \tfrac{1}{2}) = \frac{1}{2}.$$

With this additional information, it is now possible to base the analysis on Bayes's Theorem.

      By (16.2.3), the posterior probability function of θ is given by

$$f(\theta|x) = c(x)\theta^x(1 - \theta)^{n-x}\frac{1}{2} \quad \text{for} \quad \theta = 0, \tfrac{1}{2}.$$

If $x > 0$, then $\theta = 0$ and $\theta = \frac{1}{2}$ have posterior probabilities 0 and 1, respectively. If $x = 0$, the posterior probabilities are

$$P(\theta = 0 \,|\, X = 0) \ = \ c/2; \qquad P(\theta = \tfrac{1}{2} \,|\, X = 0) \ = \ c/2^{n+1}.$$

Since the sum of these must be 1, we find that $c = 2^{n+1}/(2^n + 1)$, and hence that

$$P(\theta = 0 \,|\, X = 0) \ = \ \frac{2^n}{2^n + 1} \ ; \qquad P(\theta = \tfrac{1}{2} \,|\, X = 0) \ = \ \frac{1}{2^n + 1}.$$

If the woman has at least one hemophilic son $(x > 0)$, she must be a carrier. If she has only normal sons $(x = 0)$, the probability that she is a carrier decreases as n increases.

Example 16.2.2. Suppose that components are received from a manufacturer in large batches, and let θ denote the proportion of defectives in a batch. A random sample of n items is chosen from the batch, and is found to contain x defectives. If n is small in comparison with the batch size, the probability of x defectives in the sample is

$$P(x; \theta) \ = \ \binom{n}{x} \theta^x (1 - \theta)^{n-x}. \qquad (16.2.6)$$

Given no additional information, inferences about θ would be based (16.2.6).

It may be that similar batches are received at regular intervals from the same manufacturer. The value of θ will vary somewhat from batch to batch. If the manufacturing process is reasonably stable, one might expect the variation in θ to be random, and introduce the assumption that θ is a random variable with probability density function g, say. Data from past samples would be used to help determine the form of the prior density function g.

An assumption which makes the mathematics easy is that θ has a beta distribution with parameters p and q,

$$g(\theta) \ = \ k\theta^{p-1}(1 - \theta)^{q-1} \quad \text{for} \quad 0 < \theta < 1, \qquad (16.2.7)$$

where $k = \Gamma(p + q)/\Gamma(p)\Gamma(q)$. Then, by (16.2.3), the posterior p.d.f. of θ given x is

$$f(\theta \,|\, x) = c(x)\theta^{x+p-1}(1 - \theta)^{n-x+q-1} \quad \text{for} \quad 0 < \theta < 1,$$

which is also a beta distribution with parameters $x + p$ and $n - x + q$. Probabilities can be computed by numerical integration, or from tables of the F-distribution; see Problem 6.9.14.

Of course, it would be unwise to assume (16.2.7) merely because it leads to simple mathematics. Data from past samples should

be used to check the adequency of (16.2.7), and to estimate the parameters p and q. As additional data accumulates, further checks of
the model can be made, and more precise estimates of p and q can
be obtained. Procedures such as this, in which data are used to give
information about both the current value of θ and the prior distribution of θ, are called empirical Bayes methods. $\Box$

In the two preceding examples, it was natural to regard the
value of θ as having been generated by a repeatable experiment.
Prior probabilities for θ-values then correspond to the relative
frequencies with which the various θ-values would be expected to a-
rise in many repetitions of the experiment. It is possible, concep-
tually at least, to verify the prior distribution empirically by ac-
tually repeating the experiment to obtain a sample $\theta_1, \theta_2, \ldots, \theta_n$ of
θ-values. These values could be compared with the assumed prior distribution using methods similar to those in Chapter 11. However, the
analysis would usually be complicated by the fact that only estimates
$\hat{\theta}_1, \hat{\theta}_2, \ldots, \hat{\theta}_n$ were available.

Applications such as these, in which the prior distribution
is the probability model of a physical process which generates the
value of 0, are not controversial. However, Bayesian methods are
sometimes advocated in situations where θ is thought of as a constant.
The prior distribution may be an objective summary of the prior state
of knowledge concerning θ, or it may be a statement of an individual's
subjective beliefs about θ. There are differences of opinion among
statisticians concerning the appropriateness of Bayesian methods in
such situations.

Fiducial Prior Distributions

It may be that the conditions for the fiducial argument were
satisfied in some previous experiment involving the same parameter
The fiducial distribution of θ from the previous experiment might
then be used as the prior distribution of θ in the current experiment.

Example 16.2.3. Suppose that, in a previous experiment, N components
with exponentially distributed lifetimes were tested until failure.
From Example 16.1.4, the fiducial distribution of the mean lifetime θ
has p.d.f.

$$g(\theta) = \frac{1}{\theta \Gamma(N)} (\frac{t}{\theta})^N e^{-t/\theta} \quad \text{for} \quad \theta > 0,$$

where t is the total of the observed lifetimes. In the current experiment, n additional components are tested simultaneously, and
testing stops after a predetermined time period T. From Section 9.6,
the likelihood function of θ based on the current experiment is

$$L(\theta) = \theta^{-m}e^{-s/\theta} \quad \text{for} \quad \theta > 0,$$

where m is the number of components which were observed to fail, and
s is the total elapsed lifetime of all n components (including those
whose failure times were censored). By (16.2.3), the p.d.f. of the
posterior distribution of θ is

$$f(\theta) = c\theta^{-m}e^{-s/\theta} \cdot \theta^{-N-1}e^{-t/\theta}$$

$$= c\theta^{-(m+N+1)}e^{-(s+t)/\theta} \quad \text{for} \quad \theta > 0.$$

It can now be shown by change of variables that $2(s + t)/\theta$ has a χ^2
distribution with $2(m + N)$ degrees of freedom. Hence tables of the
χ^2 distribution may be used to obtain the posterior probabilities of
statements about θ.

Note that it would not be possible to derive a fiducial distribution for θ on the basis of the current experiment, or on the
basis of the previous and current experiments combined. In each case
the minimally sufficient statistic is two-dimensional, and there exists
no ancillary statistic.

If there were no censoring in the second experiment, the two
experiments could be combined to give a single experiment in which
N + n components were tested to failure. A fiducial distribution for
θ could then be derived as in Example 16.1.4. The same result would
be obtained by taking the fiducial distribution of θ from the previous experiment as the prior distribution in Bayes's Theorem. However,
the latter procedure seems inappropriate because it violates the symmetry between the two experiments, and it may lead to unacceptable results in more complicated situations. For further discussion, see
D.A. Sprott, "Necessary restrictions for distributions a posteriori",
Journal of the Royal Statistical Society, B, 22(1960), pages 312-318.

Prior Distributions which Represent Ignorance

Various attempts have been made to formulate prior probability distributions which represent a state of total ignorance about
the parameter (see H. Jeffreys, *Theory of Probability,* 3rd edition. Ox-
ford: Clarendon Press, 1961). These are generally derived from arguments of mathematical symmetry and invariance.

293

Let us consider the simplest case, in which nothing is known about a parameter θ except that it must take one of a finite set of values $\{1, 2, \ldots, N\}$. It might be argued that, since there is no reason to prefer one of these values over another, they should be assigned equal probabilities (Laplace's Principle of Insufficient Reason). The statement that the N possible parameter values are equally probable is then supposed to represent a complete lack of knowledge of θ .

The above argument implicitly assumes that there exists some probability distribution which appropriately represents total ignorance. If this assumption is granted, then the assignment of equal probabilities seems inevitable. However, the assumption itself is questionable. It would seem more reasonable to represent prior ignorance by equally likely, rather than equally probable, parameter values. If the N parameter values are equally probable, then $P(\theta \neq 1) = (N - 1)/N$, and this would seem to be an informative statement. However, no such statement is possible if they are assumed to be equally likely, because likelihoods are not additive (Section 9.7).

Now consider a parameter θ which can take values in a real interval $0 < \theta < 1$, say. Great difficulties arise in trying to formulate a probability distribution of θ which represents total ignorance. If one assumes that the distribution of θ is uniform, then one-to-one functions of θ will generally not have uniform distributions because of the Jacobian involved in continuous change of variables. If θ is totally unknown, then presumably θ^3 is also totally unknown, but it is impossible to have a uniform distribution on both of them. This problem does not arise if prior ignorance is represented by equally likely parameter values, because likelihoods are invariant under one-to-one parameter transformations.

For further discussion, see Chapter 1 of *Statistical Methods and Scientific Inference* by R.A. Fisher (2nd edition, New York: Hafner, 1959).

Subjective Prior Distributions

In yet another approach to the use of Bayes's Theorem, the prior distribution is taken to be a summary of an individual's prior belief about θ . See, for example, H. Raiffa and R. Schlaifer, *Applied Statistical Decision Theory*, Boston: Harvard Univ. Graduate School of Bus. Admin., 1961; and L.J. Savage, *The Foundations of Statistical Inference*, London: Methuen, 1962. According to the advocates of this approach, the prior distribution for θ is to be determined by introspection, and is a measure of personal opinion concerning what the value of θ is likely to be. Bayes's Theorem is then used to modify opinion on the basis of the experimental data.

Any statistical analysis involves some elements of subjective judgement - for instance, in the choice of the probability model. Nevertheless, this subjective input is open to public scrutiny and possible modification if poor agreement with the data is obtained. The same is not true of a subjective prior distribution, which is entirely a personal matter. A subjective prior distribution may be based on nothing more than hunches and feelings, and it seems a mistake to give it the same weight in the analysis as information obtained from the experimental data. The subjective Bayesian approach may prove to be valuable in personal decision problems, but it does not seem appropriate for problems of scientific inference.

16.3 Prediction

Suppose that we wish to predict the value of a random variable Y whose probability distribution depends upon a parameter θ. We assume that θ is unknown, but that a previous set of data gives some information about the value of θ. In predicting Y, we have two types of uncertainty to contend with: uncertainty due to random variation in Y, and uncertainty due to lack of knowledge of θ. We wish to make statements about Y which incorporate both types of uncertainty.

For example, suppose that the lifetimes of a certain type of rocket component are exponentially distributed with mean θ. We have tested n components, and have observed their lifetimes $x_1, x_2, \ldots, x_n$. We wish to predict the lifetime of another component, or perhaps the lifetime of a system made up of several such components. Even if we knew θ, we could not make exact predictions because lifetimes are subject to random variation; that is, components run under identical conditions will generally have different lifetimes. The problem is further complicated by the fact that we do not know the value of θ, but have only limited information obtained from the n components tested. Both the randomness of Y and the uncertainty about θ will influence predictive statements about Y.

Throughout the discussion, we assume that the probability model is appropriate. Mathematical models are only approximate descriptions of reality, and predictions based on them may be wildly in error if they are poor approximations. Errors of this kind are potentially the most serious, and in many situations it is difficult to estimate how large they are likely to be. Although we can and should

check the agreement of the model with the past data, we cannot check
the agreement with the future values which we are trying to predict.

Prediction problems have tidy solutions in the special case
where all of the available information about θ can be summarized in
the form of a probability distribution for θ (fiducial or Bayesian
posterior). Suppose that θ has probability density function f,
and that Y has p.d.f. $g(y;\theta)$ depending upon θ. If we interpret
the latter as the conditional p.d.f. of Y given θ, the joint p.d.f.
of Y and θ is $g(y;\theta)f(\theta)$. We then integrate out θ to obtain
the marginal p.d.f. of Y,

$$p(y) = \int_{-\infty}^{\infty} g(y;\theta)f(\theta)d\theta. \qquad (16.3.1)$$

This distribution combines uncertainty due to random variation in Y
with uncertainty due to lack of knowledge of θ, and is called the
predictive distribution of Y.

Prediction problems are more difficult when there is no probability distribution for θ. A procedure which is sometimes useful
in this situation will be discussed in Section 16.5.

Predicting an (n+1)st observation from an exponential distribution

Suppose that n independent values are observed from an
exponential distribution with unknown mean θ. We wish to predict the
value of Y, an (n+1)st observation to be taken from the same exponential distribution.

The fiducial argument is applicable in this case. From
Example 16.1.4, the fiducial p.d.f. of θ based on the observed sample is

$$f(\theta) = \frac{1}{\theta\Gamma(n)}(\frac{t}{\theta})^n e^{-t/\theta} \quad \text{for} \quad \theta > 0,$$

where $t = \sum x_i$ is the observed sample total. Given θ, the p.d.f.
of Y is

$$g(y;\theta) = \frac{1}{\theta} e^{-y/\theta} \quad \text{for} \quad y > 0.$$

By (16.3.1), the p.d.f. of the predictive distribution of Y is

$$p(y) = \int_0^{\infty} \frac{1}{\theta} e^{-y/\theta} \cdot \frac{1}{\theta\Gamma(n)}(\frac{t}{\theta})^n e^{-t/\theta} d\theta \quad \text{for} \quad y > 0.$$

Upon substituting $u = (y + t)/\theta$ and simplifying, we obtain

$$p(y) = \frac{t^n}{\Gamma(n)(t+y)^{n+1}} \int_0^\infty u^n e^{-u} du.$$

The integral on the right equals $\Gamma(n+1)$, and hence by (2.1.14),

$$p(y) = \frac{t^n}{(t+y)^{n+1}} \cdot \frac{\Gamma(n+1)}{\Gamma(n)} = \frac{nt^n}{(t+y)^{n+1}} \quad \text{for} \quad y > 0.$$

Integrating with respect to y now gives

$$P(Y \le y) = \int_0^y p(v)dv = 1 - \left(\frac{t}{t+y}\right)^n \quad \text{for} \quad y > 0,$$

and probabilities of statements about Y can easily be obtained. These probabilities take into account both the random variation of Y and the available information about θ.

In Example 9.5.1 we considered $n = 10$ observed lifetimes with total $t = 288$, and in this case

$$P(Y \le y) = 1 - \left(\frac{288}{288+y}\right)^{10} \quad \text{for} \quad y > 0.$$

We use this to make predictive statements about the lifetime Y of another component of the same type. For instance, we obtain

$$P(Y \le 5) = 0.158, \quad P(Y \ge 75) = 0.099$$

and so on. Also, we find that

$$P(Y \le 1.48) = P(Y \ge 100.6) = 0.05.$$

The interval $1.48 \le Y \le 100.6$ is called a <u>90% predictive interval</u> for Y. As one might expect, the interval is quite wide, indicating that we cannot predict the lifetime of a single component Y with much precision.

It is of some interest to compare the above results with what we could obtain if we knew the value of θ. If we assume that θ is equal to its maximum likelihood estimate, we have

$$P(Y \le y \mid \theta = 28.8) = 1 - e^{-y/28.8} \quad \text{for} \quad y > 0.$$

From this we obtain

$$P(Y \le 1.48) = P(Y \ge 86.3) = 0.05.$$

The central 90% interval is $1.48 \leq Y \leq 86.3$, which is not much narrower than the 90% predictive interval. This indicates that most of the uncertainty in predicting Y is due to the random variation of Y rather than to lack of information about the value of θ.

Predicting a future value from a normal distribution

Suppose that we wish to predict a future value of Y, where $Y \sim N(\phi, c_1)$ with c_1 known. Suppose further that ϕ is unknown, but that all available information concerning ϕ is summarized in the (fiducial or Bayesian) distribution $\phi \sim N(\hat{\phi}, c_2)$ where $\hat{\phi}$ and c_2 are known. Then by (16.3.1), the predictive distribution of Y has p.d.f.

$$p(y) = \frac{1}{2\pi\sqrt{c_1 c_2}} \int_{-\infty}^{\infty} \exp\{-\frac{1}{2c_1}(y - \phi)^2 - \frac{1}{2c_2}(\phi - \hat{\phi})^2\}d\phi.$$

This integral may be evaluated by completing the square in the exponent to produce a normal integral. After a bit of algebra, we find that $p(y)$ is the p.d.f. of a normal distribution with mean $\hat{\phi}$ and variance $c_1 + c_2$. Hence the predictive distribution is

$$Y \sim N(\hat{\phi}, c_1 + c_2).$$

An easier way to obtain this result is to write $Y \equiv \phi + \sqrt{c_1} Z_1$ where $Z_1 \sim N(0,1)$, and $\phi \equiv \hat{\phi} + \sqrt{c_2} Z_2$ where $Z_2 \sim N(0,1)$, independently of Z_1. Combining these gives

$$Y \equiv \hat{\phi} + \sqrt{c_1} Z_1 + \sqrt{c_2} Z_2$$

where $\hat{\phi}, c_1$ and c_2 are known constants. Now (6.6.6) and (6.6.7) give $Y \sim N(\hat{\phi}, c_1 + c_2)$ as before.

Example 16.3.1. Suppose that we have already observed n independent measurements $x_1, x_2, \ldots, x_n$ from $N(\mu, \sigma^2)$ with σ known, and that we wish to predict the average value $\overline{Y}$ of m future observations from the same distribution. From Example 16.1.3, the fiducial distribution of μ based on the x_i's is $\mu \sim N(\overline{x}, \sigma^2/n)$. The sampling distribution of $\overline{Y}$ is $\overline{Y} \sim N(\mu, \sigma^2/m)$. Hence by the discussion above, the predictive distribution is

$$\overline{Y} \sim N(\overline{x}, \sigma^2(\frac{1}{n} + \frac{1}{m})).$$

This distribution combines uncertainty due to lack of knowledge of μ with uncertainty due to random variation in $\overline{Y}$. If $n \to \infty$, then $\overline{x} \approx \mu$. The uncertainty due to lack of knowledge of μ is then negligible, and the predictive distribution becomes the sampling distribution of $\overline{Y}$. On the other hand, if $m \to \infty$, then uncertainty due to random variation in $\overline{Y}$ becomes negligible, and the predictive distribution becomes the fiducial distribution of μ.

If σ is also unknown, we can integrate over its fiducial distribution as well to obtain

$$\frac{\overline{Y} - \overline{x}}{\sqrt{s^2(\frac{1}{n} + \frac{1}{m})}} \sim t_{(n-1)}$$

where $s^2 = \frac{1}{n-1} \sum(x_i - \overline{x})^2$; see Section 16.4.

Example 16.3.2. Suppose that the normal linear regression model (15.1.1) has been fitted to n observed pairs (x_i, y_i), $i = 1, 2, \ldots, n$. We now wish to predict the value Y of the dependent variable when the independent variable has value x. For instance, in Example 15.2.1 we might wish to predict the systolic blood pressure Y of a particular woman aged $x = 50$ years.

If σ is known, the argument preceding the last example may be applied. The sampling distribution of Y is $N(\mu, \sigma^2)$ where $\mu = \alpha + \beta x$. One can argue that $\hat{\mu} = \hat{\alpha} + \hat{\beta}x$ carries all of the relevant information about μ. From Section 15.3, we have $\hat{\mu} \sim N(\mu, c\sigma^2)$ where

$$c = \frac{1}{n} + (x - \overline{x})^2/S_{xx}.$$

Hence, from Example 16.1.3, the fiducial distribution of μ is $N(\hat{\mu}, c\sigma^2)$. It now follows that the predictive distribution is

$$Y \sim N(\hat{\mu}, (1 + c)\sigma^2).$$

If σ is unknown, we replace σ^2 by $s^2 = \frac{1}{n-2} \sum \hat{\epsilon}_i^2$ to get

$$T \equiv \frac{Y - \hat{\mu}}{\sqrt{(1 + c)s^2}} \sim t_{(n-2)}.$$

A central 95% predictive interval for Y is then

$$Y \in \hat{\mu} \pm a \sqrt{s^2[1 + \frac{1}{n} + (x - \overline{x})^2/S_{xx}]}$$

where $P\{|t_{(n-2)}| \le a\} = 0.95$.

For instance, in Example 15.2.1, the central 95% predictive interval for the blood pressure of an individual woman aged 50 years is

$$Y \in 137.68 \pm 16.30.$$

From Section 15.3, a 95% confidence interval for the mean blood pressure of all women aged 50 years is

$$\mu \in 137.68 \pm 4.61.$$

The interval for Y is much wider than the interval for μ, because there is considerable variability in systolic blood pressure among women of the same age. Even if we knew μ exactly, we could not predict the value of Y very precisely.

16.4 Inferences from Predictive Distributions

Consider the situation described in Section 14.1. There are two unknown parameters, ϕ and σ, with ϕ being the parameter of primary interest. Two independent statistics $\hat{\phi}$ and V carry the relevant information about ϕ and σ; their sampling distributions are

$$\hat{\phi} \sim N(\phi, c\sigma^2); \quad U \equiv V/\sigma^2 \sim \chi^2_{(\nu)}.$$

If σ were known, $\hat{\phi}$ would be a sufficient statistic for ϕ, and inferences about ϕ would be based on the above sampling distribution of $\hat{\phi}$. If σ is unknown, this sampling distribution can no longer be used because it depends upon σ. However, we can obtain a predictive distribution for ϕ which does not depend upon σ, and then use this predictive distribution for inferences about ϕ.

The statistic V carries all of the information about σ when ϕ is unknown, and the pivotal quantity U satisfies the conditions for the fiducial argument. To obtain the fiducial distribution of σ, we replace V by its observed value νs^2, giving

$$\sigma^2 \equiv \nu s^2/U \quad \text{where} \quad U \sim \chi^2_{(\nu)}.$$

Now, by (16.3.1), the predictive distribution of $\hat{\phi}$ given s is

$$p(\hat{\phi}; \phi, s) = \int_0^\infty g(\hat{\phi}; \phi, \sigma) f(\sigma; s) d\sigma.$$

We can avoid formal integration in this case by using (6.9.11). We have

$$\hat{\phi} \equiv \phi + Z\sqrt{c\sigma^2} \quad \text{where} \quad Z \sim N(0,1)$$

and substituting for σ^2 gives

$$\hat{\phi} \equiv \phi + Z\sqrt{cvs^2/U} \equiv \phi + T\sqrt{cs^2}$$

where $T \equiv Z \div \sqrt{U/v} \sim t_{(v)}$ by (6.9.11). Hence predictive statements for $\hat{\phi}$ given s are obtained from

$$T \equiv \frac{\hat{\phi} - \phi}{\sqrt{cs^2}} \sim t_{(v)}. \qquad (16.4.1)$$

This is the same as (14.1.5) except that now S^2 has been replaced by its observed value s^2.

In this problem, S^2 plays the role of an ancillary statistic. Since its distribution does not depend upon ϕ, S^2 gives no direct information about the magnitude of ϕ. However, its observed value indicates the informativeness or precision of the experiment with respect to ϕ. By the arguments of Section 12.8, S^2 should be held fixed at its observed value s^2 in making inferences about ϕ. Thus it would seem appropriate to base inferences about ϕ on the predictive distribution (16.4.1), rather than on the sampling distribution (14.1.5).

In fact, one will obtain the same numerical values for significance levels and confidence intervals whether one uses (16.4.1) or (14.1.5), and so the distinction does not matter in this case. It does matter in more complicated situations, such as the Behrens-Fisher problem to be considered below.

Note that the pivotal quantity T in (16.4.1) satisfies the conditions set out in Section 16.1, and could be used to obtain a fiducial distribution for ϕ when σ is unknown.

Behrens-Fisher problem

The problem of comparing two normal means when the variance ratio is unknown is called the Behrens-Fisher problem; see Section 14.4.

Suppose that we wish to make inferences about $\phi_1 - \phi_2$ where $\hat{\phi}_1 \sim N(\phi_1, c_1\sigma_1^2)$ and $\hat{\phi}_2 \sim N(\phi_2, c_2\sigma_2^2)$. If σ_1 and σ_2 are known, we can base inferences for $\phi_1 - \phi_2$ on the sampling distribution of $\hat{\phi}_1 - \hat{\phi}_2$:

$$\hat{\phi}_1 - \hat{\phi}_2 \sim N(\phi_1 - \phi_2, c_1\sigma_1^2 + c_2\sigma_2^2).$$

Now suppose that σ_1 and σ_2 are unknown, and that all the available information concerning σ_1 and σ_2 is carried by the statistics V_1 and V_2. Furthermore, suppose that $\hat{\phi}_1, \hat{\phi}_2, V_1$ and V_2 are all independent, with

$$U_1 \equiv V_1/\sigma_1^2 \sim \chi^2_{(\nu_1)}; \qquad U_2 \equiv V_2/\sigma_2^2 \sim \chi^2_{(\nu_2)}.$$

The fiducial distributions of σ_1 and σ_2 are given by

$$\sigma_1^2 \equiv \nu_1 s_1^2/U_1; \qquad \sigma_2^2 \equiv \nu_2 s_2^2/U_2$$

where $s_1^2 = v_1/\nu_1$ and $s_2^2 = v_2/\nu_2$. Integrating over these fiducial distributions will give a predictive distribution for $\hat{\phi}_1 - \hat{\phi}_2$ which depends only on $\phi_1 - \phi_2$.

The following is an equivalent derivation which yields the same result but in a more convenient form. The predictive distribution for $\hat{\phi}_1$ given s_1 is given by

$$\hat{\phi}_1 \equiv \phi_1 + T_1\sqrt{c_1 s_1^2} \quad \text{where} \quad T_1 \sim t_{(\nu_1)}$$

and similarly, the predictive distribution of T_2 given s_2 is

$$\hat{\phi}_2 \equiv \phi_2 + T_2\sqrt{c_2 s_2^2} \quad \text{where} \quad T_2 \sim t_{(\nu_2)}.$$

Thus the predictive distribution of $\hat{\phi}_1 - \hat{\phi}_2$ is given by

$$\hat{\phi}_1 - \hat{\phi}_2 \equiv \phi_1 - \phi_2 + T_1\sqrt{c_1 s_1^2} - T_2\sqrt{c_2 s_2^2}$$

$$\equiv \phi_1 - \phi_2 + T\sqrt{c_1 s_1^2 + c_2 s_2^2}$$

where T is a linear combination of T_1 and T_2:

$$T \equiv T_1\sqrt{\frac{c_1 s_1^2}{c_1 s_1^2 + c_2 s_2^2}} - T_2\sqrt{\frac{c_2 s_2^2}{c_1 s_1^2 + c_2 s_2^2}}$$

The distribution of a linear combination

$$T \equiv T_1\cos\theta - T_2\sin\theta ,$$

where $T_1 \sim t_{(\nu_1)}$ and $T_2 \sim t_{(\nu_2)}$ are independent, is called the Behrens-Fisher distribution. It is tabulated in the Fisher and Yates

Statistical Tables for Biological, Agricultural and Medical Research.   In this case we have

$$\tan \theta = \frac{\sin \theta}{\cos \theta} = \sqrt{\frac{c_1 s_2^2}{c_2 s_1^2}}$$

so that θ is a function of the observed variance ratio s_2^2/s_1^2.

When σ_1 and σ_2 are unknown, inferences about $\phi_1 - \phi_2$ may be based on the pivotal quantity

$$T \equiv \frac{(\hat{\phi}_1 - \hat{\phi}_2) - (\phi_1 - \phi_2)}{\sqrt{c_1 s_1^2 + c_2 s_2^2}}$$

which is referred to tables of the Behrens-Fisher with parameters ν_1, ν_2, and θ, where

$$\theta = \operatorname{Tan}^{-1} \sqrt{\frac{c_1 s_1^2}{c_2 s_2^2}} \quad .$$

A similar result can be derived for inferences about any linear combination of ϕ_1 and ϕ_2.

Note that the Behrens-Fisher distribution arises in connection with the <u>predictive</u> distribution of $\hat{\phi}_1 - \hat{\phi}_2$, in which s_1 and s_2 are held fixed at their observed values. Alternatively, one might consider the <u>sampling</u> distribution of

$$T' \equiv \frac{(\hat{\phi}_1 - \hat{\phi}_2) - (\phi_1 - \phi_2)}{\sqrt{c_1 S_1^2 + c_2 S_2^2}}$$

where now S_1^2 and S_2^2 are random variables. The distribution of T' in repeated samples is <u>not</u> Behrens-Fisher, and it will depend upon the unknown parameter $\lambda = \sigma_1^2/\sigma_2^2$.

Many statisticians are of the opinion that inferences should always be based on sampling distributions, so that probabilities can be interpreted directly as relative frequencies in repetitions of the experiment. As a result, they do not accept the above solution based on the predictive distribution of $\hat{\phi}_1 - \hat{\phi}_2$. On the other hand, it seems appropriate that S_1^2 and S_2^2 should be treated as ancillary statistics and held fixed at their observed values as is the case in the above solution. If one insists that S_1^2 and S_2^2 be treated as ancillary statistics, then inferences cannot be based on a sampling

303

distribution.

16.5 Testing a True Hypothesis

Sometimes one can generate intervals of "reasonable" values
for an unknown quantity y by the device of testing a true hypothesis
H. One assumes a value for y, carries out a test of significance,
and finds the significance level SL(y). A small significance level
indicates an inconsistency, and if H is known to be true, doubt is
cast on the value assumed for y. One can define a 95% interval or
region as the set of all values of y such that SL(y) ≥ 0.05.
Several examples will be given to illustrate this procedure.

Example 16.5.1. Suppose that $\bar{X} \sim N(\mu_1, \sigma^2/n)$ and $\bar{Y} \sim N(\mu_2, \sigma^2/m)$
independently of $\bar{X}$, where σ is known. Given observed values of
$\bar{X}$ and $\bar{Y}$, we can test $H:\mu_1 - \mu_2$ using the result that
$\bar{X} - \bar{Y} \sim N(0, \sigma^2(\frac{1}{n} + \frac{1}{m}))$ under the hypothesis. The significance level
will be 5% or more if and only if

$$-1.96 \le \frac{\bar{x} - \bar{y}}{\sqrt{\sigma^2(\frac{1}{n} + \frac{1}{m})}} \le 1.96. \tag{16.5.1}$$

Now suppose that we don't know $\bar{y}$ but we do know that $\mu_1 = \mu_2$. This
would be the case if $\bar{Y}$ were the average of m future observations
to be taken from the same $N(\mu, \sigma^2)$ distribution as the original sample $x_1, x_2, \ldots, x_n$; see Example 16.3.1. Now (16.5.1) yields a 95%
interval for $\bar{y}$:

$$\bar{y} \in \bar{x} \pm 1.96 \sqrt{\sigma^2(\frac{1}{n} + \frac{1}{m})}.$$

This interval consists of the values of $\bar{Y}$ such that a test of the
true hypothesis $\mu_1 = \mu_2$ would produce a significance level of 5% or
more. The same interval can be obtained as the central 95% interval
in the predictive distribution of Y; see Example 16.3.1.

Example 16.5.2. Suppose that X and Y are independent variates
with X ∼ binomial (n, p_1) and Y ∼ binomial (m, p_2). Given observed
values of X and Y, we can test the hypothesis $H:p_1 = p_2$ as in
Section 12.2. If we don't know Y but we do know that $p_1 = p_2$, this
test can be used to produce an interval of values for Y. With small
samples the procedure would be to select a value for Y, carry out an
exact test based on the hypergeometric distribution as in Section 12.2,
and determine the exact significance level. This could be repeated

for all possible values of Y, and the 95% interval would consist of those values of Y for which the significance level was 5% or more.

Under conditions described in Section 6.8, the hypergeometric distribution (12.2.2) can be approximated by a normal distribution with mean and variance

$$\mu = \frac{nt}{n+m} \; ; \qquad \sigma^2 = \frac{ntm(n+m-t)}{(n+m)^2(n+m-1)}$$

where $t = x + y$. The approximate significance level will be 5% or more for $-1.96 \le \frac{x-\mu}{\sigma} \le 1.96$, or equivalently, for $(x-\mu)^2 \le 3.841\sigma^2$. Substituting for μ and σ^2 gives a quadratic inequality which can be solved for t and hence for $y = t - x$.

For instance, suppose that we previously observed 12 successes in 20 trials, and want to predict y, the number of successes in 30 future trials with the same success probability. Taking $n = 20$, $x = 12$ and $m = 30$ in the above gives the inequality

$$(12 - \frac{20t}{50})^2 \le 3.841 \frac{600t(50-t)}{2500(50-1)}$$

which is satisfied for $22 \le t \le 37$. Hence the 95% interval for $y = t - x$ is $10 \le y \le 25$. If we select any value of y outside this interval and then test the true hypothesis $H : p_1 = p_2$, we will get a significance level less than 5%.

Example 16.5.3. Suppose that a lake contains n tagged fish and m untagged fish, where n is known but m is unknown. A sample of fish is caught, and is found to contain x tagged fish and y untagged fish. What can be concluded about m?

We assume $n + m$ independent trials, and let the probability that a particular fish is caught be p_1 for tagged fish, and p_2 for untagged fish. Given m, we could test $H : p_1 = p_2$ as in the preceding example. If we assume that H is true, then a small significance level casts doubt on the value chosen for m. As in the preceding example, an approximate 95% interval for m can be obtained as the set of values of m for which $(x - \mu)^2 \le 3.841\sigma^2$.

For instance, suppose that there are 110 tagged fish in the lake, and that the sample contains 20 tagged and 478 untagged fish. Then $n = 110$, $x = 20$, $y = 478$ and $t = x + y = 498$, and the inequality becomes

$$(20 - \frac{54780}{110 + m})^2 \leq 3.841 \, \frac{54780m(m-388)}{(110+m)^2(109+m)}.$$

This inequality is satified for $1794 \leq m \leq 3986$. For any value of m outside this interval, the observed values $x = 20$, $y = 478$ would give a significance level less than 5% in a test of $H: p_1 = p_2$.

Since $n = 110$, an approximate 95% interval for the total population $n + m$ is given by $1904 \leq n + m \leq 4096$.

Note that the inference for m depends on the assumption that $p_1 = p_2$. This assumption is perhaps questionable, because a fish that has once been caught and tagged may have a greater (or smaller) probability of being caught subsequently.

Example 16.5.4. Suppose that the normal linear regression model (15.1.1) has been fitted to n observed pairs (x_i, y_i), $i = 1, 2, \ldots, n$. Now we observe an additional value y, for which the corresponding value of the independent variable x is unknown. The problem is to make inferences about x.

Suppose that a value is given for x. Then from the normal linear model, the estimated mean value of Y is

$$\hat{\mu} \equiv \hat{\alpha} + \hat{\beta}x \sim N(\mu, c\sigma^2)$$

where $c = c(x) = \frac{1}{n} + (x - \bar{x})^2/S_{xx}$. We take $Y \sim N(\mu_1, \sigma^2)$ and test $H: \mu_1 = \mu$. Under H, we have

$$Y - \hat{\mu} \sim N(0, (1 + c)\sigma^2)$$

and the significance level is 5% or more if and only if

$$(y - \hat{\mu})^2 \leq 3.841(1 + c)\sigma^2.$$

Substitution for $\hat{\mu}$ and c gives

$$[y - \hat{\alpha} - \hat{\beta}x]^2 \leq 3.841\sigma^2 [1 + \frac{1}{n} + (x - \bar{x})^2/S_{xx}].$$

The 95% interval for x is the set of x-values for which this quadratic inequality is satisfied. For any value of x outside this interval, a test of the hypothesis $E(Y) = \alpha + \beta x$ will give a significance level less than 5%.

For σ unknown, we replace $3.841\sigma^2$ above by $a^2 s^2$, where $s^2 = \frac{1}{n-2} \sum \hat{\epsilon}_i^2$ and $P\{|t_{(n-2)}| \geq a\} = 0.05$, giving

$$[y - \hat{\alpha} - \hat{\beta}x]^2 \leq a^2 s^2 [1 + \frac{1}{n} + (x - \bar{x})^2 / S_{xx}]. \qquad (16.5.2)$$

If we take x as known and y as unknown in this inequality, we get the central 95% predictive interval for a future observation Y; see Example 16.3.2. For y known but x unknown, we get a 95% interval for x. This interval consists of all values of x such that the observed value of the future observation Y belongs to its 95% predictive interval.

If the slope β were zero, then an observed y would not determine a value of x. Thus one might anticipate difficulties when the estimated slope $\hat{\beta}$ is not significantly different from zero. A test of H: β = 0 is based on

$$\frac{\hat{\beta} - 0}{\sqrt{s^2 / S_{xx}}} \sim t_{(n-2)},$$

and the condition for $\hat{\beta}$ to be significantly different from zero at the 5% level is

$$\hat{\beta}^2 > a^2 s^2 / S_{xx}$$

where $P\{|t_{(n-2)}| \geq a\} = 0.05$. If this condition is not satisfied, the 95% interval for x will be either the entire real line, $-\infty < x < \infty$, or else the entire real line with a finite interval removed. These results can be derived by examining the discriminant and the sign of the second-degree term in the quadratic inequality (16.5.2).

The problem of estimating the independent variable x given a value of the dependent variable y is called the calibration problem. It arises when the quantity x which is of interest cannot be measured directly, but we must make do with a measurement y which is related to x. The equation relating E(Y) to x is called the calibration curve. In this case we have assumed a linear calibration curve $E(Y) = \alpha + \beta x$ with independent $N(0, \sigma^2)$ errors.

APPENDIX A. ANSWERS TO SELECTED PROBLEMS

9.1.1 $\hat{\lambda} = \sum x_i/n = 0.25$

9.1.4 $\hat{\theta} = (2x_1 + x_2)/2n = 0.55$;   exp. freq.   30.25, 49.50, 20.25

9.1.7 $L(N) = N^{-n}$ for $N \geq$ largest sample value $x_{(n)}$, and $L(N) = 0$ otherwise. $L(N)$ decreases as N increases, so $\hat{N} = x_{(n)}$.

9.1.9 $L(\theta) = \Pi p_i^{f_i}$ where $p_i = \theta^{i-1}(1-\theta)$ for $i = 1,2,3$ and $p_4 = \theta^3$.
$\hat{\theta} = (f_2 + 2f_3 + 3f_4)/(f_1 + 2f_2 + 3f_3 + 4f_4) = 0.5$
Exp. freq. 100,50,25,25; poor agreement with obs. freq.

9.3.2 $L(\mu) = \Pi p_i^{f_i}$ where $p_i = \mu^i e^{-\mu}/i!(1-e^{-\mu})$ for $i = 1,2,\ldots$.
$\ell'(\mu) = 0$ gives equation for $\hat{\mu}$; $\hat{\mu} = 3.048$.

9.3.4 $L(\theta) = \theta^m(1-\theta)^{M-m}\theta^{2n}(1-\theta^2)^{N-n}$
$\ell'(\theta) = 0$ gives $(M+2N)\theta^2 + (M-m)\theta - (m+2n) = 0$.

9.4.1   50% LI (0.168,0.355);   10% LI (0.116,0.460)

9.4.4 $\ell(\theta) = 54 \log (\frac{1}{6} + \theta) + 46 \log (\frac{1}{6} - 2\theta)$; $\ell(\hat{\theta}) = -175.74$
$r(0) = \ell(0) - \ell(\hat{\theta}) = -3.44$; $R(0) = 0.032$.   It is quite unlikely that $\theta = 0$.

9.4.7 $R(b) = 1$ for $b = 9,10$ ($\hat{b}$ is not unique)
$R(b) \geq .5$ for $b = 6,7,\ldots,17$; $R(b) \geq .1$ for $b = 5,6,\ldots,32$.

9.5.1 $L(\theta) = \theta^{-2n}\exp(-\sum x_i/\theta)$; $\hat{\theta} = \frac{1}{2n} \sum x_i$; $R(\theta) = L(\theta)/L(\hat{\theta})$.

9.5.5 $L(c) = e^{nc}$ for $0 < c \leq$ smallest sample value; $L(c)$ increases with c.
$R(c) = \exp\{n(c - x_{(1)})\}$ for $0 < c \leq x_{(1)}$;
$100p\%$ LI: $x_{(1)} + \frac{1}{n} \log p \leq c \leq x_{(1)}$.

9.5.7 $L(\theta) = \theta^{-n}$ for $\theta \leq x_1, x_2, \ldots, x_n \leq 2\theta$; i.e. for $\theta \leq x_{(1)}$ and $x_{(n)} \leq 2\theta$. Since $L(\theta)$ is decreasing, $\hat{\theta} = \frac{1}{2} x_{(n)}$; $R(\theta) = (\hat{\theta}/\theta)^n$ for $\hat{\theta} \leq \theta \leq x_{(1)}$.

9.6.2 $L(\theta) = p^k(1-p)^{n-k}$ where $p = 1 - e^{-T/\theta}$; $\hat{\theta} = T/\log(\frac{n}{n-k})$;
$R(\theta) = L(\theta)/L(\hat{\theta})$.

9.6.4 $f(x) = 4x\theta^{-2}e^{-2x/\theta}$; $L(\theta) = \theta^{-2m}e^{-2\sum x_i/\theta}(1 + \frac{2T}{\theta})^{n-m}e^{-2T(n-m)/\theta}$
$\ell'(\theta) = 0$ gives quadratic equation for $\hat{\theta}$.
$\hat{\theta} = 1813.42$;   10% LI (1382,2449)

9.7.3 $P(T \leq 100t) = 1 - e^{-100t/\theta} = 1 - \beta^t$
$L(\beta) = p_1^{29}p_2^{22}\ldots p_7^8 = \beta^{234}(1 - \beta)^{92}(1+\beta)^{19}$ for $0 < \beta < 1$.
$\ell'(\beta) = 0$ gives quadratic equation for $\hat{\beta}$.
$\hat{\beta} = 0.7245$ and $\hat{\theta} = -100/\log\hat{\beta} = 310.3$
10% LI $0.670 \leq \beta \leq 0.775 \longleftrightarrow 249 \leq \theta \leq 392$

50% LI $0.695 \leq \beta \leq 0.753 \longleftrightarrow 275 \leq \theta \leq 352$.

9.8.3　$L(\lambda) = \lambda^n \exp\{-\lambda t\}$　for　$\lambda > 0$　where　$T \equiv \sum X_i^2$.

P.d.f. of　$Y \equiv 2\lambda X^2$　is　$f(x) \cdot \left|\dfrac{dx}{dy}\right| = \dfrac{1}{2} e^{-y/2}$　for　$y > 0$, which is　$\chi_{(2)}^2$.

Now　$Z \equiv 2\lambda T \equiv \sum 2\lambda X_i^2 \sim \chi_{(2n)}^2$　by　(6.9.3).

P.d.f. of　T　is　$g(z) \cdot \left|\dfrac{dz}{dt}\right| = k(2\lambda t)^{n-1} e^{-\lambda t} \cdot 2\lambda$.

9.8.6　$f(x_1, x_2, \ldots, x_n) = \left(\dfrac{1}{2\theta}\right)^n$　for　$-\theta < x_1, x_2, \ldots, x_n < \theta$.

$L(\theta) = \theta^{-n}$　for　$\theta > t$,　$L(\theta) = 0$　otherwise, where

$T \equiv \max\{X_{(n)}, -X_{(1)}\}$.　T　is sufficient statistic for　θ.

9.8.10　$\log \dfrac{\theta}{1-\theta}$,　$\log \mu$,　$\dfrac{1}{\theta}$　(or any constant multiples of them)

9.9.2　$T \equiv \sum a_i X_i$;　minimizing　$\sum a_i^2 \sigma_i^2$　subject to　$\sum a_i = 1$　gives

$a_j = \sigma_j^{-2} / \sum \sigma_i^{-2}$.

10.1.2　$L(\mu) = \exp\{-n(\overline{x} - \mu)^2/2\sigma^2 - m(\overline{y} - \mu)^2/2k\sigma^2\}$

$\qquad = c \exp\{-[(nk + m)\mu^2 - 2(kn\overline{x} + m\overline{y})]/2k\sigma^2\}$

which depends on the　X_i's　and　Y_i's　only through　$kn\overline{X} + m\overline{Y}$.

$T \sim N((kn + m)\mu, \; k(kn + m)\sigma^2)$　by　(6.6.8)　and　(6.6.7).

10.1.6　$\hat{\mu}_1 = \dfrac{1}{3}(2x_1 - x_2 + x_3)$,　$\hat{\mu}_2 = \dfrac{1}{3}(2x_2 - x_1 + x_3)$.

10.1.8　$L(\lambda, c) = \lambda^n \exp\{-\lambda \sum x_i + n\lambda c\}$　for　$c \leq x_{(1)}$　and　$0 < \lambda < \infty$

$\hat{c} = x_{(1)}$;　$\hat{\lambda} = (\overline{x} - x_{(1)})^{-1}$

$R(\lambda, c) = (\lambda/\hat{\lambda})^n \exp\{n - n\lambda(\overline{x} - c)\}$　for　$\lambda > 0$, $0 \leq c \leq x_{(1)}$.

$\overline{X}, X_{(1)}$　is a pair of sufficient statistics for　(λ, c).

10.1.11　$\ell(\alpha, \beta) = \sum x_j \log \alpha + \sum j x_j \log \beta - \alpha \sum \beta^j$.

ML equations　$\sum x_j - \hat{\alpha} \sum \hat{\beta}^j = 0$;　$\sum j x_j - \hat{\alpha} \sum_j \hat{\beta}^j = 0$.

Substitute　$\hat{\alpha} = \sum x_j / \sum \hat{\beta}^j$　in 2nd equation; solve numerically for　$\hat{\beta}$.

10.3.4　ML eqns　$\sum(n_i - a_i) = 0$,　$\sum(n_i - a_i) d_i = 0$　where　$a_i = (n_i - x_i)/(1 - p_i)$

2nd derivatives are　$-\sum b_i, -\sum b_i d_i, -\sum b_i d_i^2$　where

$b_i = a_i p_i / (1 - p_i)$.

$\hat{\alpha} = -0.0002126$, $\hat{\beta} = -0.03297$

$-0.0354 \leq \beta \leq -0.0306$;　for　β　outside this interval, R < 0.1 for all　α.

$0.99896 \leq e^\alpha \leq 0.999992$;　for　e^α　outside this interval; R < 0.1 for all　β.

10.5.1　$R_{max}(\lambda) = (\lambda/\hat{\lambda})^n \exp\{n - n\lambda/\hat{\lambda}\}$　for　$0 < \lambda < \infty$;

$R_{max}(c) = (\overline{x} - x_{(1)})^n (\overline{x} - c)^{-n}$　for　$0 < c \leq x_{(1)}$.

10.5.4　$L_{max}(\lambda) = \lambda^m (1 - e^{-\lambda T})^m \exp\{-\lambda \sum t_i\}$;　$R_{max}(\lambda) = L_{max}(\lambda)/L_{max}(\hat{\lambda})$.

11.2.2　(a)　$\hat{\mu} = 1.383$, exp. freq. $15.04, 20.81, 14.39, 9.75 (\geq 3)$

$\qquad$ SL $\approx P(\chi_{(2)}^2 \geq 1.14) \approx 0.6$;　model is consistent with data.

(b)　$\hat{\mu} = 2.267$, exp. freq. $6.22, 14.10, 15.98, 12.07, 6.84, 4.79 (\geq 5)$

$$SL \approx P(\chi^2_{(4)} \geq 29.35) < 0.001.$$ Injuries tend to occur in groups rather than singly.

11.2.6 $\binom{4}{2}/\binom{8}{2}$;   exp. freq. 6,6,10; $SL \approx P(\chi^2_{(2)} \geq 3.58) = 0.17$.
Results are consistent with random selection of seats.

11.3.2   Define $X \equiv$ # of times benzedrine gives more rapid recovery.
If drugs are equally good, X has binomial $(20, \frac{1}{2})$ distribution.
$SL = P(D \geq 3.2) = P(X \leq 6) + P(X \geq 14) = 0.115$.
Results do not strongly contradict the hypothesis.

11.3.5   Exp. freq. 80 good and 20 defective in 100 seeds.
1st customer: $d = 3.06$, SL ≈ 0.08.   Weak evidence against
$H:p = .8$.
    No customer has a strong case against merchant.
Exp. freq. 320 good and 80 defective in 400 seeds.
Obs. freq. 300 and 100 giving $d = 5.33$, SL ≈ 0.022.
Combined results give strong evidence that $p \neq .8$.

11.3.8   Proceed as in Example 11.3.2, but with $p_0 = p_2 = \frac{3}{14}$, $p_1 = \frac{8}{14}$.
Exp. freq. 2.14,5.71,2.14;   obs. freq. 1,5,4; $d = 2.31$
There are 11 sets of freq. with $D < 2.31$,   and their total
probability is   0.651,   giving   SL = 0.349.

11.5.2   Required $N(0,1)$ quantiles are $\pm 2.20, \pm 1.73, \pm 1.48, \ldots, \pm 0.035$.
Quantile plot shows that the actual mark distn is skewed to
the left.   Lowest marks are more spread out than one would expect in a normal distn, while highest marks are less spread out.

11.5.5 $F(x) = \frac{1}{2}(1 - \cos x)$ for $0 < x < \pi$; $Q_p = \text{Cos}^{-1}(1 - 2p)$.
Required quantiles are 0.64,1.16,1.56,1.98,2.50.
Points are scattered about a straight line through the origin
with unit slope, showing good agreement with the model.

11.5.8   Plot ordered sample values versus $\frac{i-.5}{40}$ - quantiles of $U \equiv \frac{2X}{\theta}$
for $i = 1,2,\ldots,25$.   C.d.f. of U is $G(u) = 1 - (1+u)e^{-u}$.
Required quantiles, which may be read from a graph of $G(u)$,
are
     0.17   0.30   0.40   0.49   ...   1.89   1.98   2.07.

The 25 points are scattered about a straight line through the
origin (slope $\approx \theta/2$), showing good agreement with model.

11.5.10   Plot of log lifetimes versus $\log[-\log(1-p)]$ for
$p = \frac{1}{36}, \frac{3}{36}, \ldots, \frac{35}{36}$ gives points scattered about a straight line;
excellent agreement with Weibull model.   Plot of log lifetimes
versus $N(0,1)$ quantiles shows poor agreement; actual distn
of log lifetimes is skewed to the left.

12.2.2 $X \equiv$ # of def in 1st process $\sim$ binomial $(4,p_1)$
$Y \equiv$ # of def in 2nd process $\sim$ binomial $(16,p_2)$
Under $H:p_1 = p_2 = p$, $T \equiv X + Y$ is suff stat for p.   Test is
based on condit distn of X given $T = 4$.   This distn is hypergeometric. Prob. ranking or distance from mean gives SL = 0.162.

12.2.3 $X \equiv$ # good before rth def in 1st process $\sim$ neg bin (r,p_1)
$Y \equiv$ # good before rth def in 2nd process $\sim$ neg bin (r,p_2)
Under $H:p_1 = p_2 = p$, $T \equiv X + Y$ is suff for p.
$T \equiv$ # good before (2r)th def $\sim$ neg bin $(2r,p)$
$$g(x|t) = \binom{x+r-1}{r-1}\binom{t-x+r-1}{r-1}/\binom{t+2r-1}{2r-1}$$

Prob ranking or distance from mean gives   SL = 0.194.

Note: $L(p_1,p_2)$ is same as in Problem 12.2.2, but   SL   depends on the stopping rule used.

12.2.8   Under $H:\mu_1 = \mu_2 = 2\mu_3$, $T \equiv X_1 + X_2 + X_3$ is suff stat for μ_3 and has Poisson distn with mean $5\mu_3$.   Condit distn of X_i's given $T = t$ is multinomial $(t;\frac{2}{5},\frac{2}{5},\frac{1}{5})$.   Exact test as in Example 11.3.2.   For   t   large, $D \approx \chi^2_{(2)}$.

12.2.11 $T \equiv X_1 + X_4$ is suff stat for   p;   T has bin (n,p)   distn. Test of $H:p = \frac{1}{2}$ as in Example 12.1.1. $\hat{p} = \frac{t}{n}$;   exp freq are $\frac{n}{2}$ times $\hat{p}, 1-\hat{p}, 1-\hat{p}, \hat{p}$ D   is approx. $\chi^2_{(2)}$;   exact   SL   computed from

$$f(x_1,\ldots,x_4 | t) = (\begin{smallmatrix} t \\ x_1 \end{smallmatrix})(\frac{1}{2})^t \cdot (\begin{smallmatrix} n-t \\ x_2 \end{smallmatrix})(\frac{1}{2})^{n-t}.$$

12.3.2   Obs. freq. 105,302,76,335;   exp. freq. under hypothesis of independence are   90.06,316.94,90.94,320.06.   SL $\approx$ 0.012; strong evidence that vitamin C does have an effect.

12.3.3   Exact: SL = 0.00054   using prob. ranking or distance from mean. Approx. 0.00031 (uncorrected), 0.00122 (corrected).

12.4.1   (a)   SL $\approx P(\chi^2_{(4)} \geq 1.86) = 0.76$;   no evidence against hypothesis

(b)   Obs. freq. 126,271,132;   exp. freq. 132.25,264.5,132.25 SL $\approx P(\chi^2_{(2)} \geq 0.46) = 0.8$;   hypothesis is consistent with data.

12.4.5   (a)   Above   9(15)   6(8)   12(9)   23(18) Below   21(15)   10(8)   6(9)   13(18)

SL $\approx P(\chi^2_{(3)} \geq 10.58) = 0.014$;   strong evidence that birth weight is not independent of parental smoking habits.

(b)

Mother Smokes	F	$\bar{F}$	Mother doesn't	F	$\bar{F}$
Above	9(9.8)	6(5.2)	Above	12(11.7)	23(23.3)
Below	21(20.2)	10(10.8)	Below	6(6.3)	13(12.7)

Given mother's smoking habits, there is no evidence that birth weight depends on father's smoking habits.

12.4.8   (a)

Litter size	1	2	3	4
No. females	12(10)	70(80)	160(144)	184(192)
No. males	8(10)	90(80)	128(144)	200(192)

SL $\approx P(\chi^2_{(3)} \geq 7.52) = 0.06$;   weak evidence that p_i's   are not equal.

(b)   If p_i's   are equal, $\hat{p} = \frac{426}{852} = \frac{1}{2}$.   Exp. freq. in ith row are computed from $(\begin{smallmatrix} i \\ x \end{smallmatrix})(\frac{1}{2})^i$.   There are 9 d.f. since each row total is fixed and   p   is estimated. SL $\approx P(\chi^2_{(9)} \geq 13.99) = 0.12$;   data are consistent with binomial model.

(c)   Exp. freq. in ith row now computed from binomial $(i, \hat{p}_i)$ distribution; $\hat{p}_1 = \frac{12}{20}$, $\hat{p}_2 = \frac{70}{160}$, $\hat{p}_3 = \frac{160}{288}$, $\hat{p}_4 = \frac{184}{384}$. Degrees of freedom   10 - 4 = 6   since p_1, p_2, p_3, p_4 are estimated.

12.5.1 $\bar{x} = 227.49$,   n = 10.   If $\mu = 226$, $z_{obs} = 3.62$ and

$SL = P(|Z| \geq 3.62) = 0.0003$. If $\mu = 229$, $z_{obs} = -3.67$, and
$SL = P(|Z| \geq 3.67) = 0.0002$. Strong evidence against hypothesized
value in each case. Since $P(|Z| \geq 1.96) = 0.05$, SL is exactly
5% when $z_{obs} = \pm 1.96$, giving $\mu_0 = \bar{x} \pm 1.96\sqrt{\sigma^2/n}$. $SL \geq 0.05$
for $226.68 \leq \mu_0 \leq 228.30$.

12.5.5 $L(\mu_1, \mu_2) = \exp\{-[(\bar{x} - \mu_1)^2 + (\bar{y} - \mu_2)^2]/2\sigma^2\}$

$\hat{\mu}_1 = \bar{x}$, $\hat{\mu}_2 = \bar{y}$; if $\mu_1 = \mu_2$, then $\tilde{\mu}_1 = \tilde{\mu}_2 = \dfrac{n\bar{x} + m\bar{y}}{n + m}$

$D \equiv \sigma^{-2}[n(\bar{X} - \tilde{\mu}_1)^2 + m(Y - \tilde{\mu}_2)^2] \equiv (\bar{X} - \bar{Y})^2/\sigma^2(\frac{1}{n} + \frac{1}{m})$.

13.1.3 $\sum(x_i - 3727)^2 = 26.103$, $n = 15$. Tables of $\chi^2_{(15)}$ give
$6.262 \leq 26.103/\sigma^2 \leq 27.49 \longleftrightarrow 0.950 \leq \sigma^2 \leq 4.168$.

13.2.1 $n = 200$, $x = 120$, $r(p) = 120 \log\frac{p}{.6} + 80 \log\frac{1-p}{.4}$.
$r(p) \geq -1.92$ for $0.531 \leq p \leq 0.666$ (95% interval)
$r(p) \geq -3.32$ for $0.509 \leq p \leq 0.686$ (99% interval)

13.3.1 Using estimated variance: $0.532 \leq p \leq 0.668$, $0.511 \leq p \leq 0.689$.
Using actual variance: $0.531 \leq p \leq 0.665$, $0.509 \leq p \leq 0.685$.

13.3.6 $\dfrac{\partial^2 \log L}{\partial \phi^2} = \dfrac{\partial^2 \log L}{\partial \theta^2}(\dfrac{d\theta}{d\phi})^2 + \dfrac{\partial \log L}{\partial \theta}\dfrac{d^2\theta}{d\phi^2}$.

At the maximum, $\dfrac{\partial \log L}{\partial \theta} = 0$, and the result follows.

14.2.2 $\bar{x} = 61.21$, $s^2 = 14.74$; 95% confidence intervals are
$\mu \in 61.21 \pm 2.201/\sqrt{14.74/12} = 61.21 \pm 2.44$;
$3.816 \leq 11s^2/\sigma^2 \leq 21.92 \longleftrightarrow 7.40 \leq \sigma^2 \leq 42.48$.

14.2.3 Assuming n to be large, 95% confidence interval for μ has
width $2(1.96)\sqrt{\sigma^2/n}$. For width 2, we require $n = (1.96)^2\sigma^2$.
Variance estimate is $s^2 = \dfrac{80}{9}$, so $n \approx 34$. Advise about 25
additional measurements.

14.3.1 Analyse differences $X_i = B_i - A_i$.
There are 9 positive and 0 negative differences. Assuming independence among patients and no difference in drugs,
$SL = P(0 \text{ or } 9 \text{ positive out of } 9) = 2^{-8} = 0.004$. Assuming differences are independent $N(\mu, \sigma^2)$, we have $n = 10$, $\bar{x} = 1.58$,
$s^2 = 1.513$, $t_{obs} = 4.062$; $SL = P(|t_{(9)}| \geq 4.062) \approx 0.005$.

14.3.4 Assuming differences are independent $N(\mu, \sigma^2)$, we have $n = 6$,
$\bar{x} = 2.5$, $s^2 = 5.10$, $t_{obs} = 2.71$, $SL \approx 0.04$. There is some
evidence that Brand A is superior. However, brand A tires were
always tested first. It would be better to test brand A first
on 3 randomly chosen cars, and brand B first on the other 3
cars.

14.4.2 Pooled var. est. $s^2 = \dfrac{10 \times 3.79 + 15 \times 4.17}{10 + 15} = 4.02$ with 25 d.f.
$\mu_1 \in \bar{x} \pm 2.060\sqrt{s^2/11} = 6.23 \pm 1.25$;
$\mu_2 \in \bar{y} \pm 2.060\sqrt{s^2/16} = 12.74 \pm 1.03$;
$\mu_1 - \mu_2 \in \bar{x} - \bar{y} \pm 2.060\sqrt{s^2(\frac{1}{11} + \frac{1}{16})} = -6.51 \pm 1.62$.

14.4.4 $\lambda = (s_1/s_2)\sqrt{\dfrac{n(n-1)}{m(m-1)}}$.

14.5.2 $\frac{1}{6.26} \leq s_1^2/\lambda s_2^2 \leq 5.19 \longleftrightarrow 0.0445 \leq \lambda \leq 1.446$

14.5.4 $V_1 \equiv \sum(X_i - 1)^2$ and $V_2 \equiv \sum(Y_i - 1)^2$ are suff. stat. for σ_1

and σ_2. $V_1/\sigma_1^2 \sim \chi_{(4)}^2$, $V_2/\sigma_2^2 \sim \chi_{(4)}^2$, independent. Ratio

of mean squares is $(V_1/4\sigma_1^2) \div (V_2/4\sigma_2^2) \equiv V_1 \div \lambda V_2 \sim F_{4,4}$.

Observed $v_1/v_2 = 50/366$.

$\frac{1}{6.39} \leq v_1/\lambda v_2 \leq 6.39 \longleftrightarrow 0.0214 \leq \lambda \leq 0.873$

14.6.1 $Q = 9.3118$ with 3 d.f., $s^2 = 0.2389$ with 19 d.f.

$F_{obs} = 12.99$; $SL = P(F_{3,19} \geq 12.99) < 0.001$ from Table B5.

Strong evidence that means are not equal.

Obs. value of (14.6.3) is 3.01; $SL \approx P(\chi_{(3)}^2 \geq 3.01) > 0.25$.

No evidence that variances are unequal.

14R2 $n = 7$, $\bar{x} = 1.486$, $s_1^2 = 2.495$; $m = 7$, $\bar{y} = 3.214$, $s_2^2 = 2.898$

$F_{obs} = s_2^2/s_1^2 = 1.16$; $SL = 2P\{F_{6,5} \geq 1.16\} > 0.2$. No evidence

of unequal variances. $s^2 = 2.696$ with 12 d.f.

$4.404 \leq 12s^2/\sigma^2 \leq 23.34 \longleftrightarrow 1.386 \leq \sigma^2 \leq 7.346$

$\mu_2 - \mu_1 \in \bar{y} - \bar{x} \pm 2.179 \sqrt{s^2(\frac{1}{n} + \frac{1}{m})} = 1.728 \pm 1.912$.

For each pair, choose one at random to get drug 1, and give

his brother drug 2. Analyse differences as in Section 14.3.

14R5 $n = 10$, $\bar{x} = 69.56$, $s^2 = 1.425$; $t_{obs} = (\bar{x} - 70)/\sqrt{s^2/n} = -1.17$.

$SL = P(|t_{(9)}| \geq 1.17) > 0.2$; data are consistent with $\mu = 70$.

$2.700 \leq 9s^2/\sigma^2 \leq 19.02 \longleftrightarrow 0.674 \leq \sigma^2 \leq 4.75$.

For n large, length of 95% interval is $2(1.96)\sqrt{\sigma^2/n}$. For

length 0.4 we require $n = (1.96)^2\sigma^2/(.2)^2$. Taking $\sigma^2 = s^2$

gives $n \approx 137$.

14R7 $Q = 2.403$ with 3 d.f., $s^2 = 0.1804$ with 32 d.f.

$F_{obs} = 4.44$; $SL = P(F_{3,32} \geq 4.44) = 0.01$. Strong evidence

against equality of means.

15.3.1 $n = 10$, $\bar{x} = 23.25$, $S_{xx} = 1568.625$, $\bar{y} = 22.9$, $S_{yy} = 1552.56$,

$S_{xy} = 1560.4$; fitted line is $y = -0.228 + 0.99476x$.

$\sum \hat{\epsilon}_i^2 = 0.3419$, $s^2 = 0.04273$ with 8 d.f.

$t_{obs} = (\hat{\beta} - 1)/\sqrt{s^2/S_{xx}} = -1.00$; $SL = P(|t_{(8)}| \geq 1.00) > 0.2$.

$t_{obs} = (\hat{\alpha} - 0)/\sqrt{s^2(\frac{1}{n} + \bar{x}^2/S_{xx})} = -1.65$; $SL = P(|t_{(8)}| \geq 1.65) > 0.1$.

Data are consistent with $H:\beta = 1$ and with $H:\alpha = 0$.

$\mu \in \hat{\mu} \pm 2.306\sqrt{s^2[\frac{1}{n} + (x - \bar{x})^2/S_{xx}]} = 22.9 \pm 0.151$.

15.3.4   (d) $n = 10$, $\sum x_i^2 = 6974.25$, $\sum x_i y_i = 6884.65$, $\sum y_i^2 = 6796.66$;

fitted line is $y = 0.9872x$; $\sum(y_i - \hat{\beta}x_i)^2 = \sum y_i^2 - \hat{\beta}\sum x_i y_i = 0.4589$;

$s^2 = 0.05099$ with 9 d.f..

$t_{obs} = (\hat{\beta} - 1)/\sqrt{s^2/\sum x_i^2} = -4.75$; $SL = P(|t_{(9)}| \geq 4.75) < 0.001$.

There is very strong evidence that $\beta \neq 1$. If we insist that line goes through the origin, slope must be less than 1 to give a reasonable fit to the data. It is reasonable to take $\beta = 1$ or to take $\alpha = 0$, but it is not satisfactory to assume both $\beta = 1$ and $\alpha = 0$.

15.4.1 Variance estimates $0.00190, 0.00310, \ldots, 0.01110$.
Graph shows no systematic change with x in variance.
$n = 33$, $\bar{x} = 0.1276$, $S_{xx} = 0.07242$, $\bar{y} = 4.3512$, $S_{yy} = 90.81075$,
$S_{xy} = 2.55998$; fitted line $y = -0.1605 + 35.348x$.
Residual SS 0.31955; $s^2 = 0.01031$ with 31 d.f.
$\beta \in \hat{\beta} \pm 2.04\sqrt{s^2/S_{xx}} = 35.348 \pm 0.770$
$\mu \in \hat{\mu} \pm 2.04\sqrt{s^2[\frac{1}{n} + (x - \bar{x})^2/S_{xx}]} = 3.374 \pm 0.042$

Pure error SS $= 0.11620$ (22 d.f.); lack of fit SS $= 0.20335$ (9 d.f.) $F_{obs} = 4.28$; SL $= P(F_{9,22} \geq 4.28) < 0.01$. Strong evidence against straight line model. Residual plot shows curvature; fit could be improved by adding quadratic term to model.

INDEX TO VOLUME II

316

25 228